The Offshore Imperative

NUMBER NINETEEN
Kenneth E. Montague Series
in Oil and Business History
Joseph A. Pratt, *General Editor*

The Offshore Imperative

Shell Oil's Search for Petroleum in Postwar America

Tyler Priest

Texas A&M University Press, *College Station*

Library of Congress Cataloging-in-Publication Data

Priest, Tyler.
   The offshore imperative : Shell Oil's search for
   petroleum in postwar America / Tyler Priest.
      p. cm. — (Kenneth E. Montague Series
      in oil and business history ; no. 19)
   Includes bibliographical references and index.
   ISBN-13: 978-1-58544-568-4 (cloth : alk. paper)
   ISBN-10: 1-58544-568-1 (cloth : alk. paper)
   1. Shell Oil Company—History.
   2. Petroleum industry and trade—
   United States. I. Title.
      HD9569.S55P74 2007
      338.7′66550973—dc22
                                    2006024954

Contents

Illustrations

Acknowledgments

This book benefits from the contributions of many individuals.
Before thanking and acknowledging them, a little background is
necessary.

In 1998, the president of Shell Oil Company, Phil Carroll,
contacted University of Houston oil historian Joseph Pratt about
commissioning a history of his company, the U.S. affiliate of the
Royal Dutch/Shell Group, then the second-largest oil enterprise in
the world. Already at work on the history of Amoco, Joe guided
the project to me. As a business historian researching the topic of
offshore oil and gas, I found this an exciting opportunity. Shell
Oil was the undisputed leader in moving oil and gas operations
into deep ocean waters, a major technological development in the
history of oil and of modern business more generally.

In early discussions, Joe and I tried to interest the company
in a book about Shell Oil's offshore achievements. Although such
a book would sacrifice details about aspects of the company,
we argued this was the most compelling story. Shell, however,
wanted a comprehensive history that included refining, chemicals, marketing, and so forth. They saw this book as a gift to the
company's employees and retirees, and it would be impolitic to
exclude whole departments and functions. This issue of audience
never fully resolved, I agreed to write a comprehensive history.

Constant corporate reorganization threatened the project from
the very beginning. A month after the project started, Phil Carroll announced he was leaving Shell Oil. Oversight of the history
from within Shell then shifted from the president's office to the
public affairs department, a weak player in the corporate hier-
archy whose impulse to turn the book into a puff piece had to
be constantly resisted. Public Affairs did resell the project to suc-
ceeding presidents, but Rob Routs, the third one in four years,
was not buying. It eventually became clear that Carroll's objective
in commissioning the history had been to preserve the legacy of a
company that was about to disappear, to capture its history before
senior people left or died, taking their knowledge with them. We
did not make it in time.

The management changes instituted in 2002, as the book was in page proofs, represented the final dissolution of the U.S. Shell Oil Company. Shell Oil ceased to exist as anything other than a collection of companies overseen by different parts of Royal Dutch/Shell. Drastic budget cuts were imposed on the U.S. operations, including the closing of the Shell Museum in Houston, a wonderful exhibit of artifacts and displays, and the termination of the history book project. The new management team came from the "downstream" side of the business—refining, marketing, and chemicals—whereas the strength of the old Shell Oil was "upstream," in finding and producing oil and gas. The new executives apparently did not want to celebrate the history of their U.S. affiliate, whose partial ownership by public shareholders prior to 1985 had produced a semi-autonomous management structure that sometimes created tension with the majority shareholding parent. By 2002, the Shell people in Houston who cared about this history were either gone or too afraid of losing their jobs to push the matter. The new management, which answered to superiors in London and The Hague, cut funding to print the book.

The news disappointed the many retirees who gave interviews and helped with the book project, especially the offshore veterans who were rightly proud of their accomplishments. Our services agreement contract with Shell Oil gave Joe Pratt and me editorial control and control over final disposition of the manuscript. But rather than try to publish the book against Royal Dutch/Shell's wishes, I decided to write an entirely new history of Shell Oil's offshore business, the kind of study we had originally proposed to the company. This independently researched and written second book was expressly permitted in the contract. The offshore veterans were enthusiastic about the idea. Most of them allowed me to use their oral histories for the new book and facilitated another round of interviews and research.

This book is indebted to those Shell Oil alumni. Beyond their careers at Shell, they have preserved the community they created there, meeting every other year at a reunion in Galveston, Texas. This enduring network helped pull together the many strands of narrative and memory into a coherent history. Their oral histories fill the book with drama and life. Most of the alumni who participated are recognized in the text and the endnotes, but three in particular deserve special recognition. First, I want to thank Mike Forrest. A great deal of Shell Oil's offshore history, from field locations and discovery dates to the unique contributions of many individuals, is contained in accurate detail in his head. He patiently tutored me in the basics of exploration geophysics, introduced me to numerous Shell colleagues, and read multiple versions of the manuscript. The late John Redmond also was an invaluable source. Even at age 93, John's recollections of Shell Oil's history going back to the late 1930s, when he joined the company, were amazingly sharp and insightful. He lugged a draft of the manuscript around with him and provided helpful comments on it just before he passed away in September 2005. John Bookout, a towering figure in Shell Oil's history, also supported the project. His perspective from the head office was critical to making sense of developments during the last several decades.

Others assisted in creating and processing the dozens of oral histories used in this study. Sam Morton and Tom Stewart, both long-time managers in Shell Oil's head office corporate communications group, helped conduct initial interviews and edit transcripts. Later interviews were conducted as part of a study sponsored by the Minerals Management Service of the U.S. Department of Interior. Diane Austin, Tom McGuire, and Andrew Gardner of the Bureau of Applied Research in Anthropology at the University of Arizona, and Jamie Christy, a graduate student in history at the University of Houston, all contributed to that effort. Suzanne Mascola transcribed the interviews. Others who provided research assistance include Joseph Stromberg, Tom Kelly, Katie Pratt, Leah Oren-Palmer, and Tom Lassiter. I am also grateful that Jim Cox, a former editor of *Shell News*, conducted roughly thirty interviews with Shell managers in the early 1970s, making audible the voices of an earlier generation, many of whom are now deceased.

My distinguished colleague Joe Pratt, the series editor for this book, deserves credit for initiating the recovery of Shell Oil's history. He also assisted with interviews, read multiple drafts, and influenced my conceptualization of the post-1945 history of the U.S. petroleum industry. On this project and others, I have been fortunate to have Joe as a mentor, collaborator, and friend. This study also benefits from suggestions made by oil historians Diana Davids Hinton and David Painter.

I appreciate the work of everyone at Texas A&M University Press who expertly shepherded this book to publication. Ron Martinez produced the excellent maps.

Finally, I want to thank my wife, Landon Storrs, and my son, Mason, who put up with Shell Oil for an unexpectedly long time.

The Offshore Imperative

Introduction

In 1956, Shell Oil geologist M. King Hubbert made a startling prediction. In a presentation before a regional meeting of the American Petroleum Institute in San Antonio, Hubbert estimated that U.S. oil production would peak in 1970. Plotting historical production statistics and his calculations of future production on a bell curve, based on ultimate discoveries of two hundred billion barrels of oil, he warned that oil output in the United States would drop at the same rate it had risen. Henceforth, companies would drill for dwindling supplies that were harder to find. Critics ridiculed and dismissed "Hubbert's peak" at the time, citing the erroneous warnings of oil shortages by earlier forecasters. But when it became clear U.S. oil production had indeed peaked in 1970, Hubbert became a prophet. "This time the wolf is here," wrote State Department oil expert James Akins in acknowledgment of this fact.[1]

The confirmation of Hubbert's prediction, however, was soon overshadowed by the continuing transition of the U.S. oil market from a domestic to an international one. Industry commentators shifted their attention from the United States to the Organization of Petroleum Exporting Countries (OPEC) and the Middle East, especially Saudi Arabia, which seized control of world oil production and prices through an OPEC embargo in 1973. Similarly, historical studies of petroleum in the United States during the post-1945 period have largely been subsumed by a predominant focus on the Middle East.[2] M. King Hubbert's famous 1956 prediction and his 1969 forecast that world oil production (based on a total endowment of 2.1 trillion barrels) would peak in 2000 have received renewed attention in recent years as economists and geologists have debated the timing of the impending peak in world production and speculated about the extent of Middle Eastern reserves.[3] But little effort has been made to assess the implications of Hubbert's accurate 1956 prediction for the historical evolution of the U.S. oil industry in the late twentieth century. This evolution revolved around the efforts of U.S. oil firms to stave off the decline in domestic production through intensive exploration and technological innovation.[4]

These efforts had several important effects. They extended U.S. reserves and production, tempered OPEC's control over world prices, revolutionized exploration and production technology, and pioneered the move into new oil and gas frontiers, namely offshore, in the United States and elsewhere. The center of gravity in oil production may have moved from the United States to the Middle East in the early 1970s, and Saudi Arabia may have replaced Texas as the swing producer in the world oil market, but the center of financial and technological dynamism has remained in the United States. Houston is still recognized as the capital of the world oil industry.

One Houston-based firm distinguished itself in the industry's race against depletion. This was M. King Hubbert's employer, Shell Oil Company. The semi-autonomous U.S. affiliate of the worldwide Royal Dutch/Shell Group, Shell Oil was long one of the most successful oil operators in the United States. Widely respected and envied throughout the industry, the company's chief strength was in the upstream business of exploring for and producing hydrocarbons, as opposed to downstream refining, transportation, and marketing, although it was strong in those areas too. Shell Oil appeared to have a knack for finding productive reserves in the United States and doing it profitably. Hubbert's bosses at Shell were initially skeptical of his peak oil prediction, but they soon started taking it seriously. The company went after unconventional deposits in difficult and risky environments as well as in older, seemingly played-out fields. In the process, Shell made a name for itself in geophysical prospecting, petroleum geology, enhanced and heavy oil recovery, pipelining, even coal mining and alternative energy.

The jewel in the company's portfolio of oil and gas interests, however, was offshore Gulf of Mexico. Shell Oil pioneered many of the early moves offshore and continued to lead the way into deeper waters and new geologic trends. For decades, the company dominated the Gulf of Mexico, discovering and producing more oil and gas than any other firm. Barring future major discoveries elsewhere in the United States, which seems unlikely, the Gulf of Mexico will have provided the most significant extensions to U.S. petroleum reserves in the post-WWII period. For years, nothing could compete with the tremendous deposits at Prudhoe Bay, Alaska, discovered in 1967. But on an oil-equivalent basis, output from the Gulf, providing close to 25 percent of U.S. oil and gas production, today already exceeds Texas and, with new "deepwater" reserves fast coming on stream, will soon surpass Alaska. Even conservative estimates see this percentage rising to one-third of the U.S. total by 2010.[5]

An in-depth study of the company that took the lead in making the Gulf of Mexico the most explored, drilled, and developed offshore petroleum province in the world is long overdue. The limited historical analysis of offshore oil and gas development in the Gulf of Mexico so far has highlighted the rise of the specialized offshore drilling and service industries.[6] Very little has been written on the oil companies who found the oil and gas and hired the drillers and platform builders. An earlier corporate history of Shell Oil covers the period through World War II, before the offshore age, and recent studies of Shell focus on the worldwide operations of the Royal Dutch/Shell parent organization. As the leading innovator offshore, Shell Oil is a logical place to begin examining the understudied topics of strategic decision-making, scientific research, management of technology, and corporate organization and culture within modern oil companies and how these activities applied to offshore development.[7]

This development over the past fifty years demonstrates the technological sophistication, capital intensiveness, and functional complexity of the modern petroleum industry. Although we have good journalistic coverage of modern technological advances in the industry, historians of business and technology have not seriously addressed innovation in extractive enterprises such as oil.[8] As this study will show, the introduction of digital computing technology had a giant impact on the evolution of seismic exploration and the engineering of marine platforms. Yet historical studies of the application of digital technology have largely ignored its groundbreaking adoption in the upstream oil business.[9] Because of the attention given to the Middle East, the political and strategic dimensions of oil have taken precedence over the technological ones, leaving studies of innovation to deal mainly with manufacturing and information-age enterprises. Historians and social scientists also privilege structural/institutional and cultural approaches to explanations of innovation.[10] Such approaches, however, still need to be grounded with firm-level studies. Industry-level analysis helps us understand broad trends and macro-level organization, as well as the process of innovation within networks that extended beyond the firm. But examination of individual companies is needed to comprehend how and why decisions were made and how firms participated in these networks. Shell Oil developed internally a capacity for innovation that was exceptional within the industry and the offshore business. Unlike some other companies who scaled back exploration in the United States and redeployed overseas in the postwar period, Shell never lost sight of its core mission to at least replace, if not augment, its petroleum reserves in the United States and never wavered in its core belief that technology could expand the realm of the possible and ultimately lower costs.

This study analyzes the sources of innovation and the development of strategic vision in Shell Oil's search for petroleum in postwar America. A major factor shaping Shell Oil's exploration and production (E&P) organization and culture was the company's unique relationship with its parent. Royal Dutch/Shell has always had an enigmatic presence in the United States. The Group constantly downplayed the fact that its U.S. subsidiaries were foreign controlled, with great effect. The "Americanization" of Shell dates back to 1922, when the British-Dutch combine, to shield its interests from anti-foreign attacks in the United States, engineered a merger of its American subsidiaries that resulted in American investors owning around 35 percent of the stock in the Group's U.S. holding company, the later-named Shell Oil Company. Antitrust concerns and the growing decentralization of Royal Dutch/Shell's worldwide management after World War II allowed Shell Oil to achieve greater operating autonomy from its majority shareholder. No longer a collection of smaller companies largely managed and coordinated from abroad, Shell Oil became a company organized and financed in the United States and led by Americans.

That fact that foreign shareholders still controlled the U.S. company could not be hidden, however. This contributed to Shell's efforts to be a "good corporate citizen" in the United States. The company was quick, relative to many of its competitors, in recognizing new social and environmental concerns, and it was never tarnished by corruption or influence-peddling scandals, in contrast to some of its rivals.[11] Foreign ownership also imposed limits on Shell Oil's operating autonomy. Most significantly, the majority shareholders and their representatives on the company's board of directors discouraged Shell Oil from operating outside the United States where the U.S. firm might have encroached on other Group companies. Except for a nominal involvement in Canada and a relatively brief move into other countries in the 1970s and 1980s (as a result of a minority shareholder lawsuit), Shell Oil restricted exploration and production to the United States and sold its oil and chemical products only in the U.S. market. Although it was not a formal edict—the minority shareholding interest prevented at least the appearance of the Group issuing direct commands to Shell Oil management—this geographic constraint was understood.

The company therefore had to hone its focus on finding domestic petroleum reserves. Because of U.S. import quotas as well as the Group's typically "crude short" supply position (i.e. its own equity crude oil supplies were not enough to meet the demands of its vast worldwide transportation, refining, and marketing system), Shell Oil could not count on the Group to supply its U.S. downstream assets and thus never relaxed its search for petroleum in the United States. Shell Oil remained essentially an E&P company, and its E&P managers usually won internal battles over the allocation of capital to various functions. As the prospects of finding large reserves in heavily explored onshore areas in the United States dwindled and as the growth of production slowed on its way to Hubbert's peak, Shell Oil's long-term profitability, and indeed survival as a relatively independent American oil company, increasingly depended on what exploration vice president Bob Nanz called the "offshore imperative."[12] Beginning in the 1950s, the continental shelf off U.S. coasts offered the most promising potential for big oil and gas finds. Shell Oil *had* to push aggressively offshore.

Of course, the offshore realm held great uncertainties and risks. The ocean was still a largely unexplored and untested frontier. Even if petroleum could be found there, producing it profitably would require fundamentally new engineering and operating approaches. In other words, even if uncertainties about the existence of hydrocarbons and marine conditions could be reduced, plunging into ever deeper waters meant risking capital, the environment, and, less frequently but sometimes unavoidably, human lives. Technological innovation offered a promising way to minimize all these kinds of risk, although not always the cheapest. Confined to the domestic oil province, Shell managers believed the only way the company could compete with larger rivals in finding and producing more petroleum was through leadership in technology.

It is one thing, however, to recognize the importance of technology and another to imagine, develop, and apply it. The Royal

Dutch/Shell Group had long been committed to research in the United States and Europe, especially on the Dutch side. Shell Oil inherited this commitment, but after World War II, the company's technological orientation took on a life of its own. Visionary leadership was essential to this. Top executives achieved their rank not only by being good managers and businessmen, but also because they were geoscientists and engineers who could understand and embrace new directions in technology. Management hired personnel very selectively and nurtured the talents of its employees, exhorting them to push the boundaries of geology, geophysics, marine drilling and production, petrophysics, reservoir engineering, and pipelining. The company created one of the most respected earth science research laboratories in the world, obtained the most talented scientists and engineers, and constructed workable interfaces between research and operations. Top managers learned to trust the wisdom of their technical people, as opposed to the wisdom of the stock market, which today increasingly narrows corporate decision-making into short-term thinking. Developing hydrocarbon reserves and advanced technologies are long-term endeavors, and Shell Oil established a reputation for thinking and planning for the long term.

Shell Oil paid both its hourly and salaried employees as well as any company in the industry, and employees all down the line responded by creating a culture bound by a sense of common purpose and loyalty. Employment stability was common among profitable U.S. businesses during the postwar period, but Shell appeared to have an unusually large number of people who remained with the company their whole careers. "I never saw anyone from Shell looking for a job," remembered John Reilly, a former exploration manager with Pennzoil, referring to the 1950s and 1960s.[13] Even in the 1970s, when the cherry picking of talent off competitors became widespread, Shell was able to keep many of its technical teams intact. Shell people practiced teamwork before it became a fashionable business school concept. The exploration and production departments, especially, combined teams from different specialities and divisions to work on difficult problems. The family feeling at Shell Oil also reflected the culture of its parent, a large organization of allied companies managed by common consent. The quality and cohesion of Shell Oil's organization was not something merely touted or manufactured by the company but was widely acknowledged within the industry. "Boy, we watched just about every move Shell made, to make sure that we were trying to keep up with them," admitted Chuck Edwards, former chief geophysicist of Chevron.[14]

This narrative, however, does not simply celebrate Shell Oil's organization and achievements. The company had flaws. Shell's independence and selectivity produced a somewhat secretive, inbred culture. Counterparts in the industry often grumbled about the arrogance of Shell people. Afflicted at times by the "not invented here" syndrome, Shell Oil plowed money into developing technology when acquiring it elsewhere might have been more practical or when technological solutions were not the most cost effective. The company's commitment to technologically intensive exploration and production in the United States, particularly in heavy oil, backfired in the late 1980s when the price of crude oil plunged. After missing out on Prudhoe Bay, Shell invested a lot of money searching for oil in Alaska and came away with virtually nothing. Its efforts to fight depletion onshore and preserve autonomy for its American managers were ultimately losing battles. In 1985, the Group bought out the minority shareholders in a bitter struggle and in the late 1990s broke apart the old Shell Oil and absorbed the remaining pieces into a new globalized organization.

The stories of Shell Oil's setbacks and failures in this history do not diminish the successes but demonstrate how real people managing a real company had to come to grips with complex market, technical, political, and environmental forces and make hard choices in real time. Too often, the oil industry is perceived as monolithic, led by an oligopoly of major firms who conspire to advance their interests against those of the larger public.[15] It also is easy to bash oil companies, albeit deservedly so at times, for their contributions to industrial pollution, global warming, and society's excessive dependence on hydrocarbons, without trying to gain perspective on the actual workings of those companies. The period after 1973 was competitive and challenging for oil firms and their managers, and they did not all respond the same way and certainly not in concert. The chaos of soaring and plunging prices called into question conventional approaches that had succeeded so well in the postwar years. Heightened environmental concerns and political fears about inadequate supplies of petroleum fed a new wave of government regulations that fundamentally altered decision-making in the petroleum industry. At the same time, price uncertainties forced managers to work diligently, with limited success, to cushion the impact of price volatility on corporate strategies. Those in charge of Shell Oil's vertically integrated operations in the United States faced the added challenge of managing the company's divided ownership with Royal Dutch/Shell. Drawing heavily on interviews with numerous retired Shell personnel, as well as some from outside the company, this study uses first-hand accounts to convey the human dimensions of broad industrial change and relates how the imagination, talent, and hard work of people throughout the company shaped its evolution.

An appreciation of the human drama makes Shell Oil's pioneering steps into deepwater Gulf of Mexico, its signature achievement, all the more notable. In the 1990s, Shell E&P set a series of stunning water-depth records for offshore drilling and production, extending operations to several thousand feet in the Gulf of Mexico with tension-leg platforms and to depths more than a mile with subsea wells. Earlier generations could not have imagined such undertakings, and with price tags of a billion dollars or more, these projects sobered even the most adventurous oil executives. Engineers did not simply step forward in the last decade and create from scratch methods for finding and producing oil and gas from such astounding depths; they built on a half century of gradual accumulation of knowledge and improvements to technical systems. From the first semi-submersible drilling rig and first subsea wellhead in the early 1960s, to the discovery of the "bright spot" method of seismic interpretation in the late 1960s, to early deployment of 3-D seismic in the 1970s, to numerous record-setting platforms in every decade since the 1950s, Shell Oil continually made oil plays in the Gulf much deeper than anyone else in the industry was willing to venture. Innovation in Shell Oil E&P was cumulative. The steady development of expertise and experience gave Shell's leaders the confidence to take risks in the 1980s that few companies contemplated and pioneer a whole new deepwater frontier. It is often tempting to focus on the dark side of technology, on its capabilities for social disruption and environmental degradation. But as the eminent historian of technology Thomas Hughes reminds us, technology is fundamentally a creative activity. Seen as such, historians have some responsibility "to sympathetically portray those who have seen technology as evidence of a divine spark," as well as "those who consider the machine a means to make a better world."[16]

Creative technology at Shell Oil was not confined to offshore exploration and development. Shell's onshore operations across the United States and in the company's few but significant explorations abroad all contributed to Shell Oil's culture of innovation. In addition to the history of marine technology, the Shell Oil E&P story encompasses enhanced and heavy oil recovery, pipelining, coal mining, and alternative energy, which all testified to the application of Shell's technical brainpower. Moreover, the company's refining and petrochemical businesses were extremely important and technologically sophisticated in their own right. All aspects of the American energy business in the post–World War II period bore the imprint of Shell Oil.

Without depreciating the achievements of the other parts of Shell's business, this study focuses on Shell Oil's offshore record, which stands as the company's chief legacy. This is the first time this history has been told in any detail from the early days through the stunning developments of recent years. Shell's story is unique, but it also illuminates the modern history of the petroleum industry. It exemplifies the postwar race against depletion, marking the rise of offshore oil and gas in the passing era of domestic petroleum abundance. It also dramatizes major technological and organizational transformations in oil. As retired Shell Oil platform engineer Peter Marshall describes it, the industry went from "intuitive engineering and an entrepreneurial spirit" in the early days to a "reliance on computers and an expectation of no surprises today."[17] Put another way, oil companies made great technological strides but may have lost some of their creative energy in the journey. The old "can-do" spirit of daring and fortitude has been replaced by a consensus-driven process less tolerant of failure. Furthermore, technological leadership has largely passed from the major oil companies to service companies and contractors. Offshore operators have achieved safer, cleaner, and more systematic approaches to offshore development but perhaps by sacrificing some of their inventiveness and role as technological pacesetters along the way.

Organizational change at Shell Oil in the 1990s also reflected broader trends. Shell Oil offers an important case study of postwar industrial change in the U.S. South and the history of foreign investment in the United States, both relatively new and vibrant subjects of inquiry in recent years.[18] The gradual but steady decline of the United States as a major oil-producing nation, despite technological advances offshore and onshore, eventually eroded justification for a nationally organized and autonomously managed Shell Oil Company. At the same time, political liberalization and deregulation on a global scale, and a merger movement in the world oil industry, radically restructured the international oil business. Undertaking a new globalization strategy in 1997, Royal Dutch/Shell began weaving parts of its U.S. operations more deeply into its other international businesses. Loss often stimulates efforts to preserve, and the following chapters are an attempt to preserve at least a part of the old Shell Oil. The lasting achievement of this company was its ability to recognize and address the offshore imperative that has now spread to other companies and regions around the world.

The Americanization of Shell Oil

The red-and-yellow pecten shell is one of the most universally recognized corporate symbols in the world. It is the emblem of the Royal Dutch/Shell Group of companies, often referred to simply as "Shell" or "the Group," the world's third-largest industrial organization behind British Petroleum and Exxon-Mobil. In 2004, Royal Dutch/Shell operated in 140 countries, employed 120,000 people, owned 10.2 billion barrels of oil-equivalent reserves, operated 46,000 service stations, and owned an interest in 53 petroleum refineries worldwide. It had an annual net income of $18.2 billion and net assets of $84.2 billion.[1] Formed by a unique merger in 1907 between Royal Dutch Petroleum and the English Shell Transport & Trading Company, Royal Dutch/Shell grew to become one of the towering giants in the petroleum industry, an industry whose prominence in the history of the twentieth century was virtually unrivaled.

A key aspect of this growth was the emergence of the Group's U.S. affiliate, Shell Oil Company, as a major oil company in its own right. During the 1940s and 1950s, Shell Oil asserted considerable autonomy from its majority shareholder. In expanding and consolidating the many parts of its business, it projected a new public image, downplaying its majority foreign ownership and declaring Shell to be an American company. The average American bought it. Even to this day, many Americans think that Shell companies around the world are subsidiaries of the Houston-headquartered Shell Oil.

More than a public relations ploy, the Americanization of Shell Oil signified a real assertion of independence by the American company from the Group after the Second World War. Political, legal, and cultural constraints, in addition to industry structure, dictated a clear administrative and operational separation between the Group's U.S. affiliate and the rest of its far-flung global empire. In an imperfectly integrated world economy consisting of many distinct national economies, and with a prospering and protected U.S. oil market, this separation suited the Group's increasingly decentralized approach to international investment and management during the postwar period. This separation also stimulated and enabled Shell Oil's ambitious search for oil and gas offshore and the cultivation of the technological capabilities to realize this ambition.

Origin of Royal Dutch/Shell

The alliance between Royal Dutch and Shell came about through an unlikely pairing of two strong personalities. Marcus Samuel, a timorous but tenacious Jewish trader from the East End of London, had inherited a small fortune from his father, who had imported from the Far East, among other things, seashells, which were used for decoration and jewelry. In the 1890s, Samuel expanded the business into kerosene, lighting fuel refined from oil. In alliance with the Paris Rothschild family of merchant bankers, who controlled oil refining at Batum on the Black Sea coast, Samuel organized a new system of distributing kerosene in the East by bulk tanker shipments. The tankers were named after seashells, thus the origin of the Shell trademark. In 1897, Samuel combined his growing Far East syndicate of businesses into the Shell Transport & Trading Company (Shell T&T). But Shell soon became locked in fierce competition with Standard Oil, the giant trust created by the legendary John D. Rockefeller, who was known for mercilessly undermining and swallowing competitors.[2]

Samuel and Rockefeller both coveted oil-producing properties in the Dutch East Indies, where the Royal Dutch Petroleum Company had established itself as a successful oil producer. Under the leadership of Henri Wilhelm August Deterding, Royal Dutch had grown into a substantial integrated oil company—combining oil exploration, production, and refining—halfway around the world from the other major oil-producing regions in the United States and Europe. Seeing mutual advantages in an alliance to resist the takeover maneuvers and price-cutting tactics of the formidable Standard Oil, Deterding and Samuel in 1906 agreed to combine their companies. Although Samuel pushed for a 50:50 deal, Deterding insisted on 60:40 and control over management. Brilliant but autocratic, with a mastery of accounting and finance, Deterding had risen to become a managing director of Royal Dutch by the young age of thirty-four. As he once wrote, "mine is a personality which does not readily submerge itself."[3] With Shell Transport & Trading's fortunes declining in the early 1900s, Samuel had no choice but to accept the Dutchman's terms.

The two companies, however, reached an unusual agreement. They united their interests but retained separate identities. They transferred all oil lands and refineries to a newly formed Dutch company, N.V. De Bataafsche Petroleum Maatschappij (the Bavarian Petroleum Company), headquartered in The Hague, while an English firm, the Anglo-Saxon Petroleum Company, located in London, took over all transportation and storage assets. Both Royal Dutch and Shell became holding companies for these new operating companies. Each of these companies and their subsidiaries was a legal entity, together comprising the Royal Dutch/Shell Group, even though the Group itself did not exist in law anywhere in the world. Five managing directors, two from Shell and three from Royal Dutch, oversaw the affairs of this unique enterprise. But for the next thirty years (until his retirement in 1936), Henri Deterding ruled Royal Dutch/Shell's expanding empire, the only one that could truly rival Standard Oil and its offspring after the trust was broken up in 1911.[4]

The amalgamation proved to be very successful and surprisingly harmonious. The British and Dutch for centuries had maintained close diplomatic ties. Tensions between the two nationalities in the Group never prevented accommodation and mutual respect. Royal Dutch Petroleum had achieved a controlling majority in the combination but did not abuse its authority or wield it at the expense of its British partners. Deterding had not desired conquest or victory over Shell but rather access to its assets and markets, its corporate presence in the world financial center of London, and its association with the British Empire. Royal Dutch/Shell became as much a British entity as a Dutch one. The name "Shell," after all, was imparted to the Group and its products around the world. Moreover, Royal Dutch and Shell shared what Deterding called the "straight-line policy" from the pre-merger era. Whereas Standard Oil purchased and refined its products in the United States and exported them around the world, both Royal Dutch and Shell, largely to combat Standard's price cutting, tried to deliver on the straightest line possible, seeking the nearest source of supply to meet local demand. This principle would guide the Group's global expansion and shape the way it built its operations in the United States.[5]

Invading Standard Oil's Backyard

Royal Dutch/Shell first entered the United States on the eve of the Great War. With the intensifying price wars between Standard and the Group in the Far East and Europe, the United States became increasingly important to the Group's global strategy. "Until we started trading in America," explained Deterding, "our American competitors controlled world prices—because . . . they could always charge up their losses in underselling us in other countries against business at home where they had a monopoly."[6] The U.S. Supreme Court's 1911 decision to dissolve the Standard Oil trust into thirty-three separate companies stiffened Deterding's resolve to "operate in Standard's backyard." He regarded the case as a clever legal gimmick that did not prevent the companies from acting in concert.

Shell started on the West Coast and then expanded into the mid-continent. In 1912, the organization established a small marketing company in the Pacific Northwest, stretched down the coast into exploration and production in California, where it made famous discoveries at Signal Hill and elsewhere, and then broadened the hunt for oil into Oklahoma. Shell's rapid expansion threatened Standard Oil and its offspring, who maligned Shell as a foreign exploiter. Burgeoning numbers of tank cars and service stations painted in Shell's emblematic yellow and red colors provoked William Randolph Hearst to dub Shell the "yellow peril," presumably at the behest of Shell's rivals.[7]

Deterding came to realize that building a large and powerful Shell organization in the United States might require placating American opinion by enlisting American capital. The Union Oil Company of Delaware presented Shell with such an opportunity. This company, which owned 26 percent of the Union Oil Company of California, had fallen on hard times and was ripe for takeover. A merger between Shell and Union in 1922 brought all companies under a holding company called the Shell Union Oil Corporation, with approximately 65 percent of the stock held by the Group (through another subsidiary called the Shell Caribbean Petroleum Company) and the remaining 35 percent held by groups of U.S. investors. Shell's move to "go public" almost backfired, however. A majority of Union California's stockholders formed a bloc dedicated to preventing Shell's interest in their company from growing.

The campaign even recruited national politicians who, preying upon anti-British sentiment leftover from the war, charged Shell with being a tool of the British government. A Senate resolution passed in June 1922 called on the Federal Trade Commission (FTC) to investigate Shell's holdings and determine whether or not Great Britain and its dominions, the Netherlands, Romania, "or other countries having oil lands," discriminated against American companies seeking oil leases. The resulting 152-page report answered critics who accused Shell of being controlled by the British government and uncovered nothing improper about Shell's worldwide or U.S. operations. By the time the report was issued, negotiations had begun to allow American companies entry into the Iraq Petroleum Company. To further insulate itself from attack, Shell Union disposed of its Union California stock, and the company soon won new public acceptance by maintaining a clean reputation during the oil scandals of the mid-1920s.[8] During that time, Shell built and acquired new refineries and pipelines, established affiliated companies in research and chemicals, and expanded marketing into the East Coast.

The depression of the 1930s, however, forced retrenchment. As relative newcomers to the U.S. market during a period of high real estate and building costs, the Shell companies had paid dearly for their properties, which now saddled the organization with large overhead and amortization charges. In the spring of 1931, Group Managing Director Frederick Godber arrived in the United States to lay down the line on cost-cutting measures. The Shell companies refinanced their debt, surrendered valuable undrilled leases, and cut production and staff. The reduction of the exploration staff and budget at a time when the Group's U.S. affiliates were short crude oil forced Shell to miss an opportunity in the mid-1930s to match the great acquisitions of crude reserves, especially along the Gulf Coast, by competitors such as Humble (Standard of New Jersey/Exxon), Magnolia (Socony/Mobil), Gulf, and Texaco. In 1938, Shell did open two new sectors in the giant Wasson field of West Texas. The period of the 1930s, nevertheless, can still be characterized as one of missed opportunity, which would have important ramifications for Shell Oil's exploration strategy after the war. In the final move in the streamlining efforts of the thirties, Shell Union consolidated its chief operating subsidiaries into a new nationwide company. In 1939, the California, Midwest, and Eastern companies merged to become Shell Oil Company, Inc., and the St. Louis (Shell Petroleum Corporation) and New York (Shell Eastern Petroleum Products) executive offices came together in New York City's RCA building.[9]

As in all aspects of American life, the Second World War was a watershed for Shell Oil. The war emergency forced Shell to post-pone further organizational restructuring and get busy supplying the Allies' voracious appetite for petroleum products. At the urg-ing of James H. "Jimmy" Doolittle—a Shell aviation fuels man-ager, world-famous aviator, and World War I ace—Shell scientists in the United States and Holland invented catalytic cracking tech-nology capable of synthesizing 100-octane gasoline needed for a new generation of aviation engines.

Of the total 100-octane produced during the war (1942–45), Shell supplied more than 13 percent. Shell Oil also was the first company to produce commercial butadiene, a critical building block of synthetic rubber. Meanwhile, Shell men and women served their country during the war. The Shell soldier of whom the world heard most was, of course, Doolittle. He rejoined the Air Corps as a major and in April 1942 led a flight of B-25 bombers on the famous air raid over Tokyo, boosting the sagging morale of a nation in the aftermath of Pearl Harbor only four months before and earning him the Medal of Honor. Two years later, the raid was immortalized in the film *Thirty Seconds over Tokyo*, in which Spencer Tracy played Doolittle.[10]

American Citizenship

The 1939 consolidation of the Shell companies into one nation-wide operating company, Shell Oil Company, Inc., was a major step in the maturation of Shell as a national business. The Group's American operations, however, still consisted of several operating companies governed by two separate organizations, one east and one west of the Rocky Mountains. The company retained dupli-cate sets of vice presidents and department heads in New York and San Francisco. The West Coast may have been the company's moneymaker in years past, but by 1940 crude oil production east of the Rockies had almost doubled West Coast production. At the end of the war, the West Coast no longer loomed as large in Shell's national organization.

Finally, in 1949, Shell Oil completed the formal centralization of the organization in New York. Under the new structure, the San Francisco office declined in relative importance to the head office. The holding-company arrangement between the Shell Union Oil Corporation and its chief operating subsidiary, Shell Oil Company, Inc., was terminated. Shell Union was renamed Shell Oil Company, acquiring all assets and assuming liabilities of the chief subsidiary, which was dissolved. Shell Oil no longer consisted of a collection of small companies largely managed and coordinated from abroad. Now it was a single entity, operating on a national scale, financed through U.S. earnings, and with U.S. managers taking more initiative in decision-making.

The simplification of Shell's corporate structure hinted at profound changes in the company. A wave of retirements cleared the way for Americans to move into all levels of management. The infusion of Americans created greater autonomy for the company within Royal Dutch/Shell. This maturation corresponded to America's new superpower status and the emergence of the U.S. economy as the most dynamic in the world. The 1950s witnessed the evolution of Shell Oil from a medium-sized, foreign-controlled company to a major industrial concern that few people regarded as anything other than American. By 1952, the company employed a stable workforce of 31,000, with 27 vice presidents, most of whom were U.S. citizens. As former president Monroe Spaght testified, before the war "one was forever apologizing and going out of his way to be sure that it was understood that we were American. With some of the senior people being non-Americans, that was a bit of a chore. This all changed after World War II."[11]

Officers of the U.S. Shell companies traditionally had been British or Dutch citizens sent over for a time before other assignments with the Group. When they came to the United States, many remained largely external to Shell U.S.A. The war, however, altered this pattern by disrupting the replenishment of Shell managers with Group men and slowing the entry of new people, foreign or American, into Shell ranks. After the consolidation in 1949, the number of Group managing directors on the Shell Oil board eventually declined from five to two. A new influx of Americans filled a great need for engineers all across the company. Like other U.S. corporations at that time, Shell hired a generation of young Americans who grew up in the Depression, many of them having served in the armed forces during the war.

The naturalization of Shell Oil's Scottish-born president, H. S. M. "Max" Burns (president from 1949 to 1961) embodied the gradual Americanization of Shell. All subsequent presidents of Shell, except one, would be American-born. Although Burns had risen to a high position within the Group, he nevertheless asserted increasing autonomy for Shell Oil within it. Burns married an American and became a naturalized citizen. "More American operator than European paternalist," wrote *Forbes,* Max Burns symbolized the new Shell Oil Company.[12] Burns possessed keen intelligence, broad vision, and tremendous self-confidence. He conversed on a wide range of subjects. He read the bridge column in the paper every morning, yet he never played a hand of bridge in his life. He was simply intrigued by its mathematics, a subject in which he held an advanced degree from Cambridge University, to go along with an advanced degree in physics from the same institution.

Fellow leaders in the industry respected Burns, which led to his being named chairman of the American Petroleum Institute's board of directors. He was the first Shell representative to serve as the head of the U.S. oil industry's most prominent trade association.[13] For a company once branded as the yellow peril, this event was rich in symbolism. Before this time, as Shell Oil chief counsel under Burns, Bill Kenney, pointed out, "there was a great deal of timidity within Shell from a standpoint of sticking our neck out and getting it chopped off by someone who might say, 'after all, you're nothing but a foreign-controlled company, what business have you to offer advice on questions of national policy?'"[14] By the mid-1950s, this perception was fading.

Max Burns also won respect for Shell from U.S. financiers on Wall Street. He opened up Shell Oil's accounting to New York's security analysts. This was unusual for a president whose corporation was only 35 percent "public" and whose predecessors, according to *Forbes,* "were wont to wire London for permission to answer financial questions." The analysts discovered a very profitable business. During 1942–52, Shell's net return on investment was 14.4 percent, equal to that of Jersey Standard and near the top of the industry.[15] Shell Oil did not depend on its parent for capital, financing its tremendous postwar expansion with U.S. dollars, largely from retained earnings and write-offs (depreciation, depletion, and amortization). During the 1945–60 period, Shell took on long-term debt only three times, the largest issue being $125 million by Shell Union in 1946 to retire existing bond indebtedness. In 1951, Shell Oil's sales revenues exceeded the $1 billion mark, climbing to more than $1.8 billion by 1960.[16]

In 1954, the first-time listing of Royal Dutch Petroleum shares on the New York Stock Exchange (NYSE) helped Shell Oil solidify ties to the U.S. financial community. After conceding to publish consolidated accounts for its hundreds of companies on a quarterly basis, Royal Dutch Petroleum, which had long resisted the transparency required by American securities markets, became the first European company to have its shares admitted to the NYSE in the postwar period. By 1956, Royal Dutch shares had doubled in price, and U.S. investors had increased their holdings to nearly 25 percent of the company's outstanding capital. Including the 35 percent minority ownership in Shell Oil, valued at $660 million, total U.S. investments in Royal Dutch/Shell stood at over $1.3 billion.[17] Even during the isolationist revival of 1950s America, Shell was increasingly perceived as an American enterprise under American control.

"Our Saving Grace"

From 1950 on, Shell Oil Company became a more autonomous and vital part of the Group's worldwide activities. In most years, it accounted for anywhere between one-quarter and one-third of the Group's global profits.[18] The Group began to view its interest in Shell Oil as a sound investment that need not, and indeed should not, be micromanaged. As an outside observer once commented: "There is no imported management at Shell. The Group does not call the signals from London or The Hague. The American managers don't carry chips on their shoulders against the Group's managing directors. They just provide healthy dividend checks every quarter."[19]

The Americanization of Shell Oil, however, was not something forced upon the Group by an upstart affiliate. The Group had always been a decentralized organization, "not a corporation, not a legal entity, but a condition," wrote one U.S. business magazine. "This condition is international business in its most highly developed sense."[20] Royal Dutch/Shell was simply a name used to designate nearly five-hundred companies owned (wholly, jointly, and indirectly) and coordinated by its three principal operating companies: the Anglo-Saxon Petroleum Company of London, the N.V. De Bataafsche Petroleum Maatschappij (B.P.M.) of The Hague, and the Asiatic Petroleum Co., Ltd., of London.

Despite the decentralized and cooperative principles underlying the arrangement, control over the entire realm rested in The Hague and London, and Europeans made the decisions. Prior to the war, the Group hired only a few foreigners and nationals to run its affiliates and then chiefly on the basis of "how closely they resembled Europeans." A classic Shell story told of an urgent cable once sent from an affiliate to London: "Lubrication oil sales dropped 5 percent. Send two more cricket blues."[21] Sir Henri Deterding, moreover, retained a large amount of power. In his later years, his rule had a detrimental effect on the organization as his autocratic tendencies and Nazi sympathies hardened, which alarmed the British directors of Shell and both the British and Dutch governments. Deterding's retirement in 1936 and death in 1939, however, signaled the end of an era for Royal Dutch/Shell. Thereafter, the chairman of the managing directors was retired at the age of sixty.[22]

Contact and decision-making between the two parent companies in the post-Deterding era was not distinctly clarified or formalized. To the extent it was, the seven managing directors became the policy-making Executive Committee, which in 1958 was formally designated as the Committee of Managing Directors (CMD), after a corporate reorganization initiated by the American management consultant McKinsey & Company. Each director assumed a specific area of responsibility. Out of the urgent necessity to recover from the devastation of the war, when the Dutch headquarters had to be relocated to Curaçao, the directors redistributed many functions and powers down through the organization.[23]

In supervising operations, the Group maintained its traditional division of labor. Administrators in London supervised finance, transportation (which included the world's largest tanker fleet), and marketing. Administrators in The Hague, meanwhile, presided over production, manufacturing, and research. As part of the McKinsey reorganization, which was designed to preserve decentralization but improve coordination, the Group set up a Coordinating Committee of eight administration heads under the Executive Committee to handle developments and problems affecting interactions between different phases of the business.[24] Of course, top management retained final authority, especially on budgets and questions involving major capital expenditures. But wherever possible, Group executives refrained from getting bogged down in managing day-to-day operations of the subsidiaries.[25]

The status of a subsidiary company depended on its structure of ownership, its financial independence, and the degree to which it was integrated. Group companies concerned with only one phase of operations required close coordination. Shell Oil, on the other hand, could operate with relative autonomy. Production, refining, and marketing were all well developed in the United States, unlike in most other nations. Shell Oil had built up fully integrated operations that were, for the most part, independent of the rest of the Group.[26]

The Group recognized the social and political appeal of employing qualified nationals in positions of authority and not just as hired hands. John Hugo Loudon, general managing director from 1952 to 1964, anticipated nationalistic trends and urged the hiring of more locals in executive positions. Once described as "a handsome man whose casual movements seem under strict control and whose most deliberate movements seem strictly casual," Loudon exhibited the refinement bred by a cosmopolitan background.[27] He had held positions around the world, spoke five languages fluently, and came from a distinguished lineage within the organization.[28] Loudon and Sir Francis Hopwood, Shell Oil's chairman from 1951 to 1956, had the vision to recognize that the best way for Shell Oil to grow and develop in the United States was by making it a truly American venture.[29]

In many ways, the independent strength of Shell Oil proved valuable to the Group during its postwar reconstruction. "Shell Oil was our saving grace," confessed Loudon.[30] The U.S. company helped raise valuable dollars to pay for expansion in the oil-producing countries of Venezuela, Colombia, Ecuador, and Canada. In 1948, Royal Dutch/Shell gained access to a quarter of a billion dollars through a complex transaction involving the Shell Caribbean Petroleum Company, the subsidiary that held most of the Group's interests in that part of the world, including the Group's interest in Shell Union Oil. With demand for equities weak, Shell Caribbean placed a $250 million bond issue (at 4 percent interest over twenty years) with ten U.S. insurance companies, pledging its 8.8 million shares (worth $350 million at the time) in Shell Union as collateral.[31] By financing expansion of oil production in the Western Hemisphere, the loan balanced out the Group's global expansion program, at the same time easing the pressure on British and Dutch cash reserves in dollars. *Fortune* magazine called this deal the "biggest private international business transaction of 1948."[32]

Shell Oil lent a wealth of experience and technical knowledge for the global rebuilding effort after World War II. The war had devastated Royal Dutch/Shell more than any other major. The

Group lost access to its Romanian oil as well as nearly half of its ships. Its Far Eastern properties were wiped out. European operations were at a standstill. The refineries were in terrible shape, if not entirely destroyed. The Group's manpower and research organizations had been severely depleted. By contrast, Shell Oil had made great progress in every facet of the business. As Americans were coming into positions of responsibility in their own company, they were also called upon to bring Group technicians up to speed on advances in such areas as seismology, refining, and marketing. Shell Oil people also helped the Group adjust to a more competitive international business environment as decolonization and the collapse of the pre-war international oil cartel exposed the Group to greater competition. A constant stream of personnel from Group companies ventured to the United States to learn about the U.S. oil business. In these efforts, Shell Oil became a feeding ground on which the Group resuscitated itself.[33] By its fiftieth anniversary in 1957, the Group had rebounded from adversity and regained international prominence. Its empire included nearly 250,000 employees spread throughout 76 countries, oil production in 17 of them, 47 refineries, and the world's largest tanker fleet (over 500 ships).[34] As the third largest industrial organization in the world, it was expanding faster than the top two, General Motors and the Group's archrival Jersey Standard.[35]

Renewing the Research Mission

The diversity and multiplicity of Royal Dutch/Shell's activities made dedication to research essential to global competitiveness. After the war decimated the Group's European research capabilities, Monroe Spaght, who at the time was vice president of Shell Development, Shell Oil's research organization, helped reconstitute Group-wide research on a jointly planned and coordinated basis. By 1957, the Group employed six thousand people in research, one for every forty-five employees. The new research arrangement depended on Shell Development as a much more vital source of innovation for the Group than before.

Because Shell Oil had come through the war in such a strong position, Shell Development received almost 50 percent of the Group's entire research budget. The Emeryville research laboratory in California, opened in 1928 with a staff of thirty-five, grew by 1946 into an operation employing over one thousand people and occupying seven acres of land. The talent, energy, and money poured into research produced results at Shell Development and throughout the company. By 1953, Shell Development had over 2,100 patents in force, and new patents were added at a rate of about 200 per year.[36]

The wide room given for innovation, however, created growing strains between basic and applied research. Many at Shell Oil and Shell Chemical who were involved with the commercial applications of research felt they lacked sufficient control over the objectives set forth in Shell Development. Consequently, research activities were dispersed. Congress spurred the proliferation of research by passing tax incentives after the war to make research less costly. Various parts of Shell Oil started opening up their own specialized laboratories. While oil and gas production research was conducted at Emeryville, exploration was completely separate. In 1945, Shell established a new E&P research organization in Houston under the direction of Dr. Harold Gershinowitz, former director of manufacturing research in the East. Two years later, in the southwest Houston suburb of Bellaire, the group moved into a new forty-four-thousand-square-foot, state-of-the-art laboratory building adjoining the existing geophysical research laboratory built in 1936. At the dedication of the Bellaire Research Center (BRC), Max Burns announced that oil shortages during World War II had given Shell a new mission "to develop more precise and reliable ways of finding new oil" and "to develop more efficient means of recovering oil once it is found."[37]

Top managers in E&P believed these two imperatives were closely related. Instead of splitting the research division into the traditional realms of exploration and production, Gershinowitz organized three departments along academic lines. They worked closely together to study the physical, chemical, and geological aspects of each problem under consideration. Laboratory manager Noyes D. Smith Jr. assumed the hectic job of coordinating work among these units. Initially numbering 150 persons, the staff increased to 500 by 1956. An engineering department added in 1955 addressed the technical demands of increasing well depths and the growing interest in marine locations. Finally, an extensive training program gave new recruits skills in exploration and production that were not being taught in the universities. The program eventually became known throughout the industry as "the Shell University."[38]

Despite the fanfare surrounding the Bellaire lab, its early days were uncertain. Many in E&P operations remained skeptical that the endeavor would yield practical solutions in the field. Indeed,

1
Entrance to Bellaire Research Center in Houston. During the 1950s and 1960s, it was one of the most respected earth science research laboratories in the world. *Photo by author.*

E&P research replicated the organizational problems of research within Shell Oil as a whole. The Technical Services Division (TSD), created within Shell Oil in 1951, provided training, gave technical assistance to the E&P area staffs, and generally acted as liaison between research and operations. While the TSD arrangement worked in some areas, such as reservoir engineering, in other areas, such as geology and petrophysics, it tended to erect barriers to communication rather than break them down.

Shell Oil's research activities obviously needed to be molded into a more effective organization. The challenge was to insure a comfortable interface between research and operations, keeping the company's technical brain focused on business considerations without short-circuiting the legitimate long-range research interests of the company. With patience and imagination, Shell Oil managers worked out two reorganizations in the 1950s to achieve this. The first step came in 1953, when they combined all basic research at Emeryville, Bellaire, and other labs in the Shell Development Company. With this reorganization, Shell Oil increased its interest in Shell Development from 50 percent to 65 percent, preparing the way for Shell Oil's buyout of the remaining 35 percent two years later in 1955. Full ownership of Shell Development permitted Shell Oil to bring all research in the United States under tighter control.

Shell Oil's purchase of the Group's independent interest in Shell Development, and indeed the entire process of Americanization at the company, must be viewed in light of the domestic political context in the United States, which was undergoing a revival of nationalism, charged by both the end of the Second World War and the chilling onset of the Cold War. The late-1940s oil deals put together by the majors in the Middle East, furthermore, produced fears of a reborn international oil cartel and set off a new round of federal government investigations that subjected Shell Oil once again to political scrutiny as a foreign enterprise.

Although no longer denounced as a yellow peril, Shell Oil still encountered obstacles to operating at both the state and federal levels. Several states still had laws and constitutions that forbade landholding by "aliens," a legal classification that included Shell Oil because of its foreign majority ownership. In 1953, the company had to lobby hard for an amendment to the state constitution of Washington that dropped a prohibition against alien landholding and paved the way for the construction of a new Shell refinery at Anacortes on Puget Sound.[39] At the federal level, prohibitions against foreign ownership of coastwise shipping and concerns about possible restrictions on the leasing of federal lands by foreign interests complicated the company's growing interest in offshore oil and gas.

The company feared that such restrictions might grow as antitrust sentiment spread across the legal and political landscape. Under the Sherman Antitrust Act of 1890, any combination or conspiracy in restraint of trade was unlawful. The Federal Trade Commission's aggressive antitrust investigation into monopolistic practices in the international oil business heightened fears that Shell might be singled out as a foreign company, as it had been in the early 1920s. In 1949, the FTC subpoenaed documents from all the major U.S. companies and produced a study, *The International Petroleum Cartel*, that uncovered extensive historical detail about the workings of the cartel and the international relationships among the firms. Although the so-called "As-Is" cartel agreement formally ended with Standard of New Jersey's 1942 consent decree and the wartime controls and regulations, the FTC and the U.S. Justice Department's Antitrust Division believed that the formation of the Arab American Oil Company (Aramco) and other

Middle East oil deals basically reconstituted the cartel.[40] In 1952, Pres. Harry Truman ordered a grand jury criminal investigation of the seven multinational oil giants. Offended by the subpoenas for documents issued to Anglo-Iranian (British Petroleum) and Shell, which appeared to represent an illegitimate assertion of extraterritoriality, the British government urged those companies not to cooperate, and the Dutch government gave similar instructions to Royal Dutch. Royal Dutch/Shell and Anglo-Iranian were eventually excluded as defendants, and Truman and Eisenhower downgraded the investigation to a civil action, exempting companies from prosecution over their deals in the Middle East. The case, however, dragged on through the 1950s with concurrent congressional investigations into possible antitrust violations in the oil industry, keeping Royal Dutch/Shell on notice about how it conducted its business in the United States.[41]

During this period, the legal relationship between a foreign company and its subsidiaries in the United States became increasingly fraught with complications relating to antitrust law and the rights of minority shareholders. In a landmark 1951 decision in the case of *United States v. Timken Roller Bearing Co.*, the U.S. Supreme Court ruled that exclusive arrangements between a parent and partly owned affiliate were liable to antitrust prosecution. A single entity could not conspire or combine to restrain trade with itself, but the *Timken* decision implied that the parent and its partly owned affiliate could be viewed as separate actors who could not lawfully agree to set prices, divide markets, or otherwise restrain trade.[42] Under the common law doctrine of corporate opportunity, furthermore, transactions between such parties could be challenged as unfair to the subsidiary's minority shareholders. One way to reduce the risk of a corporate opportunity lawsuit was to limit the exposure of a subsidiary's business opportunities to personnel from the parent company. Another was to confine the companies' operations to mutually exclusive geographical areas. However, under the *Timken* decision, such an agreement not to compete in a given area would itself be construed as a conspiracy to restrain trade and therefore subject to antitrust prosecution. If such an agreement were seen to be prejudicial to the minority shareholding interest in the subsidiary, those shareholders could also sue the parent company on the grounds of usurpation of corporate opportunity. The contradictions in the legal strictures governing the relationship between a parent and its partly owned subsidiary were stark. Nevertheless, foreign companies, no matter how small their activity within the United States, found themselves beginning in the early 1950s brought more closely under the purview of American courts.[43]

The surest way to avoid legal problems of antitrust or corporate opportunity was for a parent to conduct relations and negotiate transactions with its subsidiary strictly at arm's length. Although there appears to have been an informal understanding that Shell Oil would refrain from operating outside the United States, the Group did not issue direct commands to its American affiliate. An awareness of new legal risks, moreover, most surely contributed to the decision to allow Shell Oil to acquire the Group's full interest in the Shell Development Company. Shell Oil and the Group updated their research services and patent sharing agreement and took great pains to ensure that the price of transfers and the sharing of technology between the two were indeed done at arm's length and according to very formalized procedures so that it would not appear that profits were being moved out of the United States or that the U.S. firm was being placed at a competitive disadvantage. The Group had partnerships and joint ventures with other American oil companies abroad. Full American ownership of Shell Development reduced the risk of technologies developed in the United States by Shell Oil being shared by its parent with overseas partners and then in turn used to help those partners compete against Shell Oil in the U.S. market.[44]

The assumption of complete control by Shell Oil over research at Shell Development not only placed Shell's American companies on safer legal ground, but it allowed Shell Oil's managers to focus research more directly on practical problems encountered in U.S. operations. In 1958, Shell management altered budgeting procedures to make money and programming for research flow through the operating divisions. This decision had far-reaching effects. The E&P Research Group and Technical Services Division merged, and E&P research flourished within this structure through improved communication and increased rotation of staff between research and operations. Objectives were more clearly defined and, contrary to the fears of some, long-term research was not only preserved but expanded.[45]

Thus, Shell Development evolved from being an independent company, jointly owned by Shell Oil with the Group, into a wholly owned research arm of Shell Oil. The development of a sophisticated, American-controlled research function represented the maturation of Shell Oil as a major American company and as a distinctive company within Royal Dutch/Shell. The larger process of Americanization at Shell Oil, in which the reform of the research organization was an important part, instilled a new spirit of initiative throughout the company, especially in the search for oil and gas offshore.

Testing the Waters

Royal Dutch/Shell emerged from World War II desperately short of crude oil. It had lost key production in Europe and Asia and failed to obtain equity in the enormous reserves of Saudi Arabia and Kuwait. As a partial remedy, the Group negotiated in 1947 the largest long-term purchase contract in the history of the industry to take one-half of Gulf Oil's Kuwait production. Still, as it recovered from the war and expanded, Royal Dutch/Shell lacked enough of its own crude to match rising sales. Aggressive exploration, therefore, became its prime directive. By the early 1950s, Group companies participated in practically every area open to exploration. They were often the first to enter frontier areas, and with their strong technological background they were not afraid to tackle those with complicated geological structures, unlike some other major oil companies that sought large, simple structures.

In the early postwar period, Shell Oil undertook one of the most successful campaigns within the worldwide Group to acquire more oil and gas sources. By 1950, petroleum had eclipsed coal as the main source of energy for Americans. Relatively cheap and easy to transport, oil and gas accounted for nearly 60 percent of national energy consumption. With a booming American economy, demand for petroleum resources seemed insatiable. Although the industry had made some major discoveries in the 1930s, American oil reserves were at a near-record low. Many U.S. oilfields came out of the war worked over and depleted; exploration for new fields had been largely postponed during the conflict. The United States was the oldest and most intensively explored producing region in the world. The most obvious oil, contained in structural traps, had already been discovered. To find new fields, new kinds of traps and oil would have to be found using new exploration concepts and technologies. Research at Shell Oil's Bellaire E&P laboratory became essential to the company's drive to replace reserves.

The race against depletion in postwar United States would prove to be as technologically challenging as anywhere in the world. The biggest challenge was the move offshore. Confined to the domestic oil province but excluded from the best onshore prospects, Shell Oil looked offshore to improve its declining competitiveness. The E&P organization adopted a strategy of gradually getting its feet wet, acquiring leases in shallow water that the company could develop without having to make great leaps in technology. The Gulf of Mexico was the perfect place to do this. Its waters were mostly calm. Its seabed sloped gently, providing a relatively flat bottom and shallow water miles out from shore. And it was near established oil centers. In fact, drilling in shallow water involved little more than an extension of marsh techniques. The company's successful experience in the 1930s and 1940s in finding and producing oil at Weeks Island and other wetland areas of south Louisiana gave it the skills and confidence to test the open water.

After the war, Shell acquired large acreage from the state in the shallow bays around the Mississippi River delta and made some major discoveries, particularly in the East Bay between the South Pass and Southwest Pass of the river. The East Bay fields, along with other major ones in the Main Pass area, changed the company. Shifting the exploration focus and the corporate E&P budget to the newly established New Orleans area positioned the company as one of the most successful early pioneers in the Gulf. As an aggressive developer of offshore seismic capabilities and one of the original sponsors of mobile drilling technology, Shell Oil vaulted ahead as a leader in the new offshore game. During the 1950s, it continued to push the boundaries of technology and water depth to expand production in this new frontier. The company took a calculated gamble, but it was one that paid sweet dividends. Most people in the industry agree that offshore Louisiana *made* Shell Oil.

Wetland Bonanzas

Oilmen began addressing the challenges of marine environments long before they began to think seriously about drilling in the open waters of the Gulf of Mexico. Exploring such environments was at first an incremental process, involving the adaptation of land-based equipment and technologies to particular locations. As early as 1896, companies had drilled in ocean waters from piers extending off the beach at Summerland, California. In 1911, Gulf Oil drilled the world's first oil well in inland waters at Caddo Lake, Louisiana—the first truly "offshore" well, detached from the shore—and subsequently built numerous structures on wood pilings there using a fleet of tugboats, barges, and floating pile drivers. Following on these precedents in the late 1920s, the Soviet Union constructed extensive trestle systems offshore from Baku for drilling in the Caspian Sea, and oil firms found a solution to Venezuela's teredo-infested Lake Maracaibo by installing platforms on reinforced concrete pilings.[1]

Southern Louisiana added another level of difficulty for even the most intrepid oilmen. Swamps, marshes, and bays, all difficult to classify strictly as land or water in many places, posed frustrating transportation and operating problems. Those who did brave the challenges, such as Shell Oil, had to make fundamental adjustments not demanded in previous marine work. To a greater extent than elsewhere, they had to tap into local knowledge of the confusing and forbidding terrain. And they had to develop new and innovative means of transportation to enable surveying and drilling in wetlands where it was too hard or expensive to establish fixed foundations.

Geophysical explorations did much of the advance work in defining the problems. Although a few salt domes had been discovered and developed in south Louisiana prior to the 1920s, serious and sustained exploration for buried salt domes in the wetlands did not get under way until the mid-1920s, after the introduction of the refraction seismograph. By the end of 1926, refraction crews were combing large parts of coastal southeast Texas and southwest Louisiana for shallow salt domes. Most notable was the geophysical campaign of 1925–30 initiated by Geophysical Research Corporation crews for Gulf Oil and Louisiana Land & Exploration/Texaco, which had obtained a huge lease position along the Louisiana coastal plain.[2] A crew would typically rent boats and hire laborers and guides in the small Cajun communities where people traditionally made their living by fishing, shrimping, crabbing, frog hunting, muskrat trapping, salt mining, or harvesting sugar, rice, tobacco, moss, or oysters. A typical Shell Oil seismic crew in the 1930s included ten specially trained seismologists and technicians and six to thirty helpers or laborers hired from the community. Crews would live on quarter boats for ten days while they were on a job and then have four days off. Residents of these insular communities initially looked with suspicion upon the outsiders who came into their midst hauling strange geophysical equipment and large magazines of explosives needed for seismograph exploration. But party chiefs offered relatively good money,

Laying cable for seismic recording at Grand Lake, La., 1949. *Photo courtesy Houston LeJeune.*

which was difficult for available hands to pass up as hardship hit rural economies such as southern Louisiana beginning in the late 1920s.[3]

Where waters were deep and open enough, the outsiders rented boats and mud scows to transport their equipment to desired locations. But in the wooded swamps and thick marsh of the Bayou country, geophysical crews turned to methods and equipment used by muskrat trappers, such as navigation by flat-bottomed pirogues (pronounced pea-rogue), a French adaptation of the canoe. Often, however, thick vegetation prohibited boat traffic, and everything had to be carried by foot after tying up the boats in a nearby inlet. With their pant legs tied tightly to protect against snakes and leeches, laborers would trudge along in waist-deep swamp water dodging cypress roots and saw-toothed palmetto leaves. "Instruments, explosives, pumps and pipe for drilling, cables, and all the other paraphernalia of the seismologist's art must be carried distances often of miles, and at a rate rarely exceeding one mile per hour," wrote a *Shell News* feature from 1939. "These are the longest miles in the U.S.A.! The number of helpers in a crew is generally measured by the difficulties to be overcome in local transportation."[4]

This was suffocating, back-breaking, and dangerous work, especially as exploration techniques changed from the torsion balance (see later in this chapter) to the seismograph and all the heavy instrumentation, equipment, and explosives it entailed. The rewards of this work outweighed the risks for many young men in the Bayou communities. Not only did it offer decent pay but also opportunities for advancement and the acquisition of new technical skills. The companies hired them as surveyors and permit men as well as "pack mules." They applied their familiarity of the local terrain and people to determine lease lines and help the companies acquire permits to explore outside the leases. The indigenous development of the "marsh buggy" in the late 1930s facilitated the penetration by geophysical crews, and behind them drilling and pipeline operations, into the marshes of southern Louisiana.[5]

Drillers faced a host of new challenges as they tried to move rigs from dry land to marshes and bays. Soft, mucky silt in these areas could not tolerate the same kinds of loads that hard-ground soils could. "In these coastal marshes," wrote F. C. Embshoff of the *Shell News*, "where the land is scarcely more than a series of floating dirt rafts insecurely anchored by vegetation, there is nothing solid upon which to build a derrick."[6] Compounding this problem, drilling objectives in southern Louisiana were located at greater depths than farther north, thus requiring more drilling pipe, casing, and heavier equipment. In the marsh, drillers resorted to constructing huge "mats" out of timber upon which they placed derricks, tanks, and boilers. In the open waters of bays and lakes, drawing on experience from places like Lake Caddo and Lake Maracaibo, drillers placed their equipment on planks supported by a foundation of numerous piles driven deep into the silt bottom. Large expenditures of time and money were directed at preparing the location and foundation, constructing heavy board roads, moving in, rigging up and tearing down the derricks and associated equipment, and then hauling them to a new location. For all but a few companies, these expenditures ruled out exploratory drilling. In 1933, however, Texaco made a breakthrough in achieving mobility in wetland drilling by using a barge, the *Giliasso*, equipped with a derrick and drilling equipment, that could be floated and submerged as a stable drilling base, thus eliminating the time and expense of fixed foundations.[7]

Other companies followed Texaco's pioneering example, and by the late 1930s dozens of "floating derricks" could be seen moving through the bayous and newly constructed canals of south

3
Example of barge drilling similar to that at Weeks Island Field in the 1940s. *Photo courtesy Kenneth E. Arnold.*

Louisiana. Shell Oil applied this technique most successfully at Weeks Island, a giant salt dome in Iberia Parish on the Louisiana coast, adjacent to a lush sugarcane plantation and surrounded by swampy marshland.[8] In 1935, Shell acquired a fifty-year lease from the Miles Salt Company on the island and in 1941 took a lease from the state covering 570 acres in the bay and marsh waters.[9] The first hole, Smith-State 1 drilled in late 1941, not only made a significant discovery but broke the record for the deepest oil production at 13,520 feet. Subsequent drilling extended into bay and marsh waters, where the company deployed its special steam-powered rigs mounted on barges. Once on location, crews flooded the ballast tanks until the barge came to rest on mats made from shells or other material on the bottom of the shallow marshes. Shell's experiment with barge drilling achieved notable success. The company placed eight such barge rigs in operation between 1936 and 1948. Rig No. 7, built in 1947, had a 187-foot-tall derrick, the largest in the world at the time.[10]

By the end of 1950, Shell had completed some forty wells that yielded large volumes of high-quality crude. Other companies also picked up leases in the area, but Shell controlled most of the acreage. The Weeks Island field, covering about 1,000 acres, eventually produced over 150 million barrels of oil.[11] One of Shell's most prolific and profitable postwar fields, it played a large role in elevating the company's annual net income from around thirty million dollars just after the war to over ninety million dollars by the early 1950s. "Weeks Island crude is versatile," wrote *Shell News*, "versatile as its source, a high knoll rising from the flat marshes of Cajun country, where salt, sugar and oil are produced for the palate and convenience of millions."[12]

There was another dimension to Weeks Island's versatility. During the Korean War, Shell workers there converted an old salt mine vault into a bomb shelter. In case of a nuclear attack, as Shell oilfield worker James Hebert explained, "the people from New Orleans, the big dogs, would come live there and try to run the plant from this shelter." It was equipped with enough beds, food, and clothing to allow ten people to live for an extended period. Amenities included a telephone, water pump, electric generator, Geiger counter, and, fancifully, showers at the entrances to wash off radioactivity. In the late 1970s, the shelter was converted to an archive for Shell Oil's historical records, some nine hundred drawers full, preserved in the dry, air-conditioned environment of the salt mine.[13]

The marsh drilling and production operations of the 1930s served as a launching pad into the open water of the Gulf. In 1937, the industry took its first tentative step offshore when a joint operation by Pure Oil and Superior Oil installed in fifteen feet of water a wooden platform in the Creole field a mile from shore off southwest Louisiana. After World War II, firms began to venture out deeper. In October 1947, a small independent oil company, Kerr-McGee Oil Industries, drilled the first producing well beyond the sight of land in the Gulf. The well, in eighteen feet of water, 10.5 miles from the Louisiana shore, came in on November 14, 1947, a date often cited as the birth of the modern offshore oil industry. At the same time, Shell Oil was casting its sights farther offshore.

Casting Sights Offshore

After World War II, Shell Oil's top managers immediately grasped the need to refocus on technology in the search for oil. Shell had a lot of ground to make up on its rivals. In the mid-1930s, budget problems had prevented it from competing strongly in the acquisitions and exploration game. In 1949, Max Burns, a geophysicist and a true "crude oil man," assumed the presidency of Shell Oil. During his administration, Burns directed the lion's share of its annual budgets upstream. By 1961, sixty-seven cents of every Shell Oil budget dollar went for exploration and production. The investment paid off handsomely in the decade after World War II, doubling the company's crude oil production and adding reserves at an even greater rate.

The risks of oil exploration and production have always been high. Nobody could tell with great certainty how much oil the earth possessed or where it was located. Time and money spent surveying and drilling often produced only dry holes. Even when discoveries were made, oil's unique physical qualities often led to economically irrational production. Multiple owners producing from the same reservoir had individual incentives to produce as much oil as rapidly as possible, which resulted in the wasting of reservoir pressure or excess production. Furthermore, because fossil fuels are finite and depletable, their production costs increase over the long run. Commercial development ultimately requires access to progressively deeper and more remote reserves.

Public policies in the United States offered tax inducements to oil production and encouraged oil companies to take risks and, simultaneously, brought a measure of order and stability. The percentage depletion allowance adopted during World War I permitted companies to deduct a percentage of the gross value of their oil and gas production from taxable income. The government also ruled that "intangible" drilling costs, such as dry holes, could be expensed against current income, rather than capitalized for recovery through depreciation. In the 1930s, to stabilize markets, conserve oil resources, and prevent excess production, oil-producing states and the federal government worked out an administrative "prorationing" system that restricted oil production among all producers in proportion to their rated capacities. Although critics charged that prorationing inflated both prices (by restricting production to demand) and costs (by reducing output of the more efficient low-cost wells and keeping in operation inefficient, high-cost wells), it nevertheless enabled the industry to expand in a rational manner.[14]

Market conditions after World War II fueled a new national hunt for oil. The lifting of price controls unleashed huge demand pent up by wartime scarcity, sent crude prices soaring, and put pressure on existing supplies. To prepare the company for the new race for reserves, Shell Oil in 1946 instituted important administrative reforms that shifted more responsibilities to the E&P divisions. With national oil demands increasing day by day, Shell was embarking on a vast undertaking that required the most efficient use of personnel.

Top management wanted to simplify communications and encourage greater decision-making by local managers. In 1946, due to the growing scope and volume of operations in the Texas-Gulf Area, A. J. Galloway, vice president for exploration and production east of the Rockies, broke this region into three new areas: Houston, Midland, and New Orleans. The three new regional units were set up like the existing structure of the Mid-Continent Area, which was renamed the Tulsa Area. In July 1947, Shell established its first office headquarters in New Orleans. A staff of two hundred occupied four floors of the Richards Building two blocks off famous Canal Street.[15] A regional staff based in Houston planned and coordinated all activities of the four east-of-the-Rockies areas.[16] Under the new system, the man in the field did not always have to wait twenty-four hours to receive approval for every decision. "Previously, before you could do anything, you had to go all the way to head office," remembered Bob Ferris, a young engineer in charge of a core drilling program in Kansas at the time. After the reform, Ferris could make his own decision on the spot about whether to run pipe into a well or plug it. The new structure saved both time and money. The head office still controlled budget allocations and major decisions, but budgets were revised every month to take into account important developments. The 1946 decentralization sparked new initiative in exploration and production.[17]

A revamped and recharged E&P organization carried Shell into many new oil frontiers, such as the Williston Basin of Montana and North Dakota, Western Canada, and the Four Corners area (where the states of Colorado, Utah, Arizona, and New Mexico meet). Of all the new frontiers, however, offshore Gulf of Mexico offered the most enticing possibilities. It also promised great risks. Variables such as high waves, hurricane-force winds, seabed soil movements, underwater corrosion, and marine transportation presented imposing obstacles to offshore operations. Engineers had no useful models for designing free-standing structures in the open sea. Even if these obstacles could be overcome technically, many doubted that offshore oil would ever be economical. In 1936, Shell's own Houston office had discouraged a seismic team led by Syd Kaufman from shooting offshore near Corpus Christi with the famous last words of "ridiculous" and "we'll never drill out there anyway."[18] Shell and other adventurous companies changed their tune in the late 1940s, hoping they could discover big enough fields to cover the large costs of developing new offshore technology.

What lay behind the decision to confront such tremendous uncertainty? Confined to domestic oil provinces but excluded from the best onshore prospects, Shell Oil looked offshore to improve its declining competitiveness. The company abided by its unofficial understanding with the Group to restrict oil exploration to the United States and not go abroad, where it would be in competition with other Group companies. After World War II, however, Shell did not have a commanding E&P position east of the Rockies. The Mid-Continent's shortage of capital during the Depression meant that Shell lost prime leases and land acquisitions in the Texas-Gulf region to other major companies. Shell had nothing to rival Humble Oil's million-acre King Ranch lease in southwest Texas, or Texaco's infamous Louisiana State Lease 340, which covered the five-island salt dome trend along the coasts of Vermilion, Iberia, St. Mary, and Terrebonne parishes and was acquired through a corrupt bargain with Huey Long's Win-or-Lose Corporation. Shell's big West Coast fields, the company's cash cow in the past, were aging. But the unexplored offshore domain offered some hope of salvation. Success there would go to the companies with the best technology and ability to control costs. On this score, Shell believed it could compete with anyone.[19]

Bellaire Leads the Way

In the early days, petroleum exploration involved a lot of guesswork. Oil seeps provided the most reliable signs of accumulations. Often, prospectors used doodlebugs or other instruments of metaphysical prognostication to hunt for oil. Some superstitions held that drilling sites be kept close to cemeteries or on the right hand forks of creeks. Over time, guesswork gradually gave way to science. Supernatural insight yielded to geological theories of how rock layers were deformed beneath the earth's surface. This knowledge enabled geologists to identify structural traps, or reservoir rock configurations that trap petroleum due to their concavity downward, such as anticlines, faults, and salt domes. In the 1920s, more sophisticated tools, such as the torsion balance and refraction seismograph, replaced the dip needle and magnetic mineral rod in exploration.

These new techniques, however, were less useful in locating petroleum structures buried deep in sedimentary formations. As shallow prospects declined, exploration soon focused on finding nonstructural, stratigraphic, or subtle traps created by variations in subsurface rock formations. Although many classic theories about petroleum geology had been worked out by the end of World War II, companies realized that they needed more sophisticated geological and geophysical approaches to finding obscure oil and gas accumulations. The Bellaire Research Center (BRC) in Houston, opened in 1947, represented Shell's commitment to this endeavor. The declining ratio of U.S. petroleum reserves to production spurred the pursuit of new exploration methods. During the 1950s, BRC scientists contributed important advances in petroleum geology, paleontology, geophysics, and production technology that placed Shell at the forefront of innovation. James Parks, a former carbonate geology researcher with Shell Oil who also worked in research for other companies, called the Bellaire facility "the first, the biggest, and, in my opinion, the best" of all industry E&P research labs.[20]

BRC recruited some of the nation's most accomplished geoscientists. Geophysicist Marion King Hubbert, the associate director of exploration, pioneered theories about hydrodynamics and established the importance of rock properties to exploration concepts. In the mid-1950s, against all prevailing wisdom, Hubbert forecast the peak of U.S. oil production to happen in 1970, an accurate prediction as it turned out (see chapter 7). An accomplished and imposing figure, he spent years battling his detractors until he was finally proven to be a prophet. When engaged in technical arguments, he could become temperamental and belligerent. This gave rise to the old cliché around the lab, "that Hubbert is a bastard, but at least he's *our* bastard."[21]

Another memorable figure at Bellaire was W. S. "Doc" Adkins. A world-famous geologist and author of the Mesozoic chapter of *The Geology of Texas,* Adkins helped develop the fundamentals of stratigraphy—the study of the origins, composition, and distribution of rock strata. He became a legend at Bellaire, both for his keen insight into geology and for his eccentric behavior. "Doc would make out his expense account with a piece of chalk on the top of the desk and then call the accounting people down to copy it," remembered R. H. (Bob) Nanz, a Ph.D. from the University of Chicago and one of the first geologists hired in the new lab. The absent-minded Adkins often lost track of time and had a habit of working late into the night on interesting problems. Locked in the research center one night after hours, he broke the plate glass door to get out, leaving a note that said: "I owe you one plate glass window," signed W. S. Adkins. "Many great ones are strange; few strange ones are great," said Nanz. "Adkins was one of the great strange ones."[22]

Carbonate geology was one area of strength in Shell E&P research. Robert Dunham, another colorful and argumentative character, was a pioneer in this subfield, which was largely born in oil company exploration labs. He extended basic concepts from work on clastic rocks to carbonate rocks and introduced a set of classification terms still in use today. "Students use published classifications," said Jerry Lucia, a colleague of Dunham's at Shell. "Real professionals make up their own." Robert Ginsburg led field research into carbonate basins, which were a major objective of Shell exploration in the 1950s, setting up a hardship post in Miami and organizing trips to the Florida Keys for Shell geologists. One star researcher in carbonate geology who came out of this group was James Lee Wilson. During a wide-ranging career in both Shell Oil and academia, Wilson did pioneering work on carbonates in Texas, New Mexico, and the Williston Basin, which not only generated interest in Williston, where Shell became a major producer in the 1950s, but in basin analysis itself. His 1975 textbook, *Carbonate Facies in Geologic History,* is still regarded as the "carbonate bible." In 2002, these contributions earned Wilson the American Association of Petroleum Geologist's (AAPG) highest award, the Sidney Powers medal.[23]

The area of research that had the greatest impact on Shell Oil's E&P operations was the geology of sedimentary rocks. S. W. "Shep" Lowman, a stratigrapher and paleontologist, brought valuable expertise in Gulf Coast Tertiary rocks. Beginning in 1939, Lowman led a major operational research project, based in the bowels of the M&M (Merchants and Manufacturers) building, the largest in downtown Houston, to document fossiliferous marine transgressions observed in wildcat drill cores and mark the significant fossil extinction events for the Miocene sedimentary zone ("M" zone) in the Gulf Coast region. This project would prove extremely valuable in Shell Oil's offshore explorations.[24]

Meanwhile, Rufus LeBlanc conducted another renowned field seminar. A self-styled "barefoot Cajun boy from Bayou Tigre, Looziana," LeBlanc studied river geology and single-handedly developed the concepts of mixed-rock sedimentation. "Just a fancy word for mud, silt, sand, and gravel," he explained. After joining Shell in 1948, LeBlanc collaborated with Bob Nanz to model how sands had been deposited and shaped. In his field seminars, LeBlanc tutored a whole generation of geologists, geophysicists, and engineers at Shell. "His head in the sand proves Shell's head is out," wrote *Forbes* magazine in 1952. The sandman's down-to-earth style endeared him to students and colleagues alike. "LeBlancisms" such as "never make the map bigger than the boss's desk" became part of Bellaire lore. In 1988 the AAPG presented LeBlanc with its Powers Award.[25]

Shell Oil's work in understanding the origin and deposition of sedimentary rocks—particularly the sandstones and carbonates that would become reservoirs for oil and gas—had a considerable impact on exploration and production geology and on the company's ability to predict the presence of reservoirs and their probable distribution. Even after sophisticated geophysical techniques were developed in the 1960s and 1970s to predict the presence of hydrocarbons, notably in the Gulf of Mexico (see chapter 5), understanding the depositional origin of the reservoir was essential to improving the quality of the prediction. In all, as Miner Long, a former chief geologist at Shell Oil, explained, "the research that Shell was doing at that time in BRC was really groundbreaking and important in training our geologists how to examine and interpret those sedimentary rocks."[26]

Once geologists determined the most promising areas to drill, geophysicists had to define those areas. Shell Oil inherited sophisticated geophysical tools and techniques from the Group. Royal Dutch/Shell was one of the earliest adopters of the torsion balance, a delicate instrument developed after World War I to map the density of rock formations close to the surface. It proved useful for detecting buried salt domes, which were often associated with oil traps. Even more useful was the refraction seismograph, introduced into the United States by Marland Oil Company in 1924. The refraction seismograph recorded the time it took for sound waves to travel between a sound source and receiver, refracting off subsurface layers. A compact formation such as a salt dome would transmit sound waves at a much faster rate than the surrounding rock, in effect refracting them much as a prism refracts a light ray. Waves thus refracted would arrive at the seismometer in a very short time. By setting off and recording explosions at various points in the same area, geophysicists could gather enough data to map large structural features beneath the surface. The refraction seismograph method proved very reliable for locating salt domes of the kind that were common on the Gulf Coast.[27] In the mid-

1920s, these new geophysical techniques helped Shell subsidiary Roxana Petroleum find oil along the Gulf Coast of Louisiana and in the Permian Basin of West Texas.[28]

To keep abreast of these significant developments in geophysics, Shell Oil in 1936 established at Bellaire a lab for geophysical research, which thrived after the war in connection with the larger BRC research center. A group of highly trained technical men assembled under Frank Goldstone, chief geophysicist, did much work on improving geophysical instruments. The workshop facilities at the lab were directed by German scientist Eugen Merten, a classic inventor who employed common materials and homemade gadgets to solve technical problems. Under Merten, Shell built its own seismometers using prototype mechanical and optical devices. Dr. Merten invented as he talked, on and off duty, whether it was adapting seismic instruments to marine operations or designing a new surgical tool for his ophthalmologist. On Halloween, ringing the doorbell at the Merten home triggered a flash of electronic lightening that dazzled trick-or-treaters. While his pranks brought levity to the lab, Merten's inventiveness helped Shell develop theories about sound waves that led to better interpretation of seismic data. Shell's seismic capabilities were among the best in the industry.[29]

By 1930, the reflection seismic method had begun to supplant the refraction method as the primary tool for geophysical exploration. Reflecting sound waves off the interface between a "slow" and "fast" subsurface rock, much like the echo off a wall in a canyon, reflection seismic obtained more detailed information on underground rock structures than the refraction method could. It allowed geophysicists to plot more accurately the shape and depth of the reflecting layers. The reflection seismograph permitted the extension of seismic work to areas other than the salt-dome fields of the Gulf Coast, at that time the only region where the refraction seismograph had produced results in the United States. Beginning in the mid-1930s, new technologies for processing seismic records were introduced. California inventor and geophysicist Frank Rieber developed the "sonograph," based on the technology used in early talking motion pictures, which recorded seismic traces as sound tracks and reproduced them in phased combinations and through filters that reduced interference noises. In the mid-1950s, several developments combined to enhance greatly the acquisition and processing of reflection seismic data. The introduction of continuous velocity well logs (sonic logs), in which sound velocity measurements within a well were correlated with rock density, gave greater insight into the origin of seismic reflections. Of major importance, in 1955, was the adoption of magnetic tape for recording seismic sound waves (previously all calculations and interpretations were made on the original paper records acquired during seismic survey). Cross sections prepared from analog magnetic-tape playback, commercially developed in 1958, could include desired "filters" to adjust for time delays caused by surface effects and path geometry. Most importantly, magnetic-tape playback provided a means for economically applying the "common depth point" (CDP) or "horizontal stacking" method of shooting.[30] CDP stacking, in which tapes of individually corrected traces were combined or "composited," vastly improved the acquisition of seismic data by enhancing primary reflections and filtering out unwanted multiple reflections or noise. Invented by Harry Mayne of Petty-Ray Geophysical, it revolutionized the shooting and processing of seismic data.[31] Although key early innovations in reflection seismic came from specialty companies outside of Shell Oil, Shell was an aggressive buyer and licenser of new geophysical technologies, and BRC would go on to pioneer conceptual and practical refinements to exploration seismology.

Research on the production side of E&P, headed initially by J. P. Murphy, made equally impressive contributions. Projects on drilling fluids, pore and capillary pressures, and porosity and permeability measurements, among others, actually began at Shell's Research Laboratory in Emeryville in the mid-1930s under the direction of Dr. Alfred Loomis. In 1935–36, Bouwe Dykstra, chief exploitation engineer in the Mid-Continent Area, established an exploitation engineering group in Tulsa that included Shell's first core analysis laboratory. Working in the Petroleum Engineering Building of Tulsa University, this group undertook Shell's first water flooding experiments on properties in northeastern Oklahoma. In the late 1930s, Dutch engineer A. F. (Toni) van Everdingen, who originally worked in Dykstra's group in Tulsa, took charge of an applied research group in Houston and expanded on his earlier work with Dykstra in an effort to develop mathematical expressions for describing the flow of hydrocarbon fluids in porous rock. Although van Everdingen's studies were hampered by the limited ability of hand calculators to solve the large equations required, they nevertheless made a monumental contribution to the early understanding of reservoir behavior.[32]

Bellaire researchers carried on van Everdingen's work and

added a degree of sophistication to the mathematics of the reservoir behavior problem. Charlie Matthews, a Ph.D. graduate of Rice University who joined Bellaire in 1948, initiated studies to analyze flows and transient pressure behavior in underground oil and gas reservoirs. The timely advent of better and faster computing equipment enabled Matthews and others to develop mathematical models of reservoir performance that became essential to supplemental recovery operations.[33] Understanding reservoirs and accurately estimating the ultimate recovery potential of any oilfield would be critical as Shell Oil moved into the more costly offshore environment where revenues would have to cover the cost of expensive production facilities.

Using newly developed well-logging tools that took downhole acoustic and electrical measurements, exploitation engineer Gus Archie pioneered ways of quantifying these measurements and thus determining the amount of oil and gas in rock formations from electric logs.[34] Archie's explanation of the physics of these measurements, first published in 1942 and verified at a test well near Elk City, Oklahoma, in 1947, eventually achieved fame as the "Archie Formula" or "Archie Equation"—although it was not a unique "equation" per se, and Archie never used the phrase himself. Even so, Archie strove to develop his ideas into a system for determining reservoir characteristics from rock properties, and he founded a new branch of petroleum technology, which he named "petrophysics," or the physics of rocks. Engineers and scientists trained or influenced by Archie went on to develop this field of study and patent well-logging techniques for the company.[35] Modest as well as ingenious, Archie was regarded by many of his peers as "Shell's all-time best engineer." Progress in other areas of production research, such as production chemistry and drilling engineering, rounded out Shell's broadening technical capabilities in exploration and production and positioned the company at the forefront of the industry in attacking the new offshore frontier.

Big Strikes in Shallow Water

The large quantities of sweet crude found at Weeks Island tempted Shell to look out into the Gulf. Shell Oil's top executives, like many others in the industry, were convinced that oil reserves did not stop at the shoreline. As the State of Louisiana began offering leases for submerged lands off its coast in early 1946, Shell anticipated the opening of a whole new frontier. Hank Bloemgarten,

Shell Oil's vice president for production, and A. J. (Alan) Galloway, vice president for exploration, traveled to London early that year to consult with Shell Oil's chairman and other inside directors about budgeting for offshore exploration. They knew that it would cost much more to find and produce oil offshore than onshore but were not certain how much. "We spent the five days on the *Queen Mary* cooking up the figures," confessed Galloway. "We really produced them out of the blue sky." They had confidence, though, that the figures were "in the ballpark." At the meeting in London, Bloemgarten and Galloway made a straightforward proposal: Shell should be prepared to spend two million dollars to find out if offshore were any good and, if it was, at least twenty million dollars to go forward. If the directors did not want to spend that kind of money, Galloway told them, then they should forget about it. "They like a clear-cut proposition like that, one that's big and imaginative, better than they do nibbling away at the little things." Group officials endorsed the proposal, which included a promise that Shell Oil would obtain seismic data before any leases were purchased.[36]

In March 1946, the company launched its first offshore seismic crew, Party 88, from the docks at Grand Isle, Louisiana. Headed by A. B. Cunningham, Party 88 consisted of three leased shrimp boats outfitted for seismic work. To convert the boats *Ramona Mae* and *Utah* for shooting, the party crew scrubbed the foul shrimp odor out of the wood interiors and built storage space for the dynamite to be used as charges. On the instrument (recording) boat, the *St. Theresa*, they built makeshift plywood cabinets to shelter the instruments from the elements. With no radar or accurate means of navigation, the crew designed a crude system to locate and position these vessels. The mast of each was equipped with a rotating, reflective sphere covered with mirrors. Spotters stationed at six-mile intervals along the beach sat atop large wooden tripods to take readings and make geometric (triangular) calculations, communicating with the boats via two-way radio.[37]

Given the state of technology at the time, it was difficult to get good seismic reflections from the soft, silty mud in the water around the mouth of the river. Distortions caused by the sound signal reverberating off the ocean floor also impaired the collection of reflection data. So crews concentrated on shooting refraction fans in search of salt domes. Before the invention of the continuous drag cable, they anchored the instrument boat and lowered the seismometers (recording devices) overboard. Meanwhile, the shooting boats would fan out to a radius of six miles and detonate dynamite charges at one-mile points along the perimeter. If the water was shallow enough, the dynamite would be dropped into pipes drilled into the seafloor. In deeper water, crews would bundle up fifty pounds of dynamite, insert a cap, toss the charge overboard, let it sink to the bottom, and then detonate it.[38]

This was dangerous business, especially when crews tried to speed up operations by making up more than one charge at a time. "When the boats are moving along at, say, six or seven knots, and you are trying to do all of this, you do not have a lot of time," remembered Aubrey Bassett, who worked on some of Shell Oil's earliest offshore seismic crews. "Two hundred feet goes by in a hurry and you are just back there wildly putting all of this stuff together." Although Shell Oil's offshore seismic operations had a good safety record, in 1948 a dynamite explosion on board a Party 88 boat killed two people.[39]

Aside from the physical dangers, offshore exploration faced other challenges. The need to triangulate from landmarks to navigate and position the boats limited how far surveying could move offshore. Rough weather frequently reduced or halted working time and ran up costs. Even moderately bad weather could shut down an entire operation. When weather permitted, recording boats often encountered difficulties with the cable snagging on debris and breaking. Another major problem was the "bubble effect" caused by the underwater explosions. An air bubble from an exploded charge would rise to the surface and then collapse and expand, transmitting distorted sounds to the seismometers. Furthermore, dynamite charges in the seabed or dropped to the bottom also ran the risk of blowing apart oyster beds in those shallow waters. No wonder fishermen were initially hostile to offshore seismic operations.[40]

Thus, early seismic prospecting offshore was slow, tedious, dangerous, and not entirely useful. But improvements came steadily. In 1947, Shell crews finally solved the bubble problem by floating the charges about four feet below the water, rather than letting them sink to the bottom. They first used driftwood and syrup cans as floats, before advancing to balloons and "turkey bags" inflated by air compressors. To address the cable problem, Dr. Eugen Merten at Shell Oil's Bellaire laboratory devised a sled-type float on which to mount the seismometers. This minimized the snagging and breaking of the cables, and it improved seismic records by reducing the noise that traveled along the sea bottom when a charge was detonated. Also in 1947, Shell upgraded its capabilities by replacing the shrimp boats with two modified World War II minesweepers, the *Ora Mae* and the *Edna S.*[41]

Shell expanded its seismic fleet by acquiring other war-surplus rescue boats. Larger boats could carry more dynamite and operate in heavier seas. At the same time, the company introduced Shoran (Short-Range Navigation), a system developed during the war in which two radio signals transmitted by a sender and returned by two radio location stations allowed for greater accuracy and flexibility in boat positioning and wider ranging navigation. Finally, Shell deployed a new seismic prospecting method called "continuous reflection profiling." By this method, an instrument boat towed a cable 3,200 feet long with sixteen sections, and at each 200-foot interval, gimbaled seismometers took readings from shots detonated from a shooting boat 100 feet away. Continuous reflection profiling allowed marine parties to keep up a continuous schedule covering as much as twenty-five miles per day, with crews working on rotating shifts of ten days on and five days off.[42] Although the cost of outfitting a marine seismic crew remained greater than a land crew, the increasingly high rate of data gathering on the water eventually yielded much lower overall costs. By the mid-1950s, it was not unusual for marine crews to cover sixty-five miles per day, which was quite impressive when compared to fifty miles of continuous coverage per *month* for a respectable dry-land seismic party working under good conditions.[43]

Of course, all these advances in offshore seismic operations came after April 1947, when the State of Louisiana offered leases in the shallow water near the mouth of the Mississippi River. The money bid for a lease constituted a cash "bonus" (or "signature bonus") offered in addition to rental and royalty payments (12.5 percent, less than the standard onshore) simply to obtain the lease. Public lease auctions at the state sales accepted only sealed bids, so it was desirable to obtain as much information as possible, both on the properties up for lease and on the competition. But because of the early problems encountered by marine crews, Shell Oil did not have all the desired seismic data to bring to the sale. A. J. Galloway and his New Orleans area manager, Ernest G. Robinson, suspected, however, that leases could be purchased rather cheaply for the untested offshore acreage.

The exploration department had some gravity (torsion balance) indications of big salt dome structures around the Mississippi delta, but they did not have good seismic data to back them

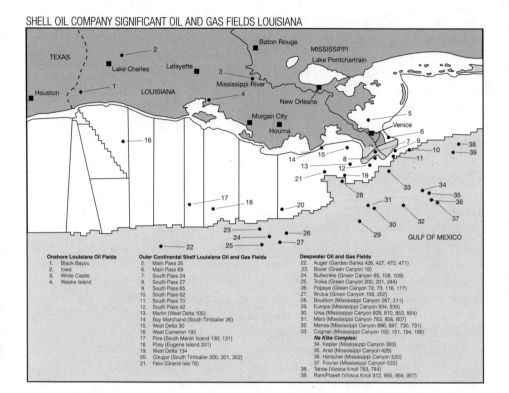

SHELL OIL COMPANY SIGNIFICANT OIL AND GAS FIELDS LOUISIANA

Onshore Louisiana Oil Fields	Outer Continental Shelf Louisiana Oil and Gas Fields	Deepwater Oil and Gas Fields
1. Black Bayou	5. Main Pass 35	22. Auger (Garden Banks 426, 427, 470, 471)
2. Iowa	6. Main Pass 69	23. Boxer (Green Canyon 19)
3. White Castle	7. South Pass 24	24. Bullwinkle (Green Canyon 65, 108, 109)
4. Weeks Island	8. South Pass 27	25. Troika (Green Canyon 200, 201, 244)
	9. South Pass 65	26. Popeye (Green Canyon 72, 73, 116, 117)
	10. South Pass 62	27. Brutus (Green Canyon 158, 202)
	11. South Pass 70	28. Bourbon (Mississippi Canyon 267, 311)
	12. South Pass 42	29. Europa (Mississippi Canyon 934, 935)
	13. Marlin (West Delta 105)	30. Ursa (Mississippi Canyon 809, 810, 853, 854)
	14. Bay Marchand (South Timbalier 26)	31. Mars (Mississippi Canyon 763, 806, 807)
	15. West Delta 30	32. Mensa (Mississippi Canyon 686, 687, 730, 731)
	16. West Cameron 192	33. Cognac (Mississippi Canyon 150, 151, 194, 195)
	17. Pine (South Marsh Island 130, 131)	**Na Kika Complex:**
	18. Posy (Eugene Island 331)	34. Kepler (Mississippi Canyon 383)
	19. West Delta 134	35. Ariel (Mississippi Canyon 429)
	20. Cougar (South Timbalier 300, 301, 302)	36. Herschel (Mississippi Canyon 520)
	21. Felix (Grand Isle 76)	37. Fourier (Mississippi Canyon 522)
		38. Tahoe (Viosca Knoll 783, 784)
		39. Ram/Powell (Viosca Knoll 912, 955, 956, 957)

4

Map of significant Shell Oil fields offshore Louisiana.

Courtesy Ron Martinez.

up. Nobody had any idea yet, as well, whether drilling would be feasible in the soft mud and currents of the Mississippi River's mouth. In violation of their pledge to the Group directors, Galloway and Robinson placed bids in the April 1947 sale anyway. They simply could not pass up the opportunity to acquire property at such a low price. In an area called East Bay or South Pass at the south tip of the Mississippi delta, Shell picked up a total of 65,000 acres, including three 5,000-acre leases covering South Pass Blocks 12, 23, and 24 for $4.27 per acre, an adjoining lease on South Pass Block 11 for $2.06 per acre, and South Pass Blocks 27 and 28. They also acquired 34,500 acres for similar prices farther east, which included Block 69 in the Main Pass area. By the end of the summer, Shell had acquired a total of 170,000 acres in leases for nearly $1.5 million in bonus bids (approximately $8.80 per acre). Both the South Pass and Main Pass areas would soon establish a lasting foundation for Shell offshore in the Gulf. "It just shows that when you're responsible for a job," explained Galloway, "every now and then you have to break the rules."[44]

In the summer of 1947, some 125 Shell Oil employees from Houston moved their families to New Orleans to help ramp up the company's new offshore enterprise. They colonized four floors of the Richards Building as well as neighboring offices while their wives, joked the Shell magazine, *The Pecten,* coped with the "inevitable 'how-can-I-ever-make-these-chintz-curtains-fit-this other-type-of-windows' problems."[45] The adjustment took some time, and the company did not begin to drill its new offshore leases until late 1948, after improvements to marine seismic operations yielded better clues as to where oil might be found and a plan had been worked out for drilling in ten-foot water depths, an unprecedented step for Shell Oil.

Main Pass 69 received less river discharge than South Pass 24, so it was the first area tested, using techniques adapted from marsh operations. The drillers filled in around the well locations with oyster shells and then sank "submergible" drilling barges, like those used at Weeks Island, on these fills. To provide protection from the waves of the open Gulf, Shell purchased old war-surplus LSMs (Landing Ship Medium, a vessel used to transport tanks and heavy equipment) to deploy as floating breakwaters that doubled as storage for mud tanks and other equipment. After completing the wells, workers refloated the barges and LSMs and moved them to the next location.[46]

The second well at Main Pass 69 discovered oil. But success came with a frightful warning about the hazards of offshore operations when, in March 1949, Main Pass Well No. 23 blew out. Caused by the fracturing of the formation under the surface pipe, the blowout sent gas and flames roaring out of the well. Fortunately, only one man was injured and none killed. "Gulf winds whipped fiery fronds through the derrick girders as the drilling crew hastily abandoned ship," reported *Shell News.*[47] Howard Shatto, a young engineer working on the company's first diesel electric rig at the time, remembered the event very clearly: "My boss, John Pittman, came running down the hall of the New Orleans office. He said, 'Get out to the airport, get the airplane and pick up the injured man!' We took off at New Orleans airport and 85 miles south we saw this pillar of smoke. We headed straight for that. The California Company had picked up the fellow who was slightly burned, not seriously." The blazing, crumpled rig even made the cover of *Life* magazine. Firefighters valiantly battled the flames for twenty-one days before the well cratered badly, completely engulfing the drilling rig and two LSM breakwaters. "After things settled down, they went in with a sounding line," said Shatto. "There was a hole just over 100 feet deep in the center, almost a perfect cone shape 1,500 feet in diameter. And there was no sign of the rig or the ships." Shifting formations below the crater eventually bridged the hole and killed the well. This calamitous ordeal taught Shell a valuable lesson about operating offshore: only the most advanced drilling techniques, equipment, and safety measures could contend with such an environment. The lesson was reinforced later that year when a small hurricane blew through the area, damaging many rigs and barges.[48]

Shell crews soldiered on to develop the field. Producing and transporting offshore oil presented new challenges. In an inland field, pipelines and rail tank cars made transporting oil relatively routine. But handling oil produced over a mile offshore was another matter. Shell engineers chose barges to carry oil from small "well jacket" or "well protector" platforms. Still, storms and rough water played havoc with barge traffic, often halting it all together. The 1949 hurricane made this painfully clear. As the number of wells and capacity for production increased, Shell found it would eventually need a more permanent method of handling the oil.

The South Pass area would be the proving ground for new production techniques. Preoccupied with the problems at Main Pass 69, improving seismic data, and studying the unique conditions of the delta, Shell did not drill its first well on the South Pass leases until the spring of 1950. These leases were in the bay between South Pass and Southwest Pass, the deepest and longest finger of the three river outlets at the tip of the delta and the main entrance to the Port of New Orleans. "There was a lot of doubt that you could get drilling rigs to work down there," recalled Frank Poorman, a Shell Oil production engineer. "Nobody knew exactly how to do that."[49]

Exploratory drilling took place following the Main Pass example using a submersible barge and floating breakwaters. The first well, No. 1 State 1008 in South Pass Block 24 completed on April 18, 1950, hit oil.[50] Then the second and third also struck pay. It became clear that cargo barges could not handle the apparently large amount of oil that could be anticipated from the field. After intense study about whether to base production facilities onshore or offshore, Shell engineers decided to build production platforms in open water and connect them by submerged pipeline to a tank farm onshore. After a year and half of drafting and planning, following a design pioneered by the California Company at

South Pass 24 field
production plat-
forms with seismic
crew in foreground,
early 1950s.
*Photo courtesy Morgan
City Archives.*

Bay Marchand, they installed three production platforms linked
to a smaller central one by cat walks and built upon standardized,
pre-stressed concrete pilings driven into the sea floor. The pilings
supported the platforms at a height of twenty-seven feet above
the water, while a network of small pipelines tied the wells to the
production platforms. Another small line, buried on the bay floor,
ran from the production platform to the land-based storage terminal two and a half miles away.[51]

Shell Oil's development of the South Pass 24 field, which
included blocks 11 and 23, was a landmark in the history of offshore Louisiana. The first 19 wells drilled hit oil. The field was a
giant. Containing 671 million barrels of original reserves, it was
the largest discovered offshore for many years and a bonanza for
the company. Not bad for leases that cost roughly four dollars per
acre. During the 1950s, after the resolution of the tidelands controversy (see below), South Pass 24 became the fastest-growing
Louisiana field, both in number of producing wells and in crude
production. Shell's well numbers climbed to 125 by mid-1956,
with daily production averaging 40,000 barrels per day (by 1968,
Shell had produced 200 million barrels from the field). "When I
went to work in production in East Bay," recalled Cliff Hernandez, a former roughneck with Shell, "as far as you could look, in
any direction, they had wells." Calco and Texaco also had wells
in the field, but Shell was the dominant producer. Said Explora-

tion Vice President A. J. Galloway about South Pass 24: "We just lucked into something big."[52]

The field was not an undeserved reward, however. Shell Oil had weighed the risk, developed its marine capabilities, and capitalized on opportunity. But the discoveries at Main Pass 69 and South Pass 24, as well as numerous others made by competitors during 1949–50, raised the stakes for the 1950 state lease auction. Shell, Calco, Kerr-McGee, and others had proven that it was possible to drill and operate in open water and actually make money out there. With results from its seismic tests in 1949, the New Orleans E&P group identified several desirable tracts on which they wanted to place bids, but they still had to convince the head office in New York of their value.

Bouwe Dykstra, who had moved to New Orleans in 1949 as general manager of the area office, led the charge. Dykstra was an experienced and well-traveled veteran of Royal Dutch/Shell around the world and a strong and dynamic character. A tall, broad, blue-eyed Dutchman "whose capacity for work is equal to one far younger, and whose sharp wit knows no master," wrote an *Oil and Gas Journal* feature in 1959, he had joined the Group in 1926 as a geologist. He worked in Romania and Sumatra before transferring to the United States where he became chief exploitation engineer first in Tulsa and then Los Angeles. He then spent six years during and after the war as technical manager with Shell of Colombia. In Bogotá, Dykstra became well-acquainted with Max Burns and upon returning to the United States in 1946 used his corporate connections and forceful personality to climb the management ranks and get himself installed in New Orleans where the hottest action was.[53]

Dykstra was a good cardplayer—gin rummy was his game— and a shrewd judge of the competition. For the 1950 lease sale, he believed Shell Oil would have to put up some "real dough" to expand its position offshore. With geophysical parties scouring the Gulf and more companies gearing up for the sale, leases could no longer be obtained for peanuts. Before the sale, Dykstra and his exploration manager, a Dutch compatriot named Freddy Oudt, went to New York to discuss lease bids with Max Burns and top management. They suggested that Shell ought to bid $650,000 for a tract in Main Pass Block 35 (Battledore Reef). The figure— which pales in comparison to sums paid later—nonetheless startled everyone in the room. After much discussion, Burns approved the proposal. Still slightly uncomfortable with the big dollar figure, however, the president switched the numbers six and five in the proposed bid to make it $560,000 instead of $650,000. As it turned out, Shell beat out Calco by just a few thousand dollars.[54]

The lease proved to be another winner, the first of many that Dykstra would acquire for the company. By the end of 1953, Shell had more than twenty wells producing in the field and had installed another stilted central production platform, similar to the one in South Pass 24, but with an even higher deck, thirty-four feet above the water level, to avoid storm waves that had been measured at twenty-six feet in that area. Main Pass 35 joined Main Pass 69 and South Pass 24 as three of the four most prolific fields in the Gulf of Mexico during this period. "Those fields were where we really made the bucks," explained Bob Ferris, a division production engineer who later became vice president of production. "I remember Dykstra, at one point there . . . said 'Oh, just drill it. Don't bother doing any geology.'" Between 1947 and 1954, only Calco surpassed Shell in production (19.6 million barrels to 7.3 million barrels), largely due to a high output from its Bay Marchand field.[55]

The unusual islands on stilts exemplified the resourceful improvisation needed to succeed in the new offshore environment. Shell and other companies designed and installed sturdy, durable, and safe structures as inexpensively as possible. But they were stepping into the unknown. They lacked knowledge and theories about the dynamics of waves, wind, and the seabed. The thirty-four-foot-high deck on the Main Pass 35 production platform was considered high for the time, but subsequent information made it look dangerously low. For engineers and E&P operators eager to study these problems, the Delta Division of the New Orleans Area Office was the place to be. New Orleans assigned about a half dozen young engineers to design the flowlines and production structures for these early platforms. They usually rushed their designs from the drawing board to the drilling rig. Keith Doig, one of those engineers, remembered working out of a little office above a pawn shop in the town of Westwego across the river from New Orleans. "At that time it was probably one of the most active places in the Shell Oil Company," said Doig.[56]

The Mississippi Delta fields transformed Shell Oil, amply rewarding its big push for new reserves and setting the company's sights in a new direction. Although the Gulf remained mostly uncharted territory for the oil industry, the discoveries in the Delta emboldened Shell and a few other intrepid oil firms to move even deeper. Before they could do so, however, they had to await resolution of a politically charged controversy over ownership of submerged lands on the U.S. continental shelf.

Taking the Plunge

The "Tidelands Controversy" originated in the late 1930s when the discovery of oil beneath the sea raised the question of whether the U.S. government or individual states had authority to lease offshore acreage. "Tidelands" was a misnomer, since nobody disputed that the states held title to shoreline terrain between high tide and low tide. At issue was control of underwater lands beyond tidal limits. Anticipating worldwide interest in ocean resources, President Truman in September 1945 proclaimed federal authority over the subsoil of the continental shelf contiguous to the land area of the United States. California, Texas, and Louisiana defied this proclamation and continued to lease offshore land. Citing historical territorial claims, Texas asserted its jurisdiction to three leagues (10.4 miles) offshore while Louisiana declared ownership of lands within 27 miles of its coastline, which the state drew as far out as possible from the marshy waters of southern Louisiana.[57]

The U.S. Justice Department responded with a series of suits against the states. U.S. Supreme Court decisions in 1947 against California and in 1950 against Louisiana and Texas ratified the Justice position, declaring that the federal government possessed "paramount rights" that transcended the states' rights of ownership. The decisions raised legal questions about much of the land already leased by the states. As royalties were impounded in Texas, California, and Louisiana, a national political drama unfolded. On December 11, 1950, the Supreme Court issued a supplemental decree prohibiting further offshore operations without the authority of the United States. The decree prompted the Department of the Interior to ban new explorations but permit wells being drilled to continue to completion. Offshore leasing and exploration came to a halt for three years, as Congress held seemingly endless rounds of hearings and the 1952 presidential candidates postured around the issue.[58]

The controversy slowed Shell Oil's plunge offshore. It did not affect Shell Oil's lease in Main Pass 35, for that was located in Breton Sound, recognized by the federal government as inland water. Development proceeded relatively rapidly there. But the tidelands conflict did affect Main Pass 69 and South Pass 24, claimed by both the United States and the State of Louisiana.

Shell Oil received special permission from the U.S. Secretary of Interior to drill wells in the South Pass field "in accord with sound conservation policies" and to undertake seismic exploration, so long as charges were detonated in the water and not on the submerged lands themselves. But exploratory drilling on Shell Oil's other leases in the area was suspended. Sympathies from Shell and most oil companies, especially the independents, lay with the states. Accustomed to dealing with state regulators, these firms were not eager to answer to less friendly officials in Washington.[59] After months of rancorous debate, Congress finally passed compromise legislation signed in May 1953 by newly elected Pres. Dwight Eisenhower. The Submerged Lands Act validated all state leases awarded before the Supreme Court decisions and reserved to the states all land within three nautical miles of their shore. The Outer Continental Shelf Lands Act (OCSLA), passed two months later, placed all offshore lands beyond the three-mile limit under federal jurisdiction and authorized the Department of Interior to lease tracts under competitive bidding.[60] Louisiana and Texas tried to block implementation of the OCSLA, which delayed several federal lease sales in the late 1950s. In 1960, the Supreme Court ruled that Louisiana's jurisdiction was limited to three nautical miles from the coastline but awarded Texas and the Gulf Coast of Florida a boundary out to three leagues based on those states' historical claims. Litigation continued for years, however, over determining the exact location of Louisiana's coastline and the division of escrowed revenues between the state and the federal government.[61]

The establishment of a legal and jurisdictional framework for leasing in 1954 nevertheless initiated a new wave of exploration and development offshore, where Shell and other companies had enjoyed incredible success in finding oil.[62] During the suspension of full-scale operations, companies such as Shell had time to collect new seismic data, review old data, and refine their understanding of prospective acreage, which left them eager to renew the quest. Shell's geological and geophysical capabilities kept it ahead of the pack. Even as they competed fiercely in exploration, however, the pioneering companies also recognized a common objective, and they cooperated in gathering data about marine operations. In 1948, twenty oil companies holding offshore leases formed the Offshore Operators Committee (OOC) to promote systematic data accumulation and sharing. The industry gained knowledge from oceanographers and meteorologists about wave dynamics and weather patterns. Beginning in 1951, the American Petroleum Institute's Project 51 conducted a program of core drilling, aerial mapping, and seismic surveys to evaluate seafloor composition and delineate the sedimentary boundaries in the Gulf.[63]

Buoyed by success and armed with a better understanding of the marine environment, Shell expanded aggressively in the Gulf. In 1953–54, production crews carried out extensions to the company's three main shallow water fields, and drillers made new discoveries in deeper Louisiana waters on state leases obtained in 1947 but held up by the Tidelands controversy. New Orleans area production manager John W. Pittman supervised the ambitious and complex drilling and production program. A former all-state football tackle from McCamey, Texas, in the days when high school teams were comprised of oil-field workers, Pittman started with Shell Oil in 1934 as a roughneck. He eventually graduated from the company's engineering training program to become one of Shell's early reservoir engineers. He left for the war a week after Pearl Harbor and won three bronze stars and a purple heart commanding a 91st Division infantry battalion through Italy. Pittman returned to Shell as a production superintendent in south Louisiana and rose to production manager in New Orleans in 1953. He exploited the skills he acquired as a military commander in mobilizing a workforce to tackle the challenge of offshore production. "People who stay in this business must conquer things," he once said.[64]

Pittman deployed his troops effectively to help create a unique and bustling work environment offshore. Gangs of roughnecks, roustabouts, derrickmen, drillers, and engineers labored intensively on twelve-hour shifts, seven-days-on and seven-days-off, on barge rigs and platforms to drill new wells and bring in offshore production. Because workers had to commute to their jobs, first in crew boats and later in helicopters, the time and costs involved led to this system of concentrated work schedules, which in other times and places was expanded to fourteen or twenty-one days on and off. Such schedules strained marriages and families. But it also permitted the geographic dispersal of worker residences often hundreds of miles from work sites and kept intact established rural communities, allowing people to continue participating in traditional activities of farming, trapping, and fishing. The pay was good, and many workers remained with the company for their whole working lives. In the mid-1950s, Keith Viater remembered receiving an attractive three dollars per hour plus benefits

6
Shell Oil's East Bay Central Facilities with Shell service station and crew boats in foreground, 1950s. *Photo courtesy Ken Arnold.*

and Provident Fund profit-sharing after a year.[65] Cliff Hernandez worked thirty-four years for Shell Oil, much of the time in drilling and production in the East Bay, and retired in 1987 with a full pension worth $750,000. For young single men, the work paid extremely well and offered opportunities for advancement and the acquisition of new technical skills. They came from all over the region—Mississippi, Alabama, Texas, Arkansas, in addition to Louisiana—to hire on offshore. Some worked for contract companies, and others were employed directly by Shell. To accommodate many of these workers, Shell set up a camp at its central facilities on the Southwest Pass channel, a thirty-mile boat journey from Venice, where the road ends in the deep-delta Louisiana parish of Plaquemines. By 1960, the facilities housed 160 employees, 80 working each week and 40 on duty at any one time. Next to the tank farm and related installations, the camp had air-conditioned sleeping barracks, mess hall, theater, warehouse, basketball court, and other facilities. "They had a TV room," remembered Hernandez. "They had one room with just telephones. Then, they had a game room where people played cards."[66]

By 1954, the first major offshore boom in the Gulf of Mexico was on. But in order to make deeper water plays, the industry needed mobility in drilling. Shell could not extend its barge-breakwater technique much beyond ten feet of water. The small jacket platform and floating tender method pioneered by Kerr-McGee and favored for exploratory drilling in the early 1950s was quite expensive. Oilfield engineer John T. Hayward first envisioned a barge that could maintain stability in deeper water and move from location to location with little set-up cost. In 1949, he designed the Barnsdall *Breton Rig 20* submersible barge, a drilling platform supported by columns on top of a barge. The *Breton Rig 20* could be sunk and refloated in twenty feet of water. A landmark engineering achievement for the industry, it paved the way for the "strange armada" of drilling vessels that soon ventured into the Gulf.[67]

The industry did not initially rush to embrace the submersible barge design. When Alden J. "Doc" Laborde, a former marine superintendent with Kerr-McGee, tried to sell his idea for a new Hayward-type barge to drill in open water, he met skepticism and ridicule. Undeterred, Laborde eventually persuaded a group of investors led by Charles Murphy Jr. of Murphy Oil to help him form the Ocean Drilling and Exploration Company (ODECO), based in New Orleans. In late 1953, ODECO started to build its $2.5 million barge, named *Mr. Charlie* (after Charles Murphy Sr.; "Mr. Charlie" was also slang for a white boss in African American blues music). But Laborde had no customer for it. Then he met Bouwe Dykstra and John Pittman, who were searching for a way to drill on Shell Oil's leases in South Pass 27 and 28, a promising area in thirty feet of water, just south of production at South Pass 24. Shell had obtained these leases in 1947, but the tidelands controversy and the prohibitive water depth had held up drilling. Dykstra immediately took a liking to Laborde and his new rig. "I was very much impressed with this fellow," said Dykstra. "He was not only a smart engineer, but he had a vision."[68]

Dykstra had trouble, however, selling the experimental concept to Shell's management in New York, who feared that a spectacular failure might bring adverse publicity. But Dykstra insisted on making the deal. "Bouwe was perfectly prepared to take responsibility for everything he did," said A. J. Galloway, who often butted heads with Dykstra but retained great respect for the Dutchman. "In the end, he put his career on the line for it and finally prevailed," Doc Laborde fondly remembered. Dykstra and Laborde drew up the offshore industry's first ever day-work contract. Shell hired *Mr. Charlie* for six thousand dollars per day, on the condition that it perform to Shell's satisfaction. Laborde agreed to permit technical oversight by Shell. As he put it, "I had to take on numerous technical people from Shell's staff, all smarter than I and all skeptical and trying to cover their backs in case things did not work out." When Bruce Collipp, a naval architect hired by Shell in 1954, inquired into the stability of the barge during submersion, he was not exactly reassured by ODECO's response that "we are only going to sink it in water depths so that if one corner

touches down, it can't turn over." After further studies, ODECO added two buoyancy tanks to improve stability and signed a five-year drilling contract with Shell.[69]

In June 1954, the odd-looking craft set off with much suspense and fanfare to drill its first exploratory well in Shell's South Pass Block 27 lease. Reporters and photographers from *Life* magazine accompanied *Mr. Charlie*'s crew, while boats filled with industry observers and competitors trailed behind. The "singularly monstrous contraption," *Life* marveled, had a 74- by 220-foot base to allow the barge to sink safely and quietly in 30 to 40 feet of water. The rig advanced the technology of mobile drilling by using larger diameter columns and pontoons to provide greater floating stability. On site, the barge hull was flooded and the rig settled into the soft mud over the well jacket as planned.

Drilling progressed satisfactorily for a few days until a heavy storm moved the drilling barge off location. It needed some minor repairs and modifications to reenter the well. When Laborde told this to Dykstra, he expected "at best a serious tongue-lashing, and more realistically, a decision that we had failed to perform and the contract was over." But before he could explain himself, Dykstra exclaimed, "That's a great rig you have there! I can see the day when you will need several more of them." Laborde, it turns out, was unaware that Shell's electric well log had revealed a major oil discovery.[70]

South Pass 27 joined Shell Oil's growing inventory of significant offshore fields. The discovery well, State Lease 1011 #1, completed in August 1954, had penetrated hydrocarbons in four sands between the depths of 4,800 and 9,200 feet, revealing nearly 200 feet of total pay (hydrocarbon-yielding stratum or zone). During the next two years, development drilling east and west established another large and widening trend of production. In 1957, Shell also found multiple reservoirs north of the discovery well in South Pass Block 24, leading to the theory that the South Pass 24 and 27 fields were one large field. However, subsequent geophysical work and delineation drilling, including nine dry holes, revealed a major fault separating the two fields, with the closest producers only one-half mile apart.[71]

Shell Oil's South Pass 27 discovery loudly announced the arrival of offshore Louisiana as a major new producing area of the United States. It affirmed that petroleum could be found in deeper waters and that the industry could create the technology to drill and produce in them. ODECO's *Mr. Charlie* made a name for the "column-stabilized submersible drilling barge" and went on to work three more years for Shell in South Pass 27 and in the Gulf for another thirty years before it was turned into an offshore museum near Morgan City. The vessel pioneered the launching of new-fangled drilling and construction vessels to support exploration and production in ever-deeper waters. During the next several years, oil companies discovered dozens of new fields.[72] Construction yards ramped up the assembly of steel jacket platforms and began to lay marine pipelines. All kinds of new companies, from helicopter services to geophysical contractors to offshore caterers, emerged as part of this rapidly expanding industry.

Shell Oil stepped forward as a leader, and eventually *the* leader, in the Gulf of Mexico. The New Orleans Area under Dykstra became the darling of the E&P organization and a growing corporate power center. In September 1954, New Orleans established a new marine division, headed by area geophysicist Gerry Burton, to help manage its growing offshore empire. That fall, the New Orleans office moved into a new building on the corner of University Place and Common Street, attached to the Roosevelt Hotel. The Roosevelt, dubbed "the Pride of the South," was a New Orleans landmark made famous, or infamous, by former governor Huey Long. The Kingfish had housed his campaign headquarters in the Roosevelt and spent more time there than in the capital of Baton Rouge, living rent-free on the tenth floor and often greeting politicians and foreign dignitaries in silk pajamas and bright blue slippers. Owned by Long's chief lieutenant and underworld associate, Seymour Weiss, the hotel was the setting for endless political intrigue. The governor often held forth in its legendary Blue Room nightclub while patrons swilled the bar's trademark "Ramos gin fizz," and shows featuring the brightest stars of the day were broadcast nationwide by radio.[73]

In the 1950s, the Roosevelt was still the place to be in New Orleans for socializing, playing politics, or negotiating business deals. With its new adjoining regional headquarters, Shell Oil plugged itself into this vital scene. The Roosevelt negotiated for seven floors of the new Shell building to expand its guest capacity. On the second floor was the huge International Room, accommodating banquets for 1,250 people and connected to the mezzanine floor of the hotel. On the third floor, sandwiched between the International Room and the fourth floor, where Shell's new marine division had its offices, was the Petroleum Club, which quickly became the hub of activity for executives of companies working offshore Louisiana. Bouwe Dykstra and his exploration manager, Freddy Oudt, often came down to the Petroleum Club from their fourteenth-floor executive suites to play gin rummy and conduct business. "I can remember going to lunch in the bar at the old Petroleum Club," said Ed Picou, a paleontologist in the marine division. "There would be people playing cards and talking, making deals and what not. It was the place to be."[74]

In the fall of 1954, Dykstra celebrated the opening of Shell's new digs by throwing a party for all New Orleans–area Shell employees with ten or more years with the company. It was a grand gala held in the International Room. The evening started with an hour of drinks and a steak dinner. The Marx Brothers, the legendary dance orchestra conducted by Ted Lewis, decked out in top hat and tails, and the incomparable Chris Owens, mistress of ceremonies, provided the entertainment. "For dessert, they had flaming Baked Alaska for one thousand people with the red-coated waiters coming in doing a jig with trays and Ted Lewis playing the music," remembered B. B. Hughson, a geophysicist in the Baton Rouge division at the time. "That was a fabulous party. That was celebrating!"[75]

Shell Oil could afford to reward its employees for the success they provided the company offshore. The money laid out for the gala paled in comparison to the amount of money Shell was now able and willing to pay out for offshore leases. Max Burns no longer hesitated to approve large bonus bids. Both Bouwe Dykstra and A. J. Galloway, Burns's longtime friend and former classmate at Cambridge, helped persuade the president to make big bets. Galloway remembered how he would notify Burns about bids for offshore leases: "I could call him up on the phone and say 'sorry, we're gonna spend a million dollars tomorrow,' and he'd say, 'fine, okay, tell me about it later.'" Legend has it, Burns would sometimes remark on a particular tract, "We can't afford to lose it, add a million dollars." While many smaller companies formed joint ventures to buy leases and spread costs, Shell elected to play alone. The costs would be greater but so would the potential payoff. Shell's most serious competitors, Humble Oil and Calco, followed the same strategy.[76] In the first federal Outer Continental Shelf (OCS) lease sale, held on October 13, 1954, Shell Oil paid $18.7 million for thirteen tracts spread across the South Pass, West Delta, Eugene Island, and Ship Shoal areas. In the July 1955 federal OCS sale, the company spent $14.3 million for numerous tracts, including ones offshore Texas in the High Island area.[77]

Occasionally, Shell would far outbid its closest competitor, leaving millions of dollars "on the table." In the 1955 federal sale, the company was the only bidder on fifteen of the tracts it purchased off Texas. Galloway's wife periodically asked in jest why her husband could not bring home just one of those millions they had left on the table. In all seriousness, though, Dykstra and Galloway usually concluded that they did not overvalue a lease as much as their rivals undervalued it. In a sealed bid auction, you could not try to second-guess the competition. Rather, you had to bid what you thought a tract was worth. "If you really wanted it, you had to bid the whole pocketbook," explained Galloway.[78] Behind this approach lay the conviction that if Shell obtained leases in deeper water, even at premium prices, the company would figure out a way to develop them profitably.

Confronting Limits Offshore

In the early 1950s, Shell Oil and other daring companies who braved the open water of the Gulf in search of petroleum reaped large rewards. A major disruption in international oil supply in August 1956, caused by Egypt's nationalization of the Suez Canal, underscored the growing importance of oil supplies close to tidewater ports for U.S. oil security and gave added momentum to exploration in the Gulf. But as the decade wore on, operators began to confront limits to extending their early finds. These finds were in relatively shallow waters of thirty feet or less, on relatively large salt domes. By the late 1950s, however, the prospects for further success offshore dimmed. Leasing was suspended in 1955 after the State of Louisiana obtained an injunction against federal sales. The state sought an explicit determination of the state-federal boundary offshore as well as a determination of rights of "ownership," as distinct from "paramount" rights, over submerged lands. A complicated "Interim Agreement" between Louisiana and the federal government finally worked out in late 1956 to divide the Gulf into zones of overlapping jurisdiction pending a court resolution allowed leasing to go forward again. However, by the time the agreement was reached, the industry's enthusiasm for leases softened. An economic recession, a damaging hurricane, drilling mishaps, declining oil finds in deeper waters, and an oversupply of crude following a quick resolution of the Suez crisis forced a slowdown in offshore exploration. Both dry holes and capital costs increased in water depths beyond sixty feet. The percentage of available drilling rigs working dropped from 100 percent in 1957 to 37 percent in 1958. "The rapid rise and correspondingly rapid decline in offshore drilling operations in the Gulf of Mexico," wrote the president of the American Association of Oil Well Drilling Contractors in 1959, "is one of the most surprising phenomena [sic] which has occurred in the oil business in many years."[79]

The sharp decline in leasing and exploratory drilling, however, did not mean the industry had lost interest offshore. Companies just needed time to assess the new challenges they faced, while finding ways to handle the production they had already developed. After a ten-year burst of exploration activity in the Gulf and all across the United States, as the economy slacked and oil supply became abundant again, it was time to take a breather and give operations time to catch up before embarking on another great search for oil.

One major challenge was high pore-fluid or formation pressures in deep wells. By the mid- to late 1950s, well costs offshore were rising to abnormally high levels as a result of attempts to drill to deep objectives. The name for the problems encountered in a deep well was "impenetrables," which were manifested in any number of ways: hot saltwater flows, oil flows, gas flows, gas-cut mud, heaving shale, sheath, lost circulation, no drilling progress, stuck pipe, or a blowout. Offshore operators were so worried about a potential blowout that they typically drilled with heavy mud weights even through sections with normal pressures. This kept most wells from blowing out, but it increased drilling times to such an extent that the costs of exploration and development drilling offshore were becoming uneconomical. Drilling contracts, furthermore, increasingly included escape clauses that relieved the contractor from further contractual obligations if the well encountered impenetrables.[80]

An enterprising drilling engineer at Shell Oil by the name of Chuck Stuart devised a solution to the problem. A former B-24 pilot from Oklahoma, Stuart spent his whole forty-three-year career with the company on the Gulf Coast. He was a beloved but unusual character. Many "Chuck stories" circulated wide and far within the New Orleans area office. One of the most memorable involved an incident at a well in the Main Pass Block 69 field. While attending a black tie party, Stuart received a call to log the well, and he arrived at the platform still in his tuxedo. Unfortunately, the well blew out, ruining the tux. He then claimed the cost of the tux on his expense account and, after some bemused deliberation, the accountants allowed the item.[81]

Stuart obviously had close-up experience with blowouts caused by impenetrables, and he worked hard on finding a remedy. In 1956, he made a significant conceptual breakthrough in recognizing the relationship of geopressures to a reduction in shale resistivity detected by an electric log. He then devised a drilling and casing program designed to recognize the onset of geopressures and maintain a higher rate of penetration. Stuart proposed that light, normal-weight mud be used to drill a well, rather than heavier weights in anticipation of encountered geopressured formations. If a mud was kicked out of the hole, the wellhead's blowout preventers would be closed. Putting the mud returns on a choke would control the kick. The mud weight would be increased to a level that allowed continued control of formation pressures. By eliminating the need to use heavy muds at all times, the practice would save time and money.[82]

During the course of four years, Stuart gradually applied the technique on an experimental basis to a fifty-eight-well drilling program on eleven Louisiana offshore prospects. At the same time, he ran a fifty-four-well drilling program that used the traditional drilling technique with heavy muds. The comparative results were astonishing. Every well in the fifty-eight-well program reached its target objective at an average cost reduction of one-sixth over the wells drilled in the fifty-four-well program, many of which did not reach their target objective and had to be abandoned. One notable field in which the practice provided convincing results was East Cameron Block 82 (acquired in 1954), the eastern extension of the giant West Cameron Block 192 gas field, located in 55 feet of water. Several gas reservoirs lay under leases owned by Shell Oil and Conoco, which at the time, in 1959, had more wells drilled and completed than Shell and was installing a gas transmission line to shore. It appeared that when the line was completed, Conoco would be able to outproduce Shell from the common reservoirs and thus drain some of Shell's reserves. Using Stuart's drilling technique, however, Shell drilled wells to a completion depth of about 9,000 feet in the same number of days it took Conoco, using heavy mud weight from shallow depth to total depth, to reach 2,000 to 2,500 feet. This allowed Shell to catch up with Conoco in drilling the common reservoirs before the gas sales outlet was available, thus preventing the draining of its reserves.[83]

Stuart never received proper recognition for his innovation. His discovery gave Shell Oil such a competitive advantage that he was not allowed to publish it outside the company. He first circulated his findings, "Geopressures," in September 1958 for limited distribution within Shell Oil in Louisiana and Texas. In June 1960, it found worldwide distribution with the Royal Dutch/Shell group. A third edition received a U.S. patent in 1968. But not until 1970 was it turned into a technical publication. By that time, however, geopressure technology had been widely adopted. Still, its significance to Shell Oil's offshore operations should not be understated. Stuart's discovery allowed Shell to continue drilling offshore in the late 1950s and early 1960s, saving the company millions of dollars, when all operators suspended drilling at one time or another. Thus, as Stuart wrote, "offshore operations were converted from an unprofitable to a profitable venture."[84]

In addition to drilling, another important function in Shell desperately playing catch-up in the offshore game was transportation. During the war, increasing amounts of crude oil traveled to Shell refineries through the extensive pipeline network developed by Shell Pipe Line. Barges, railroad tank cars, and ocean tankers also became exceedingly important. After the war, and before the large-diameter pipelines had been built, Shell pressed these other forms of transportation into emergency service to supplement the smaller, overtaxed pipelines. Even with the larger pipelines, water was still the cheapest method of transportation.

During 1947–52, Shell's requirements for transportation on inland waterways doubled, and those for seagoing carriers grew by 50 percent. In 1950, still prevented from owning coastwise going vessels, Shell Oil renewed and expanded its chartering of tankers. Deep Sea Tankers Limited, a Canadian subsidiary of Shell Oil, operated a number of large T2 tankers, which transported oil from overseas to supplement East Coast supplies. To accommodate expanding water movements, the company added over five hundred thousand barrels of tankage for crude storage at its marine terminals.[85] Growing crude production from offshore Gulf of Mexico further elevated the importance of water transport. Shell's Norco refinery received 85 percent of its crude oil intake by barge, increasingly from offshore Louisiana. Similarly, the Houston refinery processed greater volumes of Louisiana crude barged through the Gulf Intercostal Canal—the one-thousand-mile, man-made, protected waterway from Brownsville, Texas, to Carrabelle, Florida. Both refineries were designed initially to be supplied by water.[86]

Yet as the company reached farther offshore, it found that charters did not provide enough flexibility for a fleet composed of various kinds of boats and barges. "We didn't have the right to own a rowboat," lamented chief counsel Bill Kenney. Hence, in 1958, Kenney and his associates lobbied Congress to pass what they considered a noncontroversial piece of legislation relaxing the anachronistic restrictions on the foreign ownership of marine vessels. Congress eventually passed the bill but with final provisions that prevented foreign ownership of vessels over five hundred tons. This legislation allowed Shell to acquire a fleet of small crew and supply boats, but barges and tankers were still off limits.[87]

Barging and storage around offshore installations were expensive, not to mention hazardous in foul weather. Given enough production and the development of new offshore construction methods, underwater pipelines afforded the ultimate solution. By the mid-1950s, Shell was beginning to see more than enough production coming from its leases in offshore Louisiana to justify the expensive laying of marine pipelines. In 1955–56, Shell laid gathering lines at South Pass to produce the company's first commercial quantities of natural gas offshore. Short pipelines also were laid between wells and manifold platforms in the same area.[88]

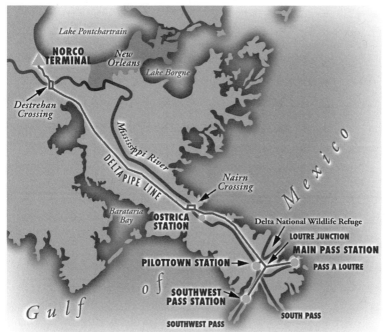

8
Map of Shell Oil's
Delta Pipeline, laid
in 1958. *Courtesy Ron
Martinez.*

Two years later, Shell Pipe Line laid 120 miles of trunk and feeder lines through Mississippi Delta marshland to connect Shell's South Pass and Main Pass facilities to the Norco refinery. Crews first dug a trench in the soggy marsh and then "pushed" the line joint by joint into the trench from a "lay barge," a method ordinarily employed on river crossings. The right-of-way also had to be routed to avoid oyster leases and muskrat dwellings. Construction superintendent C. L. Jarrett called it "the biggest little line I've ever worked on."[89] The Delta Pipe Line, as it was called, reduced the long-term cost of transporting oil from the shallow part of the Gulf and assured the delivery of crude to Norco under all weather conditions.

Another matter to be dealt with, more by Royal Dutch/Shell officials than by Shell Oil's American managers, concerned Shell's rights as a foreign-controlled company to acquire leases under the OCSLA. As detailed in chapter one, antitrust investigations by the FTC and Congress into the oil industry during the 1950s raised concerns that Shell might be singled out because of its foreign majority ownership. Although neither President Truman's Proclamation nor the OCSLA restricted foreign enterprises from acquiring offshore leases, Section 8 of the OCSLA allowed for the "nonacceptance of any bid, as may be appropriate to prevent any situation inconsistent with the antitrust laws." Shell directors feared the company's ownership structure could be seen to violate these laws. One impetus behind the assertion of national jurisdiction over submerged lands, after all, had been to protect them against foreign encroachment: Pres. Franklin Roosevelt's foremost concern in the late 1930s was that, in the absence of federal control, a European nation might try exploiting the Gulf of Mexico for oil.[90]

If Shell Oil became a target of an aggressive antitrust crusade, it was vulnerable to being cut off from federal offshore leases. To redress this, Group managers pushed for a special provision in the negotiations over a Treaty of Friendship, Commerce, and Navigation between the United States and the Netherlands. C. H. Bogaardt of Bataafsche Petroleum Maatschappij, the principal Dutch holding company in Royal Dutch/Shell, was a key negotiator for the Dutch in the preparation of the treaty, and he lobbied diligently for explicit language protecting the property rights of Dutch corporations in the United States as well as against language, as a U.S. State Department official put it, that would "lay open to review their legislation or policies on 'cartels,' or subject them to representations from the U.S."[91] Although U.S. officials found it odd that a commercial representative would take such an active role in diplomatic negotiations and that the language he pressed regarding "controlled corporations" was unusual, Bogaardt and the Dutch got what they wanted in the final treaty. Article 9, Section 5 of the treaty, signed at The Hague on March 27, 1956, provided that: "with respect to the acquisition, ownership, use and disposition of property of all kinds within the territories of either Party, companies constituted under the laws of that Party, which are controlled by nationals and companies of the other Party, shall be accorded treatment no less favourable than that accorded within such territories to companies of such other Party. . . ."[92] Such a provision still might not have fully protected Shell Oil's right to offshore leases in the event of a political inquisition, but at the very least it provided some diplomatic insurance about the security of the company's offshore leases.

Indeed, insurance of all kinds was what the offshore industry needed as it ventured into deeper waters. Based on the success of *Mr. Charlie*, medium- and large-sized shipyards along the Gulf Coast soon received orders for mobile drilling vessels. With no clear idea of what was needed for deeper water, designers quickly moved in several directions. ODECO, Global Marine, and the Offshore Company emerged as leaders in the market for various sorts of vessels. Given short time horizons for lease development, they scrambled to invent mobile units with increased depth capabilities. Each ponderous innovation pushed the industry a little deeper into the Gulf. Submersible pontoons and hulls increased dramatically in size to allow drilling at greater depths. In 1957, ODECO built and leased to Shell the world's largest submersible, *Margaret*, named after Doc Laborde's wife. With a deck supported by ten cylindrical "bottles" and a giant catamaran hull measuring the size of two football fields, the barge could work in 65 feet of water.[93]

To operate beyond this depth, however, submersible drilling units would have to be built to mammoth and impractical proportions. In the 75- to 150-foot depth range, the self-elevating "jack-up" vessel promised better success. Ungainly looking sea monsters, these barges elevated their platforms out of the water by extending, or jacking, long cylindrical legs to the bottom. One of the earliest prototypes was Glasscock Drilling Company's three-million-dollar super-rig, *Mr. Gus*, the first mobile unit designed to drill in 100 feet of water under hurricane conditions. A combination submersible/jack-up, it used a barge-type hull to transport the drilling platform in an elevated position. Once on location, the legs were jacked into the soil and the lower portion of the barge was lowered to the bottom to give additional support. The Houston Area Office of Shell Oil enthusiastically supported this design as an alternative to the column-stabilized submersible. For a while, considerable rivalry existed within the Shell E&P organization as to which was the best approach. In 1955–56, *Mr. Gus* set out to prove its worth by drilling a well for Shell sixty-three miles off Galveston, Texas, farther offshore than any company had yet attempted to drill.[94]

That well turned out to be Shell's first one-million-dollar dry hole. The twenty-thousand-dollars-per-day cost of operating *Mr. Gus* demonstrated the vast sums of money at play farther offshore. The one-million-dollar figure did not even include the cost of seismic surveys, soil boring, and other incidental operations prior to drilling, nor did it include the cost of field development if oil were struck. In 1955, as an alternative to rented mobile units, Shell experimented with a prefabricated fixed platform—the company's first permanent one installed offshore—for exploratory drilling in 72 feet of water in South Pass Block 42. Shell engineers felt that a fixed structure secured by piles driven deeply

into the seabed would hold more steady in the soft mud bottom
than a mobile platform. The location stretched the depth limits for
submersibles, and early jack-up designs still appeared somewhat
shaky. In fact, the next several years witnessed a number of mishaps and capsizings by jack-up rigs in the Gulf, including *Mr. Gus*,
which tipped over while preparing to move off location in 1957.[95]
But as Shell discovered in South Pass 42, trying to salvage fixed
platforms in case of a dry hole incurred even greater costs than a
mobile rig. At such depths, fixed platforms made sense only for
producing oil from established fields, not for wildcat exploration.

In the late 1950s, an economic recession and a growing oversupply of crude discouraged all exploration and lowered the
allowable production rate for offshore Louisiana wells, dramatically slowing activity in the Gulf. Compounding the problem, a
string of harsh weather besieged the northern Gulf. Even though
advance warnings and preparations protected many structures

and vessels from disaster, the 105-mph winds and 15-foot waves of Hurricane Audrey (listed as category 4 strength) in July 1957 inflicted heavy damage and destroyed Cameron, Louisiana, an offshore support center on the Gulf Coast. An estimated five hundred people perished in the storm, and losses throughout the region amounted to seventy-five million dollars, approximately sixteen million dollars by the oil industry. Thanks to well-conceived plans to protect employees and equipment, Shell reported no deaths or catastrophic losses, though the company suffered an estimated one million dollars in damage.[96]

The time had arrived to reassess the challenges facing the offshore industry. As in earlier oil booms, the race to develop oil offshore lost fortunes as well as made them. The downturn of the late 1950s severely exposed unsecured investments, speculative leases, and careless construction. Hardest hit by the downturn were drilling contractors and service companies. With two-thirds of marine drilling vessels idled, many of these companies either went out of business or were acquired by competitors. The frequency of accidents increased with bad weather, and budgetary cutbacks too often sacrificed safety. Consequently, insurance premiums for offshore operations soared.[97]

As drilling on the unproductive offshore Texas and South Pass 42 leases proved, Shell Oil was not immune to setbacks offshore. To make matters worse, one of the company's top production officials, John Pittman, spent a year in the hospital and lost an eye after a mid-1957 auto accident nearly took his life. Fortunately, he eventually returned to work and regained his old position in 1959. Despite setbacks, the company was still in a good position. Through a mixture of foresight, determination, and providence, Shell had acquired and developed some of the best properties yet discovered in the Gulf, and thanks to Chuck Stuart's geopressure technology, its offshore drilling program continued apace. By 1960, Shell had become the leading producer of offshore oil in the Gulf of Mexico (64,000 barrels per day out of industry total of 213,000 b/d) and had major interests in the four largest fields in the Gulf: West Delta 30, South Pass 24, South Pass 27, and Main Pass 69. The South Pass Block 24 and 27 fields, Shell's most prolific ones, combined for a total of 418 producing wells. Bill Kenney recalled his awestruck legal counterparts at Gulf Oil, who once remarked to him: "My God, Shell has stolen the whole offshore."[98]

While Shell's record in the Gulf may have demoralized some competitors, the company's dedication to technological innovation became a beacon for the offshore industry. Continued innovation was imperative more than ever if the industry hoped to offset rising costs and seek new opportunities in deeper water. Shell and other offshore pioneers had come too far to give up on this "billion dollar adventure in applied science."[99]

Betting on Technology

In the late 1950s, U.S. oil companies weathered intense competition and mounting difficulties to find oil. The industry was drilling a higher ratio of dry holes to total wells. The average size of new oil fields was decreasing. So was the number of barrels discovered per foot drilled. The search for crude in the United States required deeper wells, as well as expansion into "frontier" locations where solutions to environmental and technical problems were expensive. Cheap oil from Venezuela and the Middle East poured into the United States at a rate that even exceeded the voracious appetite of Americans with their ever-larger, ever-more-powerful, gasoline-guzzling automobiles to consume it. "No longer was the growing abundance of cheap foreign oil something to be encouraged and applauded as a way to relieve the pressure on U.S. reserves," writes Daniel Yergin. "Rather, the rising flood of imported oil was seen, at least among the independent American producers, as a dangerous threat that was undercutting domestic prices and undermining the domestic industry itself."[1]

Relief came in March 1959, when, after a decade of arguments and failed efforts to encourage "voluntary" import quotas, Senators Lyndon Johnson (TX) and Robert Kerr (OK) and other oil-state congressional leaders prevailed upon Pres. Dwight Eisenhower to impose mandatory quotas on imported oil. Despite all the exemptions built into it and administrative problems that arose from it, the quota program did serve its purpose in shielding domestic oil from foreign competition. One of the most significant domestic policies implemented in the postwar period, mandatory quotas expanded the market for more expensive domestic oil, such as that produced offshore or from enhanced recovery operations in old fields, both of which became Shell Oil specialties. With the end of the severe recession of 1958, furthermore, the U.S. economy once again moved into a growth mode, and the hunt for domestic oil resumed.

Still, the ability of oil companies to expand production did not depend solely on policy and market conditions. Competition was still keen within the protected U.S. market. To achieve success, oil companies would have to find ways to minimize risks and improve methods of exploration and production. Shell Oil confronted these challenges by betting on technology as the key to profitability. "Finding and producing oil at a profit today," stated Shell Oil Executive Vice President Denis Kemball-Cook in 1958, "depends most of all on how well we can improve our technological skills and apply them boldly and efficiently."[2]

The commitment to technology permeated Shell. The company had built up outstanding research capabilities, and senior management vigorously pushed technology as a way of gaining a competitive advantage. At times, disagreement flared over where and how far to push it, especially on the question of moving into the deeper waters of the Gulf of Mexico. But for the most part, middle managers and technical staff were protected from the internal politics of strategic decision-making and budgeting. They were simply told to find and develop hydrocarbons. Combining imagination, experience, and sheer brainpower with computers that enhanced data processing, Shell engineers and geoscientists pioneered a stunning range of new technologies in the 1960s that significantly boosted the production of oil and gas from established fields onshore and took offshore exploration and production into record water depths.

Deepwater Visions

Shell's march into so-called "deepwater" (the definition has changed since; see chapter 6) started in the mid-1950s, when offshore oil exploration and production reached what many considered to be a threshold water depth of about sixty feet. Shell enjoyed immense success with its shallow-water South Pass and Main Pass fields, located in depths less than thirty feet. But working much deeper than that, in sixty feet or more, seemed almost impossible. Submersible drilling units were impractical and unstable in deeper water. Early jack-up units designed for greater depths were prone to capsizing. And, as Shell found out in South Pass Block 42, exploratory drilling from a prefabricated fixed platform was extremely costly.

Bouwe Dykstra, vice president of the New Orleans Area, believed that offshore development had reached its limits. Although Dykstra had been a driving force behind Shell's early moves into shallow water, he now argued that going deeper than sixty feet would require special equipment, which might not be possible to develop. Even if it were, he insisted, well costs would be pro-hibitive. In a paper presented before the Fifth World Petroleum Congress in 1959, Dykstra admitted that "the industry has always been able to make important reductions in well costs once the op-erations have been systematized in an area where a large number of wells have to be drilled." He noted that one-third of the wells drilled in less than sixty feet of water had been dry holes, com-pared to a dry-hole rate of 40 percent in waters deeper than sixty feet. The normal reduction of well costs, he observed, "has not been possible in the areas beyond the 60-foot depth mark." This led him to the conclusion that deep water was too expensive and that companies would be better off sticking to production in the shallow water.[3]

Others were more sanguine about deepwater. Ronald E. "Mac" McAdams trusted in technological innovation to make it com-mercially viable. McAdams had joined Shell as a geologist in 1936 at Shreveport, Louisiana, and rose to become senior geologist in Houston after World War II and later to exploration manager in the Tulsa Area. In 1955, he was named manager for exploration at the head office, and two years later he was elected vice presi-dent for exploration, a new office created when A. J. Galloway moved up to become executive vice president. Tall and bald, Mac cut an imposing figure. His physical presence and brusqueness, along with his penchant for making snap decisions, intimidated people in exploration. He rewarded talent and innovation but ostracized and banished those who rubbed him the wrong way. "They would quake when he would make visits to the division and review what was going on," said Don Russell, reservoir en-gineer at Bellaire during this time. McAdams was an excellent geologist himself and would grill his technical people with very pointed questions. Said retired Shell Oil geologist Jerry O'Brien: "Boy, when you came in to see Mr. McAdams, you'd better have thought about all the potential questions before he asked them!"[4]

McAdams took his job so seriously that he had very little so-cial life outside of the company. But he engendered fierce loyalty from those close to him. He styled himself after legendary general George Patton, under whom he had served as a lieutenant colo-nel during World War II. When the feature film *Patton*, starring George C. Scott, premiered in 1970, associates were startled by how closely Scott's characterization resembled McAdams. "Like Patton, McAdams had many flaws," said Marlan Downey, an ex-ploration geologist at the time. "But he created morale, *espirit de corps,* and he would charge a semi with a pitchfork if he thought it would find oil for Shell." McAdams had an uncanny ability to recognize the value of new technology before other companies in the industry or anybody else in Shell exploration. "Don't let the shoreline stop you," he counseled. "The geology is good. The fields are good. And if the volumes are there, some smart engineer will always help us figure out how to make money in deeper water."[5]

James E. "Ned" Clark, vice president for production, supported McAdams's views about deepwater. The son of a cattle rancher and U.S. marshal from Southern California, Clark was a rugged outdoorsman and a jack-of-all-trades. As a youth, he had worked as a cowpuncher, horse breaker, fruit picker, chauffeur, body-guard, bulldozer operator, waiter, bouncer, and paper deliverer, among other occupations. "I worked for wages before I was old enough to go to school," he said.[6] In high school his day started at 3:30 A.M. to fit in all the jobs he held. He also found time for organized sports such as football, basketball, track and field, ten-nis, wrestling, and rodeo. His boundless energy and multiple talents served him well at Shell. Clark joined the company in 1927 as a summer laborer in Ventura. After taking an engineer-ing degree from Stanford, he advanced to various positions in the Pacific Coast and Denver Areas before becoming vice pres-ident in 1958. Self-effacing and publicity shy, Ned Clark main-tained a low profile among the company's top management. His genial, low-pressure management style contrasted distinctly with McAdams's. But that made him no less effective as a leader and offshore visionary.

Quietly but persuasively, Clark argued that offshore was the only place left in the United States where Shell could stake a com-petitive position. The other majors had tied up the best undevel-oped properties onshore. Soon after becoming executive vice pres-ident, he expanded the small offshore task group that had been established in the Technical Services Division in 1954 and created a special program, initially headed by Bob Carter, to investigate ways to develop a complete system for drilling and producing in water depths ranging out to six hundred feet. Clark maintained that if Shell beat its competitors in finding the technology for deepwater exploration, the company could obtain leases at very reasonable prices for tracts that nobody else was prepared to de-velop. "That is why you guys are on the payroll," he told the deep-water group. "We will fund you. We will support you. And we will keep you very secret."[7] The group operated under the utmost secrecy, not only externally but within Shell. When TSD merged with Shell Development in 1958, the senior managers in research were not even privileged to know what the offshore group was doing.

The creation of the E&P Economics Department in 1958 strengthened the head office's commitment to deeper water. The campaign to revise Shell's method of evaluating the profitability of potential investments originated with John Redmond, a talented chemical engineer who had joined the Tulsa office in 1936. Working under Toni van Everdingen, Redmond embraced his mentor's efforts to bring greater mathematical and technological sophistication to all of Shell's operations. "Redmond had as much individual capacity as a senior manager to handle complicated mathematical problems as anyone I have known in Shell Oil," said John Bookout. Transferred to New York in 1955 to become manager of the exploitation engineering department, Redmond inquired into something he had wondered about for a long time: the basis behind the company's profitability criteria, which stated that projects had to pay out in two years with 100 percent undeferred profit, assuming that prices and costs were held constant forever. "This had always bothered me," he said. "Mostly because I did not understand it—not because I had any better ideas at the moment."[8]

He searched in vain for a formal economic analysis behind this "payback period" policy. "It was just something that had been age old in the company," said Redmond. "But it was strict and there were very few exceptions to it."[9] Redmond convinced A. J. Galloway that Shell needed more logical and quantitative methods of measuring the economic value of any particular opportunity. Galloway agreed. Collaborating with van Everdingen and a young engineer named Gene Bankston, both from the newly organized production-economics department, and Jimmy Lyon, reservoir engineer in the Calgary Area office, Redmond developed new profitability measures using the "net present-value" approach. This approach basically factors into rate-of-return calculations the value of money over time allocated to alternative investments. The concept of net present value was not new. For the first time, however, Redmond's group tackled the problem of how to use present value in decision-making for projects and wells in the field. Petroleum companies were just arriving at this idea, spurred on by the management consulting firm of McKinsey & Company. The old standard did not accept a perfectly viable project if it did not pay out until four years down the road. Clearly, and particularly for deepwater prospects with large front-end investments, the net present-value approach offered a better understanding of the impact of time on overall profitability. It placed a premium on exploring for new reserves, whose net present value of future production might be greater than production in older fields, allowing companies to undertake more aggressive drilling.[10]

Some senior managers, especially those with vested interests in the older producing fields or areas, resisted applying the new approach in drafting the E&P budget. However, Denis Kemball-Cook, who replaced the retiring A. J. Galloway in 1958, made it a priority. An amiable and soft-spoken Englishman, Kemball-Cook had immigrated to the United States at age 22 to "take a crack at his fortune." He started at the bottom, finding work in the midst of the depression as an attendant in a New Hampshire Shell station. But unlike most grease monkeys, Kemball-Cook was a product of Oxford University, where he had majored in economics. He was quickly recognized as management material, and held various posts in St. Louis, Tulsa, and Houston on his way to becoming a director of Shell de Venezuela, responsible for economics, refining, and supply. Upon returning to Shell Oil, the self-described "vice president in charge of vice presidents" sought better ways to coordinate exploration and production. As an economist, he immediately recognized the usefulness of the new profitability tool.[11]

Kemball-Cook spent most of his first six months in office organizing a new E&P Economics Department, managed first by John Redmond. The new department, whose staff included representatives from both exploration and production, bundled together forecasting, reserve determination, and budget management with overall profitability studies. The department subsequently became a key place for rising young managers to gain a broad perspective on the workings of the company. Many of Shell Oil's high-ranking executives in the 1960s and 1970s would take a turn heading E&P Economics. In 1959, Kemball-Cook and Redmond decided to prepare the annual E&P budget on the new net present-value basis for presentation to the board of directors in London and The Hague. At the end of the presentation, Jan Brouwer, then Group E&P Coordinator in The Hague, asked only one question: "Would the present value method expand or restrict investment opportunities?" When Redmond's answer was "expand," Brouwer simply said, "Go ahead."[12]

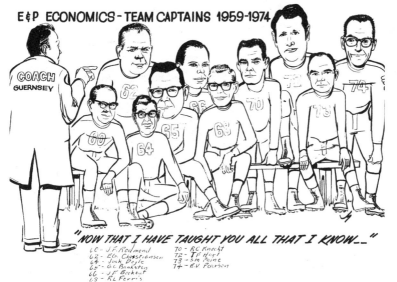

E & P ECONOMICS - TEAM CAPTAINS 1959-1974

COACH
GUERNSEY

"NOW THAT I HAVE TAUGHT YOU ALL THAT I KNOW..."

60 - J.F. Redmond
62 - E.G. Christiansen
64 - Jack Doyle
65 - G.C. Bankston
66 - J.F. Bookout
68 - R.L. Ferris

70 - R.C. Knecht
72 - T.F. Hart
73 - S.M. Paine
74 - E.V. Pearson

10

As this caricature suggests, many rising managers at Shell Oil spent time as head of E&P Economics. Pictured here: 60—John Redmond; 62—Ed Christianson; 64—Jack Doyle; 65—Gene Bankston; 66—John Bookout; 68—Bob Ferris; 70—Ronnie Knecht; 72—Tom Hart; 73—Sam Paine; 74—E.V. Pearson. *Caricature courtesy Bob Ferris.*

Even when confronted with the new approach to valuing properties and setting budgets, Bouwe Dykstra still opposed moving into deeper water. As an engineer, he had well-founded concerns that the risk of maintaining deepwater facilities for the entire producing life of a field was too great. Nobody at the time had a comfortable understanding of the design criteria or experience with safely operating such facilities in the open Gulf. The loss of a platform halfway through its producing life could mean the loss of the entire investment—the remaining reserves would probably not justify the cost of replacing it. "He recognized that we were leaving the economic safety of single-well development in comparatively shallow waters for development with much greater risks," said John Redmond. "This bothered him a lot."[13]

Dykstra was also bothered by the escalating costs of production, as he pointed out in his presentation to the World Petroleum Congress, as well as the rising costs of obtaining leases. The real eye-opener occurred at the federal "drainage sale" held for offshore Louisiana on August 11, 1959. Under the state-federal agreement, leases in the disputed zone were only offered if they were subject to drainage from productive leases nearby. John Rankin, who had taken over as regional manager of the Bureau of Land Management's New Orleans OCS office in January, remembered the sale very well. "First, it was held on my birthday," he said. "Second, my youngest daughter was born the night before. And third, to my consternation, I opened a bid from Shell Oil Company and didn't know whether I could handle such a figure." Shell Oil had bid *$26 million* for a single, 2,500-acre, half-block tract on South Pass 28, adjoining the company's producing leases in the South Pass 27 field, outbidding Texaco by $7 million. "I gulped twice and read that historic bid which was the record high price per acre bid [$10,442] for many a year [until 1964]." Shell Oil also won a 1,370-acre tract in the southern portion of South Pass 27 for $6.75 million, the third highest per-acre bid in the sale. The company paid a total of more than $50 million for ten tracts.[14]

What had happened was that in 1958 Shell had drilled some deeper exploratory wells around the South Pass 27 field and discovered new oil pay zones. Wells drilled along the south line of the company's state leases on South Pass 27 and 28—the three-mile state-federal boundary ran right through those leases—confirmed gravity data about the depth of salt and gave sure indications that production would extend into the federal parts of those two blocks. "We would drill as close as we could get to this other acreage that was going to be up for sale," remembered Bob Ferris. "And talk about tight holes!" Marine Division Manager Gerry Burton and Division Petrophysicist Sam Paine personally went out to the drilling vessels to gather the well logs without even allowing Schlumberger, the company that ran the logs, to see them. "There were state regulations requiring copies of the logs be sent to the state, and we were able to talk them into letting us hold back on those for quite some time, which was a good trick in Louisiana," said Ferris.[15]

Assured of the oil potential on the federal side of the lease line, Shell Oil desperately wanted that acreage and was willing to "bid the whole pocketbook" to acquire it. Although most per-acre bids did not run nearly as high as Shell's winning bid, the 1959 drainage sale revealed the high stakes now involved offshore in the Gulf. Numerous high bids indicated that the offshore industry was ready to get moving again. The U.S. economy was back in a growth mode. The high success rate of those offshore drilling operations that continued during the recession gave cause for new optimism. And innovations in seismic prospecting—especially the introduction of magnetic tape for recording and the improvements in data processing it afforded—stimulated renewed offshore exploration. Even though the federal-state dispute over the submerged lands still awaited a Supreme Court decision, both

Louisiana and the federal government were ready and willing to implement the Interim Agreement and begin holding general lease sales again.

Shell's top executives were prepared to place large bets in this high-stakes game. In their view, this was not a reckless gamble but a calculated risk that offered promising rewards. For Mac Mc-Adams and Ned Clark, the new profitability criteria formulated by E&P economics reinforced their conviction that deepwater development was not only technologically possible but could be made safe and economical. They had faith in their technical staff and believed in learning curves. Costs would eventually come down with improved techniques, an infrastructural base, and economies of scale. Through dogged determination and some unpleasant meetings, McAdams, Clark, and Kemball-Cook carried the day on the new profitability criteria and on the company's strategy to continue pushing the water-depth horizon offshore.

Shell was more eager than any other company to get into deeper water. The main reason for this was that it had an ace in the hole, a floating drilling vessel being secretly designed that could take exploratory drilling into three-hundred-feet-plus water depths. Upon publication of the initial Call for Nominations for the 1960 federal lease sale, the head of Shell's New Orleans legal department, George Schoenberger, convinced the Department of Interior to withdraw the call and issue a new set of leasing maps with deeper acreage out to and beyond the three-hundred-foot depth contour. With Shell's assistance, the BLM redrew the maps with "south additions" to all the old original blocks off Louisiana and issued a new call for nominations. On February 26, 1960, the BLM offered 1.17 million acres offshore Louisiana and 437,000 acres offshore Texas. Once again, offshore operators spent big— $285 million in high bids ($249 million for the tracts off Louisiana)—more than double the amount spent in any previous sale. Shell Oil spent $23.1 million for 126,720 acres offshore Texas and $13.3 million for 51,431 acres offshore Louisiana, including a number of tracts in the Grand Isle Area South Addition, where the company planned to drill, unbeknownst to anyone else at the time, with its revolutionary new vessel.[16]

A Giant Step

On August 14, 1962, Shell Oil dramatically unveiled a "floating drilling platform." This converted submersible, the *Blue Water 1*, was equipped to operate in six hundred feet of water without resting on the bottom. Announcing successful drilling from the platform, Shell reported technical progress in completing ocean-bottom wells by remote control from the surface. The unveiling of the *Blue Water 1* ended two years of speculation in the industry about what Shell had been doing with its mystery rig and captured national headlines heralding a double technical breakthrough. "Oilmen can now find and produce petroleum from the open sea regard-less of depth of water or distance from land," reported the *Wall Street Journal.* Overnight, Shell's *Blue Water I* changed the mind-set of the entire industry and opened a new era for offshore oil.

This overnight sensation, however, was many years in the making. Oil companies had been investigating the possibilities for floating drilling as early as 1948, when Shell joined Continental, Union, and Superior in a four-company consortium called CUSS to undertake a drilling program off California. The shelf drops off steeply on the West Coast, so companies could not wade gradually into deeper water with drilling units that sat on the ocean floor, as they did in the Gulf. With little experience or designs to go on, the CUSS group naturally turned to traditional marine practices. They drilled from a ship, developing techniques for taking surface cores and experimenting with a drilling rig mounted on the side of a converted former Navy patrol boat called the *Submarex.* Other floating drilling contractors soon emerged off California and in the Gulf of Mexico using similar concepts. In 1954, CUSS converted a much larger ex-Navy barge into a "drillship." The *CUSS I* drilled through an opening in the center of its hull, called the "moon pool," which gave the rig greater stability than those with derricks mounted over the side. The vessel also featured a revolutionary underwater drilling system using guidelines, underwater blowout preventers, and a marine conductor for well pipe.[17] In 1961, Project Mohole, the National Science Foundation's ambitious but eventually aborted endeavor to drill to the earth's mantle, selected the *CUSS I* to drill its test holes in the Pacific Ocean. But even the most stable drillships such as the *CUSS I* were still only reliable for working in protected or calm waters, not for all-weather drilling in open seas.

As companies explored modifications to traditional ship designs in the mid-1950s, Bruce Collipp, a bright young engineer with Shell Oil, took a different approach. In 1953, Collipp wrote his master's thesis in naval architecture at MIT on the stability requirements for increased mobility with submersible drilling vessels. "Everybody else was writing theses on towing ships," said

Collipp. "That seemed like a boring subject." He had read some articles about the submersibles in the Gulf of Mexico and discovered that the industry had not yet figured out how to measure the stability of these funny-looking things. Naval architects concentrated on reducing the force on a slender body moving through water and gave little thought to the forces on something designed to remain at a fixed location. Collipp decided to pursue this intriguing question and imagine what he would build if he were going to drill in deep water.[18] His thesis, "The Design of an Offshore Oil Drilling Rig," examined the energy relationship between waves and structures, and it proposed a novel type of mobile rig that would jack its legs and base unit to the ocean floor for support. Others in the industry were pursuing similar ideas, and in April 1954 construction started on the first jack-up rig.

Collipp's unique and timely approach to a real-world problem earned him a position as a naval architect in Shell Oil's Technical Services Division in Houston. However, it seemed to Collipp that Shell people did not really know what a naval architect was or did. Nobody told him to invent a better floating drilling unit. His first task was merely to study and help make improvements to the kinds of vessels Shell already had in operation. He went to New Orleans in 1954 to review ODECO's designs for the *Mr. Charlie* and *Margaret* submersibles and then traveled to California to inspect the *CUSS I*. "Everybody on these vessels was quite a few years older than me, and I thought they must know what is going on," said Collipp. "But I discovered that wasn't necessarily true!" While aboard the *CUSS I* in the Santa Barbara channel, Collipp observed that as long as the ship was aligned with the bow headed into the sea swells it rode pretty well. But when a storm came out of the hills and hit the *CUSS I* broadside, the ship rolled violently. At that moment, he realized that "if Shell was going to operate in deep water, we were going to have to have something other than a ship to withstand the environmental forces and to maintain a relatively stable platform that men could work on."[19]

The young engineer tried to impress this on his bosses but made little headway. "I kept saying there was a better way, and from time to time they would say, 'So what?'" In 1956, he returned to TSD to work on improving anchoring designs, an assignment that did not exactly thrill him. "Cavemen tied a rock on a piece of rope," he remembered thinking. "Here is a thing that has been around for thousands of years, and I'm supposed to improve it." He decided to take the initiative and used the $20,000 budgeted for anchor research to look into new shapes for drilling vessels. After all, he reasoned, part of the anchor design is the thing that is being anchored. In searching for a new design, Collipp recalled how the *Margaret* had handled during a rough storm. The crew had ballasted the hull underwater so the rig was riding on the columns waiting for the waves to subside. He noted that the barge did not move around as much in this position as it did during the tow out.[20]

Collecting his observations and applying new engineering concepts relating to the design of marine vessels, Collipp designed and built scale models of two unique floating drilling platforms. One was a space-frame structure consisting of three large columns and a submerged hull. Most of the vessel's mass would be beneath the surface for improved stability, and the columnar structure would minimize wave forces. He called the contraption the "Trident," after the three-pronged spear that Poseidon, the Greek god of the sea, brandished to pierce the sea bottom. Another was a circular-shaped "donut." Collipp carried his models around for "show and tell" and took them out to the University of California–Berkeley to run wave tests in the university's special tank. "I just charged ahead," he explained. "There was no one in Shell who was trained in this area at all, so no one said: 'Do you know what you are doing?'" He filmed the experiments with high-speed cameras and presented the results to Joe Chalmers, the director of TSD. They demonstrated the Trident's superior stability compared to alternative designs.[21]

Impressed by Collipp's model, Chalmers passed the idea along to the head office. Although Shell's senior management did not really understand the science behind the Trident, they gradually accepted it. McAdams and Clark had become increasingly anxious for an innovation that could take exploration into deeper waters. With enthusiastic backing from Bert Eastin, manager of production research, work progressed on all fronts of the deepwater development program, including underwater completion technology along with floating drilling. Collipp spent most of 1958 running computer-simulated tests on his model, a time-consuming task given the relatively slow speed of punch-card computers in those days. "Bellaire had this old IBM 650 computer," remembered Don Russell, "and I used to get teed off because I never could get on it because the civil engineering group had it tied up morning, noon, night, weekends and holidays doing all these analyses of waves and forces."[22] Although it had been thoroughly tested, the floating drilling idea remained in the development stage for another year, until the February 1960 sale, when Shell acquired acreage deeper than anyone. As the race to develop new offshore acreage resumed, Shell was determined to be first in deep water.

The head office ordered up a floating drilling platform to evaluate its new deepwater leases. Time was short, however. The exploration managers wanted to drill and obtain information on these leases before the next scheduled lease sale in March 1962, which was going to be even bigger than the last. Building a new rig from scratch based on the Trident design would take too long and cost too much. Fortunately, Collipp knew of an existing unit that could be converted into a floating platform. It was the *Blue Water 1*, a three-year-old, bottom-sitting, submersible owned by a consortium of four small producing companies called the Blue Water Drilling Corporation. With four stabilizing columns resembling monstrous milk bottles, the *Blue Water 1* had the desired hydrodynamic properties. The hull pontoons could be filled partway, making them buoyant enough to keep the vessel afloat but heavy enough to sink below the lash of the waves. Collipp offered to lease the unit from Blue Water Drilling for five years, under the conditions that Shell could make changes to it and that Blue Water had to keep those changes strictly secret. Blue Water's president, Sam Lloyd, readily agreed. The *Blue Water 1* had been idle for about six months, and a five-year contract in those days was unheard of. "Do whatever you want with it," Lloyd told Collipp. "Paint it red and yellow if you have to." [23]

During 1961–62, under a shroud of secrecy, Shell converted the vessel in the Ingalls Shipyard in Mississippi, right next to where the first nuclear attack submarine was being built. "The levels of secrecy were about equal, except we were up higher and could look down on what they were doing," Collipp remembered. He equipped the *Blue Water 1* with a special mooring system using eight anchors, each larger than any ever built at that time. Departing from standard practice, he used wire line rather than anchor chain, making this massive system easier to handle and quicker to install. The system was designed to overcome the biggest problem of drilling from a floating vessel: keeping the drilling platform directly above the well. Collipp supervised the development of instruments to help keep the vessel on the right spot, including a "Tiltmeter" to measure the angle between the vertical axis of the wellhead and the rotary table on the platform. Automatic ballasting controls stabilized the columns and trimmed the

11

The *Blue Water 1* semi-submersible on its maiden voyage, 1960. *Photo courtesy Ronald L. Geer.*

vessel. The $1.5 million conversion took a rig originally built to operate in sixty feet of water and increased that depth to six hundred feet and possibly beyond.[24]

As the conversion neared completion, Collipp met with a U.S. Coast Guard official, Captain McPhall, in Mobile, Alabama, to have the rig certified. But there were no classification guidelines for something like the *Blue Water 1*. Ships were supposed to be long and pointy and go from point A to point B. But this vessel was roughly square, did not transport cargo, and did not transport people. Therefore, it was not a ship.

When McPhall unrolled the blueprints, he asked Collipp, "What is it?"

"It's like an iceberg," Collipp explained. "You see, all the hulls are underwater."

"I don't know what you are describing to me," replied the captain, "but this thing is going to sink if all the hulls are underwater."

"Well, no," said Collipp, trying his best to describe in simple terms how the vessel would remain afloat. After more discussion, McPhall agreed to register the rig, but he needed to know how to classify it.

"It is kind of a semi-submerged thing," Collipp told him. The term just popped into his head.

"All right," said the captain. "We will list it as a Super Manned Barge-Semi-submersible."

The name stuck, and Collipp would henceforth be known as the "father of the semi-submersible."

The semi-submersible vessel itself was only one piece of the deepwater puzzle, albeit the major one. "I'm just the guy who gives you some place to stand," Collipp said.[25] Other bright minds recruited into Shell's deepwater program attacked the problem of how the oil would be drilled and produced. The effort began in 1955 with the offshore task force set up at Bellaire to study the whole field of offshore operations. By 1958, the special development program for the Gulf of Mexico had come to focus on floating drilling *and* underwater wellhead completion. The semi-submersible increased the depth of exploratory drilling, but new methods would be needed to produce oil from those depths. In conventional offshore producing operations, the deck of a fixed platform housed the wellhead equipment—the blowout preventer and the assemblage of control valves, pressure gauges, and chokes known as the "Christmas tree"—above the water surface. A conductor connected all this equipment to the bottom. Because platform designs had not even approached the water depths (300 to 600 feet) contemplated by the deepwater program, Shell engineers investigated the possibility of installing, completing, and maintaining a wellhead on the sea floor. And because the practical limit of diving at the time was only about 150 feet, everything had to be done by remote control. Like the semi-submersible, an underwater well represented a giant conceptual leap. But, as Collipp noted, "We were all of that age where we didn't know we couldn't do it. It was kind of like why do 18-year-olds make better soldiers than 40-year-olds? Charge what hill?" Or as Ron Geer, a mechanical engineer who directed the design of the wellhead system, remembered: "We were limited only by our imaginations."[26]

In 1958, Geer joined about a dozen other young engineers, some fresh out of school, at the Bellaire Research Center to develop the underwater completion technology that would accompany floating drilling in the Gulf. Having grown up in West Palm Beach, Florida, and served in the merchant marine during the war, Geer's love of the ocean was the main reason he had chosen to work for Shell. His first assignment, however, was in the bone-dry Permian Basin. When given the chance to get into marine work, he threw himself into developing the production wellhead and flowlines for the new system. What he and others ultimately created was complex, consisting of many specialized subsystems and components. Frank Poorman was responsible for the special blowout preventers. John Haeber helped put all this together with the wellhead suspension system. Lloyd Otteman and John Lacey designed the structure for guiding all the various components into alignment. Ed Lagucki devised an innovative "through-the-flowline" system of maintaining well production. Bill Foster came up with the instrumentation, and Ray Perner handled dimension control. Bob Carter, Keith Doig, and Art Williams were the overall project managers at BRC during the system and equipment development phase. The engineers invented many kinds of patented tools and equipment for the project. All the work progressed under tight security wraps. "We had cloak and dagger types of devises," Geer recalled. In order to preserve secrecy, Shell farmed out work on individual components to various manufacturers. "Except for team members themselves, no one had an integrated understanding of the project," said Geer.[27]

Douwe "Dee" DeVries, a mechanical engineer from Holland with broad experience in both refining and oil production, developed all the innovative equipment-handling systems that connected the wellhead to the *Blue Water 1.* Assisting Collipp in the conversion, DeVries applied the "spider deck assembly" concept, devised by Bill Craig at BRC, under the drilling derrick and developed the riser equipment and controls for the blowout-preventer stack. He also designed, built, and installed the first telescoping joint and buoyancy chamber, which became standard in floating drilling. For motion compensation, he devised an elevator system with two buckets that served as counterweights. "It was a terrific opportunity because money was no problem," recalled DeVries. "But the money we spent was worth it. We didn't just develop existing concepts. We did all this stuff from scratch."[28]

In 1959, Ron Geer and others assembled and tested the manifold system on dry land at Weeks Island and then tested it again in fifty-five feet of water off Louisiana, from a small platform at a drilling site where no oil or gas was likely to be discovered. The two tests revealed that some equipment modifications were needed. Back in Houston, at a site on Gasmer Drive adjoining Shell Pipe Line's Technical Laboratory, Shell Development built a test facility to simulate an offshore drilling platform for further work on the system. The construction of the strange-looking

12
Scale mode of
Blue Water 1 and
RUDAC under-
water wellhead
system. *Photo courtesy
Ronald L. Geer.*

mock-up rig lent mystery to the project. Passers-by speculated that the structure was everything from a drive-in movie screen to a rocket launching pad. In December 1960, after making the necessary improvements, the engineers took the equipment aboard the *Blue Water 1* (before its conversion to a semi-submersible) to a well site thirty-five miles off Cameron, Louisiana. There, in fifty-six feet of water, they lowered the guidance and alignment structure to the bottom. Five guide lines connected the columns of the guide structure to the barge platform and served as a means for landing casing, blowout preventers, television cameras, and production wellhead equipment. The subsea valves were electrohydraulically controlled from the surface and hydraulically operated through the high-pressure manifold system. On December 19, Shell's deepwater team made the world's first diverless underwater completion.[29] The well produced 288 barrels per day and continued to produce for another twenty years. After the successful test, Geer, along with Bruce Collipp, Dee DeVries, and John Lacey, prepared this Remote Underwater Drilling and Completion (RUDAC) system for operation with the converted *Blue Water 1* in deepwater. Max Clayton was assigned to be the overall project manager, and Geer was the project engineer.

After some delays in late 1961 caused by bad weather, beginning with Hurricane Carla in September, the *Blue Water 1*/RUDAC was ready for its real test. Fearing continued bad weather but pressed for time to demonstrate the worthiness of the concept before the March lease sale, the decision was made to go ahead. On a foggy night in January 1962, tug boats quietly pulled the new semi-submersible out of the Ingalls shipyard, attracting as little attention as possible. Collipp was the only Shell person aboard the rig, but other Shell engineers flew out by helicopter to meet him. "The weather was bad," remembered Geer, who was on that helicopter. "You couldn't see from the deck to the ocean surface." The rig proceeded on to its first well location forty miles southeast of Grand Isle, Louisiana, in three hundred feet of water. Over the next nine days, the crew carefully set the mooring system. The weather worsened, bringing driving rain, wind, and high seas. Although the vessel rolled and pitched, it remained relatively stable. "You walked down the deck grasping the hand rails and with your clothes flying around," said Collipp, "but this darned thing didn't move much." Contract crews who came out to service some of the equipment refused to believe the unit was floating. On one occasion, several welders cut off a piece of plate that fell off the side into the water. Standing nearby, Collipp told them that the plate had just fallen almost four hundred feet. "I don't think so," replied one welder. "We're sitting on the bottom." When told that they were actually floating in three hundred feet of water, both of them quit. Nobody could figure out how the *Blue Water 1* could float nearly motionless out on the open seas. The mooring system was hidden underneath, except for the crown buoys. The New Orleans papers speculated at length about what was happening. Rumors spread that it was sitting on a coral reef. Curious competitors spied from helicopters and work boats. One helicopter even brazenly tried to land on the rig's heliport before it was waved off. Oklahoma senator Robert Kerr, whose company Kerr-McGee was heavily committed to offshore, personally circled the rig for many days, first in an airplane and then in a boat. But like all the others, he came away with few clues.[30]

The structural integrity of the space-frame vessel, not its stability, worried Collipp the most. Living on the *Blue Water 1* for several weeks, he would lie in bed listening apprehensively to its creaks and groans. Geer and DeVries went many days without sleep, checking and monitoring everything. "I learned to sleep standing up like a horse," recalled DeVries. But the unit remained intact, and the systems worked. Rig engineers Norm Montgomery and Fines Martin relieved Geer and DeVries, after they had spent thirty-six straight days on the vessel, to supervise the drilling. In late January 1962, the *Blue Water 1* spudded the deepest offshore well ever, in 297 feet of water. As the crews sounded the bottom for the official depth of the first well, someone called out "297." "Damn!" Collipp remarked. "Why can't we just move it over a little to 300?" But that was quibbling. The well set a depth record by a long shot. During the spring and summer, the *Blue Water 1* went on to drill six more exploratory wells. Even though most of them did not encounter oil, after seven years and $7 million of research, Shell Oil had finally proven the viability of drilling and producing oil from depths previously unthinkable. When the company loosened the secrecy around the project and announced its findings in August 1962, the offshore industry began to adopt a new way of thinking. As one Shell representative told reporters who visited the rig: "We're looking now at geology first, and then water depths." [31]

Attacking the 1962 Lease Sale

The successful debut of the RUDAC *Blue Water 1* program gave Shell the confidence to bid on deepwater leases off Louisiana in the large federal lease sale of March 1962. In this sale, the government had decided to speed up the pace of development. Some twenty operators nominated 3.67 million acres, 6,000 square miles, most of which was off the coast of Louisiana (30 tracts were nominated off Texas) and in water deeper than 100 feet. Not all of this was prospective acreage, of course. A company typically nominated three tracts for every one on which it planned to bid, in order to conceal its true desires and misdirect competitors. Nevertheless, offshore operators clearly signaled their collective intention to raise the ante on the multibillion-dollar offshore game. The BLM decided to call their hands. It put up for lease everything nominated by industry, all 3.67 million acres comprised of 781 tracts, more than all four previous federal sales combined.

Slightly more than half lay in water less than 120 feet deep. The rest ranged out to as far as 600 feet, the edge of the continental shelf, which was far too deep for drilling units then operating in the Gulf.[32]

Shell's marine exploration division was itching for the sale. The leader in making preparations was division exploration manager Ronnie Knecht, another hard-charging geoscientist who took his mission to find oil for Shell very seriously. Sometimes, when his wells struck pay, Knecht was known to run down the halls of the office excitably announcing the discovery. A paleontologist with a degree from Louisiana State University (where, incidentally, he won the high jump event at the SEC track and field championships in 1946 and 1948), Knecht had studied in LSU's renowned geology department and worked in Shell's Baton Rouge exploration division. He had a firm grasp of Gulf Coast stratigraphy, and he was also a sharp technical manager, able to draw on the talents of his technical staff, who shared his belief that the geology offshore was just as promising as it was onshore and that salt domes offshore would have excellent traps, seals, and reservoir sands like those onshore. He was convinced that the application of emerging technology would overcome the difficulties sometimes encountered in finding pays around complex salt domes. Upon arriving at his new post in 1958, Knecht discovered that his boss, New Orleans exploration manager Freddy Oudt, was not high on salt domes. An old-time explorationist who preferred structures with simpler traps, Oudt had been burned by one south Louisiana dome in particular, White Castle, which at the time had not yielded much production but which later became a very good field. If Knecht or anyone else proposed a salt dome prospect, Oudt would invariably respond in his thick Dutch accent: "You don't vant another Vite Castle, do you?"[33]

Knecht was determined to make a case for attractive prospects offshore, including large salt domes. In 1959, he established a "Special Studies" team of geologists led by Jerry O'Brien to undertake a major quantitative study of all the known fields in south Louisiana. Knecht and his team hoped this study could determine the richness of fields by type and indicate if rich fields had specific characteristics that could be determined in advance of drilling using the technology of the time. According to O'Brien, "the idea was to look at all the field data available and arrive at a prospect evaluation which was based on systematic analysis rather than unsubstantiated theories." The team looked at more than four hundred fields and categorized them as salt domes (of all types), shallow non-salt features, or deep uplifts. They found that salt domes as a group were especially rich and that field richness was closely related to a number of characteristics, some of which could be determined by current exploration methods prior to leasing and drilling. Knecht and division geologists, geophysicists, and paleontologists agreed that the key characteristics that could be determined in advance were the area of structural uplift (big was best) and the thickness and quality of objectives (thick neritic sand and shale were best). The team finalized the study using these features and developed a prospect evaluation scheme.[34]

Shell also had paleontologists looking at the fossil record and estimating the age and environment of deposition in order to help predict which prospects would likely have producing sands offshore. The company had long done the most rigorous paleontological work on the Gulf Coast. In putting together the onshore salt dome package, Knecht drew heavily on his paleontologists, who were managed in the offshore division by Jim Lampton. The key was getting the correlations into the offshore on a regional scale, no small task. "It was a real challenge to do the detailed microscope work and synthesize the information into dependable time horizons for correlations from well to well," explained Ed Picou, who helped work up the data. "In a real sense, it was applied research in an operating division." Said O'Brien: "The paleontologists deserve a lot of credit. That was a big thing."[35]

The results of the team field study and prospect evaluation scheme significantly shaped Shell's approach to the 1962 lease sale. Not least, it won over Knecht's exploration bosses. When Knecht laid out the study and scheme for New Orleans Area Exploration Manager Freddy Oudt, Dr. Oudt looked at the data and was silent for a while. Then he finally sighed: "Vell Gotdam!"[36]

With agreement reached on the exploration objectives, the next step entailed collecting the geophysical data. The seven-man offshore geophysical group went on a six-day week for months with no extra compensation. They evaluated each area of interest that had been indicated by four-by-four-mile reconnaissance surveys. "We shot like mad," said B. B. Hughson, head of marine geophysics. By then, Shell's offshore crews had moved beyond the mine sweepers and acquired a more modern seismic fleet composed of larger steel boats built specifically for the open Gulf. Party 88, working under the New Orleans exploration office, had started using long-distance refraction, taking profiles up to 80,000 feet in length. While they were shooting their data, the crews would also gather intelligence and scout the coordinates of other companies' seismic operations, which was useful when it came time to develop a bid strategy.[37]

In addition to the refraction data, Shell's offshore crews also began to acquire better seismic reflection data than they had ever had offshore. Paper recording went out in the offshore in the mid-1950s. By the late 1950s, recordings were all made on photographic film, optically processed on the Reynolds machine and displayed in line sections. In 1960, Party 88 started shooting reflection seismic with a "stacked" line, using multiple sources and receivers (see discussion of stacking in chapter 2). Between 1960 and 1962, the most technologically advanced companies in the industry, including Shell, licensed the technology. Still the main signal-to-noise enhancing technique today, CDP stacking was a watershed that divided previous seismic exploration from all subsequent innovations. The first stacking operation by Shell's Party 88 used a cable 4,800 feet long with forty-eight hydrophones 100 feet apart. Although refraction seismic was still the main source of data used in the 1962 sale, reflection seismic was having a new impact.[38] With the abundance of new data gathered for the sale, the marine group's geophysicists produced detailed reflection maps, refraction interpretations on all shallow features, and gravity maps.

Once all the geophysical and geological data were collected and synthesized, and the maps drafted, Mac McAdams and Ronnie Knecht together reviewed every prospect. At one point, geologists in the marine division lined up outside Knecht's office, each clutching a large stack of maps. After the best prospects were narrowed down, the next step was formulating bid values. In the 1962 sale, the more advanced companies began to develop their bids, for the first time really, with rigorous and quantitative studies of reserve estimates, trap analysis, risk discounting, rates of return, and bidding tendencies of competitors. In previous sales, a lot of guesswork and hunches had gone into formulating "back-of-the-elbow" bids. But by 1962, the more advanced companies began to arrive at bids that contained more concrete numbers. Gene Bankston, who was with Shell Oil's E&P economics department at the time, explained: "a typical block would have some part of a potential oil field underlying it, and we would have to look at the probability of certain amounts of oil or gas, or both; and then, we had to provide a development scheme that showed how they would be developed and produced, and based on this calculate a value we could afford to bid, with the proper discounting for risk."[39]

In approaching the sale, Shell and other companies maintained obsessive secrecy about their bids and their methods of attaching dollar figures to tracts. A penny could theoretically decide who won or lost a particular tract. It is not clear how much money Shell Oil decided to expose at the 1962 sale, but it was a large amount for the time, probably something approaching $100 million. In fact, it was more than the company could handle on its own and certainly more than Bouwe Dykstra wanted to take responsibility for. Ronnie Knecht convinced McAdams and top management to borrow money from Royal Dutch/Shell so they could afford to bid what they wanted. Knecht traveled to Europe himself to present the data and make the pitch. So protective was he of the sensitive material, he kept the large folio with the maps and seismic profiles handcuffed to his wrist.[40]

The March 13 and March 16, 1962, lease sales became legendary in the industry. Everyone from that era remembers "the sale so large it took two days to read the bids." It was in reality one large sale split over two days. It pried open the Gulf of Mexico to a broader range of players: forty companies or combinations of companies bid successfully in the sale. Putting so much acreage up for sale not only provided more leases for a larger number of companies to choose from, but it also drove down the price of cash bonuses, allowing smaller companies to acquire a piece of the action. Still, the majors retained a commanding lead in exploration, especially in the deeper waters. All the hard work put in by Shell Oil's exploration people paid off, to the immense satisfaction of Ronnie Knecht. Out of the $446 million total paid by the industry in bonus bids, Shell laid out $45.5 million in winning bids for leases on fifty-seven tracts totaling 278,565 acres. The company was the third highest bidder in money, but it acquired the largest number of tracts.[41] The success in this sale led to the acquisition of significant leases in 300-plus feet of water. With this acreage and its deepwater drilling program, Shell was better prepared than any company in the industry to take the next step offshore.

Redefining the Possible

It was too prepared for its own good, however. On some of the deepwater tracts, Shell was the only bidder. "No one could operate in those depths," explained Ron Geer. "We were five to seven years ahead of the rest of our competitors." Because Shell's bids on deepwater leases did not have any competitors, the government did not honor them. This forced Shell E&P to reevaluate its secretive approach to deepwater research. Senior management concluded that there had to be competition, both to enable Shell to acquire the deepwater acreage and to stimulate the commercialization of the technology. Shell had pioneered a whole new frontier in offshore drilling, but it could not go at it alone. The costs and risks were too high. Other oil companies, as well as suppliers, manufacturers, and construction firms, could progress into deeper waters only together as an industry. The real commercial advantage for oil companies lay in figuring out where the oil was, not in figuring out how to produce it. "We realized that the only way we could ever have access to those frontier areas was to share our knowledge with the rest of the industry, to give them a base of technology from which they could expand," said Geer. "We'd still be ahead of the competition because of our hands-on experience and besides, we owned 160 patents on the technology."[42]

Thus, in January and February 1963, Shell held an unprecedented three-week "school" on offshore technology for representatives from industry and government. The company charged tuition of $100,000 per company, a sizeable amount of money for the time. Seven companies, along with the U.S. Geological Service, signed up for the series on all facets of Shell's deepwater programs. Collipp conducted seminars on floating drilling, and Geer led classes on the RUDAC system and new producing methods. Yet, as the participants discovered, these were not the only technologies Shell had developed. Howard Shatto, a division mechanical engineer from Shell's Pacific Coast Area office, presented other Shell innovations that had been achieved through a parallel program of research and development for the West Coast.

Initiated in 1959, this program sought ways of operating in the rough seas and deep waters off California. While the Gulf Coast group had been working on the RUDAC *Blue Water 1* system, Bill Bates, vice president for the Pacific Coast Area, had conceived another approach. He assigned Shatto's group to develop a different method of completing underwater wellheads and work on dynamic vessel positioning. Bates had urged Ned Clark to establish the second research program in order to increase Shell's chances of finding a workable solution to the deepwater dilemma. "The

13
Shell Oil engineer
Ron Geer speaking
at Shell's "school
for industry"
held in January–
February 1963.
*Photo courtesy
Ronald L. Geer.*

stakes are so high in offshore oil production that we had to be sure we had at least one system which would work successfully," Clark explained.[43]

An important innovation of the Pacific Coast deepwater program was a system developed by Shatto for keeping a floating vessel in position dynamically without anchors or mooring lines. In March 1961, the *CUSS 1* first successfully tested the concept of dynamic positioning using manually controlled thrusters to keep the vessel on location. Concurrently, Shell was preparing to outfit a small core drilling ship, the *Eureka*, with a manually controlled system. But Shatto felt that such a system was too difficult for a human operator to coordinate, and he convinced his bosses to equip the *Eureka* with an automatic system that he designed. It featured two thrusters and three analog controllers, which evaluated the three basic degrees of motion on the vessel—surge, sway, and yaw—and transmitted its drift-correcting orders to the thrusters.

With the work being done in the basement of the Shell office in Los Angeles, secrecy was as tight as or tighter than in New Orleans. "We put all of the things we were working on in grocery shopping carts at night, and wheeled them into a bank vault," remembered Shatto. "They took our names out of the telephone book. We just disappeared all together." In May 1961, two months after the *CUSS 1* tests, the *Eureka* began core drilling with great success. Anchored coring vessels of its size took a day or two to set anchors, drill a core hole, and recover anchors. Their water depth was limited to 200–300 feet. By comparison, the *Eureka* drilled as many as nine core holes in one day and in water as deep as 4,000 feet. Beyond its success in core drilling, it validated the concept of automatic dynamic positioning and initiated a new trend in the evolution of drillships. Since then, the industry has built about fifteen hundred dynamically positioned vessels around the world for applications from pipe laying to cruise ships.[44]

The underwater completion system developed by Shatto and other Pacific Coast engineers addressed the perceived need for diverless operations in a unique way. Code-named MO, for "manipulator operated," the system featured the use of a free-swimming remote-controlled robot "diver" designed by Hughes Tool, which had a mechanical arm capable of turning lock screws, operating valves, and attaching control hoses and guidelines. Driven by propellers and guided by sonar and a television camera, the so-called "Mobot" could be lowered by a wire cable and attached to the wellhead equipment. It then rode around the wellhead on a circular track to perform its tasks. "It was basically a swimming socket wrench," said Bill Petersen, a Shell mechanical engineer who worked on the system. The idea was to have a very simple, land-type wellhead, in contrast to RUDAC, but with a rather complex support tool in the Mobot, which could be easily recovered to the surface for repair. Other innovative engineering features of the system included flexible joints that could bend with the motion of the drilling ship and a remote-control device for disconnecting the riser from the wellhead equipment. In October 1962, Shell successfully tested the system with the Mobot by drilling a wildcat in 250 feet of water near Santa Barbara.[45]

The dazzling array of new technologies demonstrated by Shell in its three-week course encouraged the industry to tackle ever-increasing water depths and more hostile environments. Semisubmersible drilling vessels led the way. In 1963, ODECO christened the *Ocean Driller,* the first purpose-built semi-submersible. That same year, Shell contracted with the Blue Water Drilling Corporation (later purchased by the Santa Fe Drilling Company) to build the *Blue Water II.* A larger vessel than its sibling, it had the drilling rig in the center and was designed to operate in the stormy seas of the northern Pacific Ocean. After drilling wells off northern California, on leases obtained in May 1963 at the first federal lease sale ever held on the Pacific Coast, the *Blue Water II* moved north to drill on federal leases off Oregon and Washington obtained by Shell in October 1964. Soon, shipyards turned out numerous semi-submersibles of varying designs that opened up deeper and rougher areas of the world to exploratory drilling. By 1968, over twenty were in operation or under construction.

Drillship designs and dynamic positioning technology also evolved. In 1962, Global Marine, the successor to the CUSS group, launched the *Glomar II*. A much larger drillship than the *CUSS I*, the *Glomar II* found oil for Shell in the Santa Barbara Channel and later drilled in the ice-infested waters of the Cook Inlet in Alaska. Although drillships could move faster across the ocean than semi-submersibles, they still experienced substantial downtime in harsh weather. Each successive generation of drillships, however, has achieved ever-greater stability.

The technologies developed by Shell during those years still had to be improved and perfected, which in some cases would take many years. Nevertheless, they laid the foundation for deepwater floating drilling, floating production, remote-operated vehicles, and subsea completions, all of which became increasingly important as offshore operations moved deeper and expanded to other parts of the world. Most importantly, they established a new learning curve, redefining what was possible at the time and helping other companies in the industry see the potential of offshore. In 1971, Shell received the Offshore Technology Conference's first annual Company Distinguished Achievement Award for its numerous advances in deepwater drilling technology from 1955 to 1967. Bruce Collipp and Dee DeVries were later awarded the prestigious Holley Medal from the American Society of Mechanical Engineers for their contributions to developing the semi-submersible. Other Holley winners include Henry Ford; Polaroid's Edwin Land; Ernest Lawrence, Nobel Prize winner and inventor of the cyclotron; and Elmer Sperry, who invented the gyroscope. Collipp also received the Blakely Smith gold medal from the Society of Naval Architects and Mechanical Engineers and the Gibbs Medal from the National Academy of Science. In 1984, 2001, and 2002, respectively, Geer, Shatto, and Collipp each received the OTC's Individual Distinguished Achievement Award for their pioneering contributions to deepwater drilling and underwater completion technology. In his closing remarks at Shell Oil's school for industry in 1963, Douglas Ragland of Humble Oil commented that never in the industry had any company made a presentation that had such an impact on the future.

In many ways, the deepwater drilling technologies pioneered

in the 1960s by Shell were way ahead of their time. They allowed the industry to explore out to 600 feet but not to produce from those depths. The new methods of completing wells on the seafloor were very costly and still faced a multitude of technical challenges that forestalled their commercial application. Underwater completions involved a lot of expensive hardware and controls under high pressure on the ocean floor, in addition to the large expense of operating floating vessels to complete, maintain, or work over a single well. Gulf of Mexico wells often produced paraffin, wax, sand, and hydrates, which plugged the flowlines and caused endless headaches and complications. Furthermore, Shell's deepwater research program had not really addressed how to handle production from great water depths. Consequently, when Shell presented its resumé of deepwater operations to the industry, the New Orleans area office did not yet have a strategy for developing the deepwater leases acquired in 1962.

The urgency to establish production from these tracts within the government's five-year time limit prompted the New Orleans production department in 1963 to conduct detailed economic analyses of the various options for drilling and producing from deep water. In the late 1950s, Shell had adopted the use of mobile drilling units (jack-ups and submersibles) to drill and complete exploratory wells. If a discovery was made, the well was left standing and a small "well protector" platform was installed. The 1963 study proposed a new kind of approach using mobile or floating vessels to drill exploratory wells, abandoning them even if a probable discovery was made, and then redrilling them from a self-contained or tender platform at a later date when sufficient reserves were proven, at little additional or even less cost.[46] Once the preconception against abandoning wells was overcome, the "expendable well approach" proved to be the most cost-efficient way to proceed, and it provided a dependable example for other companies to follow.

In 1963–64, Shell's exploratory drilling program, employing the *Blue Water 1,* the *Ocean Driller,* and other vessels, made major discoveries in depths of 200–300 feet, most notably on the Marlin prospect in the West Delta area of the Gulf. Meanwhile, Bellaire researchers and Shell's Offshore Construction Design Group in New Orleans performed intensive technical work—engineering calculations, design analysis, oceanographic research, and economic studies—for new and larger fixed platforms. In 1965, the two leading offshore construction firms, Brown & Root of

Houston and J. Ray McDermott from Morgan City, Louisiana, built and installed five giant platforms for Shell's multimillion-dollar "Marlin System." Standing in waters ranging from 223 to 285 feet deep, these platforms were taller than all others in the Gulf. Each contained everything to drill and complete twelve wells, along with crew quarters, kitchen, dining hall, and recreation room. Brown & Root also laid a forty-mile crude pipeline to connect the platforms to Shell's Delta Pipe Line, which linked the company's shallow-water fields to the Norco Refinery near New Orleans. Shell Pipe Line's patented design for an "articulated stinger," a long flexible pontoon that eased pipe from the pipelaying barge to the bottom without buckling, facilitated the laying of what was then the world's deepest marine pipeline. In 1996, Carl Langner, senior staff research engineer, won the OTC Individual Distinguished Achievement award for his work on the articulated stinger and other pipelaying technologies, on which he acquired numerous patents.[47]

By the end of 1966, Shell's gross oil and condensate production from federal leases acquired since 1960 amounted to 33,000 barrels per day. This represented an investment of $60 million since 1963, largely for the construction of twenty-five multi-well drilling and production platforms, nine of which stood in water more than 200 feet deep. As John W. Pittman, executive assistant to the New Orleans Area vice president, remarked in 1967: "Shell has spent more money, drilled more holes, and produced more oil offshore Louisiana than anyone else."[48]

Rapid advances in platform and pipeline designs were critical to this development. The digital computer, introduced in the early 1960s, revolutionized design techniques and the analysis of wave forces on marine structures. "In a little over seven years, engineers went from electrical/mechanical desk calculators to finite element analyses of complex tubular joints and complete three-dimensional computer models of 400-foot structures," wrote Pat Dunn, one of Shell's top offshore design engineers.[49] As Shell and other companies moved gradually into deeper water, economics consistently favored fixed platforms over subsea completions. "Every time we did an analysis, we found that the cost of doing this underwater system went up at the same time our knowledge and actual experience with platforms also went up," said Gene Bankston, head of E&P Economics in the early 1960s.[50] Better-quality steel and welding practices improved the strength of joints on platform jackets, reinforcing their resistance to large waves.

Construction equipment grew in size, and giant launch barges were built to assist in platform installation. Divers aided offshore construction by finding ways to go deeper on mixed-gas air and in diving chambers.

The industry also learned valuable lessons from two "one hundred year" hurricanes in the mid-1960s that battered offshore installations and caused damages in the hundreds of millions of dollars. In October 1964, Hurricane Hilda ripped through the Eugene Island–Ship Shoal area spawning six tornadoes and destroying fourteen production platforms, most designed only to withstand a twenty-five-year storm. A year later, Hurricane Betsy swept across the offshore oil breadbasket in the central and eastern gulf generating wave heights estimated at over seventy feet, demolishing more structures and causing widespread damage and flooding in the city of New Orleans. By studying these hurricanes, engineers gathered data about meteorology, oceanography, soil movements, and hydrodynamics, which allowed them to make crucial design improvements to platforms.[51]

Shell's facilities offshore Louisiana fared better than most during these storms, thanks to a policy of designing and constructing structures to withstand high waves and winds. But the company did incur damages and extended shutdowns. The *Blue Water 1*, which was operating thirty miles offshore when Hurricane Hilda arrived, suffered a strange and catastrophic misfortune. Its crew escaped to shore ahead of the storm, but upon returning they found the vessel floating upside down, still in its anchor pattern. Apparently, one of the tornadoes produced by the hurricane had flipped it over. After several failed attempts to right the *Blue Water 1*, Shell decided to abandon and sink it. At the last minute, oilman John W. Mecom purchased the vessel with the hope of salvaging it. He moved it to shallow water but still had not saved it when Hurricane Betsy passed through and severed its mooring. The capsized semi-submersible then drifted fourteen miles across the Gulf and slammed into one of Shell's newly installed Marlin System platforms at West Delta Block 134, shearing away four of the platform's eight legs. Nobody actually saw what happened. But the *Blue Water 1* left evidence. One of its columns was found inside the platform. The improbable irony of the incident was almost too astounding to believe.[52]

The company recovered from these setbacks and forged ahead. The big offshore development for Shell in the mid- to late 1960s

14
Shell Oil's West Delta 133-A platform after being hit by the *Blue Water 1* during Hurricane Betsy, 1965. *Photo courtesy Peter Marshall.*

happened at South Pass Block 62 in the Gulf of Mexico. Acquired in the 1962 federal lease sale, this tract lay beyond known territory in excess of 300 feet of water. Shell bought South Pass 62 for a measly $188,000. "We didn't know if there would be any sands over there or not," explained Billy Flowers, who was division exploration manager in New Orleans at the time. "But we took a chance." Exploratory drilling on South Pass 62 was delayed several years as Shell appealed to the Coast Guard and other government agencies to have a shipping fairway over it moved. When the *Ocean Driller* semi-submersible finally drilled the tract in 1966, it discovered a field that turned out to contain over 100 million barrels of oil, which put it in the "giant" category. The next year, Shell designed and installed a platform in 340 feet of water in South Pass 62. With this mammoth structure capable of drilling eighteen directional wells, Shell set yet another world record for the tallest fixed platform.[53]

This and subsequent platforms installed in South Pass 62 were landmarks of deepwater development in the 1960s. But as Shell developed this field, its engineers already were studying the possibility of moving deeper. The Gulf Coast and West Coast groups that had worked separately on floating drilling and underwater completions merged in 1964 to become the Marine Technology

Group (MTG). Headed by Ron Geer, the MTG looked to extend floating drilling and subsea wellheads to 1,000-foot depths. At the same time, other engineers at Shell analyzed designs for deepwater fixed platforms. Technological innovation had brought them this far, and they believed it could take them even further.

The Trials and Triumphs of Exploration

Compared with any other energy source in the 1960s, oil was cheap. Throughout the decade, the price per barrel of crude in the United States remained in the range of two to three dollars. Import quotas limited the entry of even less expensive foreign oil into the United States and raised relative costs to consumers, but domestic oil producers filled the breach left by reduced imports, and the buyer's market endured. As the increasing scale of production fed swelling consumption, oil companies battled to carve out new market share by building refineries and service stations to take expanded throughput and engaging in periodic gasoline price wars. A wide array of new technologies allowed them to increase the refinery yield of high-value products from a given barrel of crude and improve the efficiency of transportation and distribution. These innovations also stimulated the growth of a vibrant petrochemical industry that produced a wide range of plastics and synthetic materials, which quickly became staples of modern society.

At Shell Oil during the 1960s, senior management placed new emphasis on refining, transportation, marketing, and chemicals as profit centers and diverted a greater share of the company's overall budget to these functions. Gone, at least for the time being, were the days when these downstream activities were regarded merely as a way to dispose of crude, the main source of profit. Although E&P was not starved for money and still commanded the fattest portion of the budget, it was not indulged in the same way it had been under Max Burns, who once reportedly complained, "The marketing people are thinking again, we pay them to peddle, not to think." Yet, at the same, E&P managers felt added pressure to find and produce oil. Through the application of enhanced recovery technology and the increasing pace of offshore operations, Shell steadily increased its production of crude and natural gas liquids from 400,000 barrels per year in 1960 to 500,000 barrels per year in 1965. By 1970, offshore Louisiana and Texas accounted for 45 percent of Shell's total crude production and 35 percent of its natural gas. But offshore Louisiana crude was more expensive than onshore, and oil and gas production lagged behind the big surge in marketing. The company could supply crude for only about 60 percent of its refinery needs and each year had to purchase larger amounts from the Group or other companies.

Shell Oil E&P, therefore, was caught in a three-way bind between cheap oil, rising costs, and a stretched capital budget. The only way to break this bind was through the development and application of innovative technology. Even with the new corporate focus on the downstream businesses, top management's support allowed E&P to make great technological strides during the 1960s, especially in exploration. Digital computers enabled a quantum leap in the amount of data that could be handled and manipulated, which led to almost continuous innovation in seismic processing and interpretation. Shell Oil remained on the leading edge of this remarkable period of innovation. Shell's geophysicists used advanced seismic technology to decipher the geology of Michigan's pinnacle reef trend and make the company's most profitable onshore oil play ever. And they pioneered the revolutionary method of directly detecting hydrocarbons—"bright spots" in Shell's parlance—which radically improved the accuracy of exploration in the Gulf of Mexico and similar environments around the world. Although pained by their failure to discover the giant Prudhoe Bay field in Alaska and a costly and unsuccessful drilling program off the coasts of Oregon and Washington, Shell's exploration team kept the company's stride in the race against depletion and its lead in the increasingly important offshore domain.

Competition for Capital

Upon Max Burns's retirement in 1961, Monroe Spaght—"Monty" as he was familiarly known—assumed the presidency of Shell Oil. A self-styled "scientist turned businessman," he was not the typical oilman who had spent his career in the field looking and drilling for oil. Tall and professorial, a research chemist with a Ph.D. from Stanford, Spaght was the first U.S.-born president of Shell and the first to come from outside exploration and production. He spent the first thirteen years of his career in various parts of Shell's manufacturing operations before serving during the war on the Aviation Gasoline Advisory Committee, the Naval Technical Mission to Europe, and the United States Strategic Bombing Survey in Japan. After the war, he helped reconstitute the Group's worldwide research organization and became president of Shell Development Company in 1949 and rose to executive vice president in 1953.[1]

Highly intelligent and self-confident, Spaght understood both the oil and chemical sides of the business, and he came into office aware of the need for dramatic changes. Royal Dutch/Shell recently had reformed its internal organization and had begun construction of its new London headquarters, a towering edifice on the South Bank of the Thames called Shell Centre. It was time for Shell Oil to change as well. The company had expanded rapidly during the postwar boom, and the downturn of the late 1950s exposed inefficiencies and excessive costs that had been bred into that expansion. Although Shell was still in sound financial condition, its business performance had weakened, and its net income had declined. The company faced stiff competition downstream in chemicals and refined products and upstream from a rising group of independent oil companies. "Both abroad and at home, streams of oil began to flow around the control of the majors," wrote John Blair in his classic study, *The Control of Oil*.[2]

Spaght responded by immediately embarking on a cost-cutting campaign that included the first significant lay-offs since World War II. He ordered some salaried employees into early retirement and terminated others, reducing the work force by more than 11 percent. He eliminated and consolidated marketing divisions, and in 1962–63 he faced down the company's unionized workforce, winning a bitterly contested and uniquely coordinated strike by three different unions at the company's three major refineries (Deer Park, Wood River, and Norco), a watershed in labor relations at Shell Oil that reasserted the power and prerogatives of management.[3]

No part of the company was spared. E&P experienced both a reduction in people and greater centralization of authority over personnel management at the head office in New York. Shell had not made a habit of firing people or retiring them early, certainly not in E&P, so it was a shock when these decisions came down. Monty Spaght did not earn the kind of reputation for compassion that Max Burns had enjoyed. Although close friends, the two had completely different styles. Burns managed with finesse and charm. Spaght was direct and ruthless. He was not concerned with his popularity ratings. "Monty was more like a calculating machine," said A. J. Galloway. "He was absolutely logical and followed the logical result down to its end which sometimes would bring him the wrong result because he was dealing with people, who are not logical." Shell's generous pension program and liberal severance arrangements, the "golden handshake," cushioned many of those who were let go, but the reductions still created morale problems. "It wasn't easy," admitted Spaght. "But I felt I couldn't sleep at night unless I got on with it."[4]

After the bloodshed, Spaght went to work reshaping the company with an eye toward balancing Shell Oil's integrated business. Seeing new profit potential downstream with cheap crude, he allocated a larger part of the budget to advertising and the expansion of retail outlets. As a result, the company dramatically increased its share of the U.S. refined products market. When Spaght was promoted within the Group to managing director in 1965 (the first American to hold the position), his successor, Richard C. McCurdy, president of Shell Chemical since 1953, carried on the work of investing downstream. Under McCurdy, Shell made huge outlays to build new service stations and expand its refineries and chemical plants. For this, however, the company needed to tap outside sources. Like most majors, Shell Oil traditionally had been very conservative about borrowing. Since World War II, it had relied mainly on internally generated funds. But beginning in 1965, Shell Oil, like its other major competitors, took on relatively large sums of debt. The firm simply could not rely on earnings to pay for expensive new projects. Over the next three years, Shell's total long-term debt soared from $267 million to nearly $720 million. But even after the first two $150 million bond issues, a cash crunch in mid-1966 forced Shell to put together a package of short-term loans from a group of New York banks. As spending continued, Shell raised a crucial $304 million in 1968 through a new share offering to its stockholders. The Group remitted more than $200 million to buy additional shares, while U.S. minority shareholders added another $95 million.[5]

Under Spaght and McCurdy, E&P was forced to compete harder for money, which led to some unpleasant budget fights. The managers of E&P argued for increased funding to support their wide-ranging exploration program, but McCurdy held firm to Spaght's policy of distributing a rising share of the total budget to manufacturing, transportation, and marketing. While E&P expenditures remained the largest of any function, the disparities were no longer as great as they had been during the early postwar period. During 1959–63, E&P consumed more than 70 percent of the company's average annual capital budget of $275 million. By 1965, this share was down to 54 percent ($262 million out of $485 million), where it remained for the rest of the decade. In the mid-1960s, having forecast growing sales of gasoline and other oil products, Shell embarked on a new wave of refinery modernization and expansion and moved aggressively into petrochemicals, especially ethylene and propylene. The company also fought a long but ultimately unsuccessful battle against local political and environmental opposition to build a new refinery on the East Coast, where fewer import quota restrictions made importing cheaper foreign oil an attractive opportunity.[6]

The new downstream focus was partly a function of E&P's success in producing rising volumes of crude, especially from offshore Gulf of Mexico. In the mid-1960s Shell had to expand its capacity to handle new volumes of oil and natural gas flowing from leases obtained in the 1960 and 1962 federal sales. In 1965, Shell completed the Marlin Pipe Line, which branched off the Delta Pipe Line servicing the East Bay fields. The Marlin line ran into deeper water to transport production from the company's West Delta fields (see chapter three). Shell Pipe Line also laid the Blue Dolphin Pipe Line, which brought gas from Shell's Buccaneer field into Freeport, Texas. The following year, the company completed two more offshore pipelines: the Bay Marchand Line running from Shell's platforms in South Timbalier Block 26 (see below) and the Central Gulf Gathering System, which connected fields in the Eugene Island and South Marsh Island areas to shore. In 1967–68, Shell led a consortium of companies in constructing Capline, a 630-mile, 40-inch-diameter crude pipeline extending from St. James on the Mississippi River in southern Louisiana to Patoka in south-central Illinois. Conceived, designed, and operated by Shell Pipe Line for itself and seven other companies, Capline had the capacity to channel more than 400,000 barrels of crude per day from the Gulf Coast bayous to the energy-hungry Upper Midwest, and it remains today a central North-South oil artery for the United States.[7]

Meanwhile, as all these pipelines were built, Shell prepared to develop a major pipeline gathering system for both oil and natural gas in the Central Gulf. The Red Snapper pipeline, an ambitious $127 million project headed by Shell Oil for 29 other companies, would have consisted of a 300-mile system for gathering oil and gas from water 100–150-feet deep. A competing consortium led by Tennessee Gas Transmission Company, however, eventually beat out Shell's Red Snapper group and won authorization from the Federal Power Commission in 1971 to lay its Blue Water system.[8]

The changed investment emphasis presented a serious dilemma for E&P managers. They needed to find ever-greater amounts of petroleum to supply these expanded downstream assets and replace record withdrawals from the company's crude reserves yet without a rising capital budget to achieve this. In 1964–66, Shell E&P battled manufacturing and chemicals over budget allocations and had to pick and choose among various exploration ventures. While the New Orleans Area focused on bringing in production from its large inventory of leases obtained in the early 1960s, the spotlight for exploration shifted to offshore West Coast and Alaska, two largely unexplored frontiers. With declining production in California and a rapidly growing population and demand for oil all along the West Coast, company leaders were growing anxious about finding a new source of supply for this region. Ideally, Shell's exploration managers would have pursued both offshore West Coast and Alaska with equal vigor, but money constraints prevented this, much to their lasting regret after what eventually transpired.

Alaska and West Coast Offshore

Shell Oil first entered Alaska in 1952, following the arrival of several other companies. Sam Bowlby, E&P vice president for the West Coast, spearheaded the campaign. An energetic exploitation engineer and protégé of Max Burns, Bowlby assumed the top job in Los Angeles in 1946 and held it until his retirement in 1964. For years, he was a prominent spokesman for the West Coast oil industry and a strong civic leader. Bowlby was also an avid fisherman who became enchanted with Alaska but not just for the excellent fishing. He believed the territory had significant oil potential. It possibly held the key to the continued functioning of Shell E&P as a going concern on the West Coast.

During 1952–56, the Northwest Division office in Seattle sent exploration parties to conduct geological and surface mapping expeditions on the Kenai and Alaska peninsulas, the Yukon Delta, and the North Slope. In 1955, the company began using helicopters to do wider-ranging reconnaissance and ran its first seismic tests in Wide Bay, along the Alaska Peninsula, which revealed some promising oil-bearing structures. Because of the unusually high costs of drilling and operating in the extreme conditions of Alaska, Shell took on Humble Oil as a partner and drilled a well on the coast of the bay. "They had over 70 days of winds of 100 miles an hour or more, a real hairy exercise," said Bowlby. "They found all the proper sediments, but they were too tight to produce oil. So nothing happened there."[9] Richfield Oil, in 1957, made the first significant strike in Alaska with the discovery of the two-hundred-million-barrel Swanson River field on federal land in the Kenai Peninsula about eighty miles southwest of Anchorage. This discovery accelerated momentum toward Alaskan statehood and attracted throngs of oilmen to the region. "Throw a snowball in the Territory these days and you're likely to hit a geologist," said one Alaskan at the time. By the end of 1957, the Bureau of Land Management had leased out nineteen million acres. A group of politically connected Anchorage businessmen, led by a former Army-Navy surplus store clerk named Locke Jacobs, acquired many of these leases and then turned them over to oil companies at a nice profit.[10]

After achieving statehood in 1959, Alaska began leasing tracts in the Cook Inlet, Anchorage's outlet to the Pacific Ocean. This is where industry's interest had turned in the wake of the Swanson River discovery. In 1959, Shell participated in a joint seismic survey with ten other companies in Cook Inlet. Two years later, the company acquired leases on nearly thirty thousand acres there in partnership with Socal and Richfield and another thirty thousand acres the following year. As operator of the partnership, Shell contracted for Global Marine's drillship, *CUSS II* (soon renamed the *Glomar II*), to drill in the extreme tides (up to 35 feet) and currents (up to 8 knots, changing direction every six hours) of the inlet. In September 1963, the vessel struck oil at a place called Middle Ground Shoal. Further tests revealed, impressively, a productive interval 550 feet thick. This was the second major discovery in Alaska and the first in the Cook Inlet, which would yield finds on five major fields (four oil and one gas) by the end of the decade, making it one of the hottest plays in the United States.[11] The rapid development of the Cook Inlet fields between 1963 and 1968 was an extraordinary engineering accomplishment for Shell Oil and other operators and contractors because of harsh environmental conditions that consisted of ice, tides, current, bitter cold, and earthquakes. The lack of design criteria for counteracting the extremely great lateral forces caused by winter ice floes made this development even more remarkable. In August 1964, Shell began producing from the field using a twenty-four-well drilling and production platform housed on an innovative four-legged structure designed to withstand the ice forces and tidal changes. At a cost several times more than in the normal offshore platform,

15
A Shell Oil ice-
breaker platform
in Alaska's Cook
Inlet at the start of
winter, late 1960s.
*Courtesy Anchorage
Museum Archives.*

the four legs were built of low-temperature steel and designed to withstand a sideways thrust of five to six thousand tons—"twice the thrust of the Saturn V rocket used in the Apollo moon project," said one engineer.[12]

After the discovery in Cook Inlet, the Pacific Coast exploration office looked to northern Alaska. The North Slope area of the Brooks Range mountains already had known petroleum deposits. In 1923, Pres. Warren Harding had set aside a twenty-five-million-acre site as a naval petroleum reserve (NPR). Geological investigations in the 1920s and again during World War II confirmed a field there containing one hundred million barrels. It was logical, then, as the petroleum hunt heated up in Alaska, that industry and government would expand the search north. During the summer seasons of 1959–61, many of the larger oil companies, including Shell, did geological field work and acquired lease acreage on the North Slope. In December 1960, the federal government established the Arctic National Wildlife Refuge (ANWAR), limiting industry entry to the area lying between the Colville River on the NPR side and the Canning River on the ANWAR side. During 1961–64, exploration programs expanded over what was left on the North Slope between these rivers. Sinclair and British Petroleum (BP) began collecting seismic data in 1962, shortly after forming a joint exploration venture. Between 1962 and 1964, BP-Sinclair greatly expanded their seismic acquisition, including that from Prudhoe Bay, while Shell did not.[13]

In the post–World War II period, Royal Dutch/Shell and BP had placed different emphasis on Alaska as an exploration area. During the war, Royal Dutch/Shell had lost significant reserves in Romania and Indonesia. Despite the 1947 discovery of oil by a Standard Oil of New Jersey affiliate at Leduc, near Edmonton, Alberta, Royal Dutch/Shell decided to bet on and invest in Venezuela and Nigeria instead of Canada and Alaska. When Shell Canada, with upstream assistance from Shell Oil, did enter Western Canada in 1949, it had missed out on most of the attractive reef plays and had to focus on the Alberta foothills. Western Canada and Alaska never occupied a position of high priority in the larger international Shell organization. BP also entered Western Canada late, but they placed a higher value on Alaska. In reviewing Sinclair Oil's Alaska database in 1959, a BP team of geologists and business people highgraded the North Slope. Their senior geologist clearly understood that structural plays, such as the ones Shell was involved with in Canada, would be less attractive than the huge regional arch of northern Alaska, which was more analogous to the kinds of oil plays BP was used to in Iran, Iraq, and Kuwait. BP geologist Roger Herrera noted that his company's concept of exploration "was radically different" from American oil companies, who trained their geologists and geophysicists to spend years accumulating detailed knowledge of specific places. BP, by contrast, said Herrera, moved its geoscientists every two years. "Get him out of there, because he ceased to look at the big picture. And, at that time, BP was interested only in the big picture." Said Bert Bally, Shell Oil's chief geologist at the time, "It is now obvious to me that their regional experience in the Middle East strongly guided BP's thinking. At Shell, neither the folks in The Hague nor the folks in New York or Los Angeles had that kind of background and experience."[14]

Thus, by 1964, BP had the "big picture" of the North Slope based on regional seismic data that showed the regional arch, or Barrow arch, culminating in two major structural arches or "highs," the Colville high and the Prudhoe Bay high. Based on seismic control across Prudhoe completed in 1964, BP-Sinclair mapped all the basic components of Prudhoe, which consisted of a regional south dip, a down-faulting dip to the north offshore, and truncation to the east. The key was the north dip, in an area where everything else was dipping south. Shell geologists, by contrast, did not see the dip and identify the structure in their data. They had mapped the anticlinal folds along the front of the range and had participated in one regional seismic line that ran north toward the coast of the Beaufort Sea. But the line revealed no structures. "It was just as flat as it could be," remembered Sam Bowlby.[15]

Bowlby and his exploration leaders, especially Seattle exploration manager Dick Story, nevertheless pushed for more seismic exploration on the North Slope to give the region another chance before the state held lease sales on North Slope acreage in December 1964 and July 1965. E&P Economics had done a feasibility study of Arctic development, however, and the results were not encouraging. Because of the distance to markets, the lack of infrastructure, and the high costs of drilling through the permafrost and transporting oil over it, the only way to make money in Alaska was to find a multibillion-barrel field, which seemed highly unlikely given the geology of the region. Bob Sneider, who worked on the December 1964 Alaska state lease sale evaluation team, remembered Bert Eastin, vice president of production, asking him, "How many 2–3 billion barrel fields do you know of?" Moreover, a 1963 Shell Oil pipeline study estimated that a twenty-four-inch pipeline running to Anchorage would cost nine hundred million dollars. Even if oil were found—and most people expected that it would be—the conclusion was that there would never be enough oil to pay for the pipeline, let alone make a profit. On top of that, the structures that Shell geologists had mapped along the front of the Brooks Range appeared to peter out toward the coast. As Bowlby explained, "we couldn't laugh off the fact that we had a seismic line that was technically good in every respect, but had no character to it."[16]

Still, in previous years, the proposal to shoot more seismic on the North Slope might have been funded. As budget battles were being waged with the downstream functions in the head office, however, Mac McAdams and Ned Clark, the top E&P officials in the company, had to choose between the North Slope and exploration off the coasts of Oregon and Washington. They could do one or the other but not both. Bill Bates, marine exploration manager for the West Coast, lobbied hard to acquire the offshore leases and won. As a result, Shell Oil participated in only a minor way at the state sale of leases at Prudhoe Bay, Alaska. In the December 1964 sale, BP-Sinclair acquired the Colville high acreage. In the July 1965 sale, a Richfield Oil–Humble Oil combination won the crestal portion of Prudhoe Bay, outbidding BP, which acquired most of the acreage on the flanks of the structure. Richfield did not have sufficient seismic control to define the faulting to the north, but they had enough information to be aware of its existence, some of which, BP explorationists later charged, was stolen from BP. The closest Shell Oil ever got to winning a lease in that sale was as the tenth highest bidder.[17]

At the time, it did not seem to matter. Shell had just demonstrated the spectacular potential of deepwater offshore drilling with its *Blue Water 1* semi-submersible in the Gulf of Mexico, and seismic tests off the narrow continental shelf from Point Conception, California, northward showed some very promising structures. Furthermore, by 1965, the Los Angeles office already was drilling there. In May 1963, at the first federal lease sale ever held on the Pacific Coast, Shell Oil paid more than eleven million dollars to obtain forty-nine of the fifty-eight tracts leased off the northern California coast. In the fall, the company tried drilling those leases with the *CUSS 1* and *Glomar II* drillships, but operations were eventually suspended because of inclement weather. "We were down 80 percent of the time," remembered Bruce Collipp. "Everybody was getting worn out, bruised and battered—so were the rigs." Consequently, the decision was made to introduce the semi-submersible concept to the West Coast. Shell contracted with the Blue Water Drilling Corporation to construct the *Blue Water II*, which was modeled on the *Blue Water 1*, but larger and specifically designed to withstand harsher conditions. Built by Kaiser Steel Company in Oakland and subsequently purchased by Santa Fe Drilling Company, the *Blue Water II* spudded its first well in August 1964, in 440 feet of water. Although it did not find oil, this successful test of the vessel's capabilities encouraged Shell's exploration managers to move on to the next federal lease sale for tracts off Oregon and Washington.[18]

The industry's interest in this sale was higher than for the previous one off northern California. In October 1964, Shell Oil acquired twenty-three tracts (132,000 acres) for $8.9 million, most of which were off Oregon (the industry paid a total of $35 million). Gerry Burton, who became Pacific Coast vice president upon Sam Bowlby's retirement in 1964, was particularly enthusiastic about the prospects. Less enamored of Alaska than Bowlby, Burton had headed the company's first marine division in New Orleans and felt the company's marine skills could be put to good use offshore in the Pacific. The geology also looked attractive. "Shell's exploration group theorized that perhaps the volcanic sequence found onshore would not be present offshore," said Burton at the time. "Out there, rocks might be found similar to those which have been so prolifically productive in oil in Southern California and, more recently, in Alaska." [19]

When the *Blue Water II* went to work out there, however, it drilled nothing but dry holes. After drilling ten wells off California in 1964–65, the vessel moved on to drill only four wells off Newport and Astoria, Oregon, before the effort was called off in the winter of 1966. Unfortunately, the large structures viewed on the seismic record were not salt but shale diapirs (shale cores that had broken through the overlying rocks of an anticlinal fold). As Shell Canada also learned offshore Vancouver Island, British Columbia, the sands were not there. This was a serious setback for the Los Angeles E&P office. After 1965, discussion of this highly touted and expensive drilling program completely ceased in Shell publications. An earlier *Shell News* article had pointed out that "Shell has a large stake in the outcome of this search." But now it was tough to dwell on the disappointment. [20]

In the spring of 1968, startling news from Alaska compounded the pain. The previous December, after a costly failed well by BP-Sinclair on its Colville lease, the joint venture between Humble and ARCO (Atlantic Refining merged with Richfield Oil in 1966) struck oil in a deep wildcat well at Prudhoe Bay. Step-out drilling the following summer confirmed that they had discovered something special. ARCO, Humble, and BP-Sinclair, the other major leaseholder in the area, found themselves sitting on a world-class oil field that held an original ten billion barrels of recoverable oil reserves, the largest oilfield ever discovered in the United States. [21]

The news devastated the Shell E&P organization. Here was this massive find, and they were not a part of it. They were one of the first companies to find oil in Alaska, and their geologists had sniffed around on the North Slope prior to the lease sales, but the decision was made not to expand geophysical explorations there. The companies that did have seismic data saw something unusual. "Shell didn't have the data," said Charlie Blackburn. "They would have been in there buying leases if they had had the data." Still, the most frustrating part was that nobody, not even the companies who discovered it, could have expected anything so massive. The stratigraphic trap with north dip did not fit any of the geologic models people had been using. Shell thought it had made the correct *economic* decision not to pursue the Prudhoe Bay play, based on their understanding of analog fields in both the United States and Canada. The problem was that the *geology* of the Prudhoe Bay field did not conform to those analogs. Said Bert Bally, Shell Oil's chief geologist at the time: "I simply could not visualize something as big as Prudhoe Bay because I had never seen anything like that in the whole United States and Canada, and I had a fair amount of exposure to U.S. geology. It was a lack of broad-minded experience and a lack of information which led us to what clearly was the wrong decision."[22]

The Pacific Coast E&P office of Shell Oil never recovered from it. After ARCO-Humble's discovery, Gerry Burton hastily tried to get a piece of the action for Shell Oil around Prudhoe Bay. In May 1968, production manager Jack Doyle swiftly organized a drill test before the tundra thawed on an old Shell lease called Lake 79. Shell combined information from this test with additional information gathered by geological and geophysical crews in preparation for the next state lease sale in September. However, this information was not very encouraging. Shell secured a few leases at the sale, at which industry paid a whopping nine hundred million dollars, but the company's exploration managers had to resign themselves to the fact that they were too late to make a significant claim.[23]

Meanwhile, the Los Angeles office got cold feet when it came to further exploration off the Pacific Coast. In 1965, Shell did purchase a 20 percent share in a submerged tract of the Long Beach Unit, an extension of the East Wilmington field, which the company and its partners famously developed with artificial islands so as not to create an eyesore for coastal residents. But Los Angeles managers shied from wildcat exploration. In February 1968, the Bureau of Land Management leased seventy-one tracts in the Santa Barbara channel for $603 million. Many of these leases were in water depths of 1,000 feet or more. As the leader in deepwater drilling, in possession of the *Blue Water II*, which had been re-equipped to drill in 1,000-foot depths, Shell Oil might have been expected to have been one of the more aggressive firms in this sale. Instead, the company came away with only a few leases. The Los Angeles office even agreed to sublease the *Blue Water II* to Humble Oil, which used the vessel to drill a significant discovery on its "Hondo" prospect in 850 feet of water, one of the first truly deepwater development projects offshore. The Union Oil Company's January 1969 blowout in the channel and resulting environmental opposition would slow and eventually halt development there. As it turned out, Shell was probably fortunate not to have spent big either in the 1968 Santa Barbara or Prudhoe Bay sales. But at the time it seemed that Shell had lost its nerve on the West Coast, the confidence of its exploration managers undermined by the failures in Alaska and offshore. Either way, the company's western crude reserves were declining fast, with little hope of replacing them through new discoveries.

Emphasis on Geophysics

More than anyone, exploration vice president Mac McAdams lamented Shell's failure to discover the Prudhoe Bay oilfield. Intensely competitive and passionate about finding oil for the company, he took it personally. But the setbacks in Alaska and off the West Coast in the mid-1960s only made him more determined to improve the company's technological capabilities. By the end of the decade, thanks to his leadership, Shell Oil was better prepared than any company to meet the challenges of exploration, especially in the Gulf of Mexico, which he felt was a place where Shell could still make a lot of money and beat its competitors.

McAdams prided himself on building the premiere exploration group in the world. He regularly gave pep talks to this effect, attempting to break down insularity and foster cooperation in all aspects of exploration. With support from Bob Nanz, who was the director of exploration research and head of the Bellaire lab in the 1960s, McAdams encouraged informal teamwork, bringing people together from the divisions and various research specialties to tackle problems. He forced people to butt heads if necessary. "Teamwork" in Shell exploration was not invented in the 1990s, when Shell Oil management suddenly proclaimed that they had changed from an unspecified, older way of doing things to a new approach based on greater teamwork and cooperation across lines of specialization. In fact, the research group at Bellaire had always been responsive to exploration needs, and local mangers very much encouraged close relationships with the lab. Under McAdams, teamwork received added impetus. Exploration managers took tours of duty in the laboratory and researchers spent time looking for oil. By the mid-1960s, Bellaire concentrated increasingly on research that could be directly applied in the field. On the exploration side, this research was in petroleum chemistry and geophysics. That meant less money for research on basic structural geology and stratigraphy. Bellaire had done marvelous work in these areas, but it had not yielded the kind of predictability desired for oil exploration. Some geologists ended up leaving Shell for academia, where there were fewer restrictions on scholarly interchange. But for those who stayed and saw the application of their research, the 1960s was an exhilarating time.

One of Shell's great breakthroughs in exploration geology was in understanding the origin and migration of oil. During the 1950s, Ted Phillipi, a Dutch geochemist working at Bellaire, and others developed a theory that hydrocarbons in the earth formed at great depth and pressure and then were essentially cooked out of their source rock and migrated to reservoir traps. This explanation flew in the face of conventional understanding, which held that oil formed in the fine-grained rocks near the reservoir sands. Geologists within Shell at first resisted the idea. "All that oil in the Gulf of Mexico can't come up from deep layers," they said. "How would it get up there? A special team coordinated by J. T. Smith evaluated the concept and attempted to put it to practical use. As these ideas proved valid and earned acceptance within the company, Bellaire researchers pioneered techniques for measuring the temperature histories of source rocks. By telling how hot they had been heated, geologists could determine whether the product from that rock would be oil, condensate, or dry gas. The reliability of these techniques, and their dissemination into the operating areas, enabled Shell to avoid expensive drilling in areas where all the source rocks had been heated so hot that they would only yield gas, whose commercial value was still relatively low compared to oil.

Regional geologic studies also strengthened Shell's exploration program. A key person in this area was Bert Bally. The son of a Swiss research botanist, Bally grew up hiking the Alps and Apennines of Europe and spent ten years at the University of Zurich assisting and learning from some of the world's most famous alpine geologists. He came to North America in 1953 through a fellowship at Lamont Geological Observatory and joined Shell Canada a year later. As chief geologist there, he co-authored the classic paper, "Structure, Seismic Data and Organic Evolution of the Southern Rocky Mountains," a landmark of North American geology that greatly advanced the understanding of folded belts on a regional scale and led to Shell Canada's discovery of the Waterton and West Jumping Pound gas fields.

Bally championed the use of long seismic profiles to better comprehend regional geologic structure. He was not the originator of the idea. Enlightened management at Shell Canada had obtained such profiles before he arrived there. And at Shell Oil, the critical combination of refraction and reflection techniques in regional studies was originally suggested by Willy Hafner, chief geophysicist in New York, and carried out by Glen Robertson, later chief geophysicist in New Orleans.[24] But Bally recognized the importance of reflection seismic records for geologists, and he contributed significantly to extending the concept within Shell. He eventually became the world's leading expert in using seismic records to interpret regional geology, especially in the area of folded belts. The combination of regional geologic studies and the understanding of oil origin and migration eventually induced geologists to think in broader terms about how oil might have been generated in a particular basin. This kind of thinking evolved in the late sixties and seventies into what came to be known as "petroleum systems" analysis, with Bellaire geochemist J. T. Smith as the recognized catalyst in this field.

In 1966, Bally became manager of geological research at Bellaire, where, according to Marlan Downey, he led his team of scientists by "intellectual force, not administrative dictum." Two years later, he moved to New York as chief geologist and trusted advisor to McAdams. In addition to his own research contributions, for which he received the American Association of Petroleum Geologists' coveted Sidney Powers Award in 1998 and the OTC Distinguished Achievement Award in 2003, Bally helped inspire and develop other exploration geologists. He was instrumental in getting geologists and geophysicists to work together more closely in Shell's exploration program. "We need to think of how the rocks sound, how they react to acoustic waves," he would exhort his geologists. "Don't tell me what color they are or how old they are, tell me how they will appear on seismic sections."[25]

These words reflected the unwavering commitment to geophysics at Shell Oil. In the late 1950s, analog magnetic recording and processing of seismic data had become routine, leading to methods of stacking or compositing data from multiple sources and receivers to achieve improved signal resolution. Yet geophysicists still struggled to correct precisely for variations in signal travel times through weathered surface layers. During the early 1960s, Shell geophysicist Gerry Pirsig solved this problem by devising a filtering technique that not only better determined near-surface travel times but measured velocities all the way down the reflection path to depth. "Gerry's filters gave us an optimum solution for all of those parameters in one calculation," explained Bill Broman, director of exploration research in the late 1960s. "It was a brilliant move. And from then on, stacking went full bore within Shell Oil." [26]

A mathematician, astronomer, and physicist, Pirsig had joined Shell in 1949, serving first in the New Orleans and Houston areas, and then with Shell Canada, before becoming the company's chief geophysicist in 1963. Unprepossessing, he was a large, rotund, and disheveled man who chain-smoked (as did many people in those days). Tom Hart, a rising star in exploration at the time, and quite a chain-smoking character himself, called Pirsig "the Buddha." Indeed, Pirsig became the geophysics guru within the company, enlightening others on the company's improving geophysical capabilities. During the early 1960s, Bellaire was making rapid strides with multidrum computers to process seismic data on magnetic tape, foreshadowing the move to digital processing and recording. Pirsig tirelessly traveled to all the area and divisional offices updating them on these advances. "Gerry was not hired because he had a degree in geophysics or geology," said Marlan Downey. "He was hired because he was one smart son-of-a-bitch. With his mathematical background he became the geophysicist's geophysicist." [27]

In 1967, Pirsig took a turn as manager of exploration in the Denver Area, where he used his filtering technique to help Shell decipher the unusual "pinnacle reef" geology of Michigan's lower peninsula. The Michigan play was a perfect example of how Shell combined the different talents of its scientists. The idea for investigating it came in 1962–63 from Pete Lucas, a carbonate specialist in the Denver office who had mapped the area and hypothesized a broad belt of Silurian pinnacle reefs—tendrils of ancient coral—interlaced by rich source rock running across the northern part of the lower peninsula. Lucas drew on analogies with Ontario pinnacle reefs to argue that, if present, the reefs in Michigan would likely be porous and would provide well-sealed oil traps. However, the known Silurian reefs were typically very small and contained trivial amounts of hydrocarbons. Furthermore, the gravel, peat, and moss in the glacial till overlaying the reefs made obtaining good seismic data difficult. As Bob Nanz described, "It was almost like shining beams through broken glass." [28]

Management hesitated to make a play based on this concept. But Reed Peterson, Shell district manager, persistently championed Lucas's concept and finally convinced Bert Bally of its merit.

Bally and Pirsig, who was also intrigued by the concept, rallied support within Shell for following up Lucas's work with a seismic program. The two men had previously worked together in Canada, and they recognized the similarity of the Michigan play with the Rainbow play of northern Alberta on which Shell Canada had missed out. Pirsig suggested that with improved seismic technology geophysicists might be able to see through the thick glacial till.

During the next several years, experimental seismic acquisition yielded what seemed to be disappointing results. First, it was very expensive, setting new records for cost per mile of data. Second, the data revealed something puzzling. Gerry Pirsig's filters did a superb job of making large corrections to the signals passing through the glacial till, but everywhere they expected to see reefs the reflections were not continuous; there was an absence of data. Geophysicist Leo Buonasera, who was working on a team to create seismic-geologic models for the play, thought there was something to those data gaps, that they might indicate places where the pinnacles had altered the continuity of the seismic waves. Pirsig agreed to test the theory and laid out a three-well drilling program. If it turned up nothing, Shell would walk away from the play. In 1967, when Downey brought Pirsig the ranking of new potential plays for Denver's 1968 budget, Michigan was ranked fifth, with an estimated production of eighty million barrels. Pirsig smiled and said, "Marlan, let's increase the potential to 100 million barrels and move it up two places. It may not look worth doing, otherwise."[29]

Shell Oil discovered a Silurian reef on its initial three-well program, and the race was on. Amoco was pursuing a similar concept and engaged in a fierce race with Shell for acreage in northern Michigan. In August 1969, Shell spent one million dollars to lease 1.3 million acres down-dip from the shelf margin where Amoco had concentrated its leasing. Rival companies began to lease down-dip like Shell, but by then Shell had locked up what turned out to be the best acreage in the pinnacle trend. The next problem involved deciding where to drill in areas where Shell had not shot seismic lines. It was impossible to pinpoint these small fields simply with seismic data. They were only about 160 acres in size, a few hundred feet high, and buried about eight thousand feet deep. Some feared that the money spent on expensive seismic acquisition might even exceed what could be made from the oil. Inspired by a paper on optimum search theory used by destroyers to strike submarines in World War II, Marlan Downey provided mathematical calculations for extrapolating from what data they did have to estimate the size and frequency of "acoustic-anomalies-that-might-be reefs" as seen on the experimental seismic grid. When he presented the results to management, someone made the comment, "a billion-barrel field in a thousand places!"[30]

The Michigan play required a new approach for Shell E&P. The company's strategy was to explore for large fields, because people believed that fields of one million barrels could not be profitably developed. When it became apparent that discoveries were going to consist of numerous, closely spaced, one-million-barrel fields, Ronnie Knecht, who had become general manager of E&P in Midland, led the exploration, production, and land team in exploring and developing the 150-mile-long trend as though it were one large field. Rapidly improving seismic techniques for mapping the reefs and dense coverage resulted in an incredibly high success ratio in the drilling program. Although each field contained only one to two million barrels of oil, the engineers kept the drilling costs remarkably low and the profits exceptionally high. By 1973, 145 exploration wells had resulted in 69 discoveries, and the next year the number of producing wells rose to over 400, just as the price of oil was soaring due to the OPEC embargo. Wrote the *Oil and Gas Journal*, "Not in the 1930s when Pure dominated drilling or in 1940 and again in 1960 when Marathon was the pacemaker, has a single company carried the ball like Shell in 1974." Over a thirty-year period, Michigan was Shell's single most profitable onshore oil play, yielding a gross recovery of over 350 million barrels of oil and 2 trillion cubic feet of gas. It was also one of its most intense plays; at times in Michigan, Shell had seven rigs drilling, thirteen seismic crews shooting, and untold numbers of landmen leasing. Only the combination of diligent geologic work, high-tech geophysics, and efficient drilling and completion made it possible.[31]

Shell increasingly emphasized geophysics, which was undergoing nothing short of a revolution, driven by new quantitative capabilities offered by rapidly improving computer technology. The introduction of digital computers in the early 1960s enabled a quantum leap in the amount of data that could be processed. At first, analog-to-digital converters digitized field-recorded magnetic tapes. Field recording directly in digital format soon followed. The number of recording channels used by Shell grew from twenty-four in the late 1950s to one hundred by the late 1960s. By then, Gerry Pirsig had encouraged the Bellaire lab to devise a system capable of handling one thousand receivers (the Kilotrace Project). Increasingly, according to exploration manager Bill Broman, the key challenge for geophysicists was "digesting this firehose of digital data in a scientific manner." Shell geophysicists started leasing time on IBM machines to process this data, but their needs quickly expanded. In 1965, the Bellaire lab addressed these needs by creating a new facility housing $75 million worth of UNIVAC digital computer equipment used strictly for processing geophysical data. "We didn't share this huge data processing center with anybody, any other function," remembered Billy Flowers, exploration manager of Shell's Marine Division.[32]

In the late 1960s, Shell's research into developing advanced computer technology led to a curious sidelight. Seeking to design a special-purpose computer for handling the mass of data generated by escalating numbers of seismic traces (the data recorded from one surface group of detectors), Alton Christensen, an electronics scientist at Bellaire, pioneered the early development of integrated circuits for random access memory microchips. Christensen took out many patents for Shell on his microelectronic devices, called MOSFETs (Metal Oxide Silicon Field Effect Transistors), which led Shell at one point to flirt with the idea of actually becoming a computer manufacturer. But it was only a flirtation. The company decided to stay focused on the oil business. The licenses sold on these patents to computer companies such as IBM, nevertheless, became the source of Shell Development's largest royalties ever. Given the subsequent explosion in the personal computer business sparked by the microchip, it is interesting to speculate about what would have happened had Shell developed the technology commercially.

Breakthroughs in the Gulf of Mexico

In the mid- to late 1960s, Shell chose to apply new seismic technology in the place where it had enjoyed the most success in the past: offshore Gulf of Mexico. The New Orleans Area office was still the hotbed for E&P, retaining many of the technical teams responsible for early offshore successes. However, some of the key people who had worked the 1960 and 1962 sales had moved. Bouwe Dykstra retired in 1962 and was replaced as vice president by James E. (Jim) Wilson, a young executive on the fast track (not to be confused with researcher James Lee Wilson; see chapter 2). A Texas A&M Aggie and a veteran of the U.S. Third Army Division that landed in Normandy during the war, Wilson had started with Shell in 1938 as a geologist. He climbed rapidly after

16
An unidentified
Gulf of Mexico
seismic vessel set-
ting off a dynamite
charge, 1950s.
*Photo courtesy Morgan
City Archives.*

the war to become director of exploration research and then vice president in Houston, the youngest executive in the company at age forty-four. He brought in his own men because Gerry Burton and Ronnie Knecht, two of the early managers of the marine division, were transferred to management positions in Los Angeles and Midland, respectively. By 1966, Wilson's exploration team in New Orleans included Jim Hohler as manager of exploration, Tom Connally as marine division manager, and Billy Flowers as marine exploration manager. Like Wilson, they were all young, technically accomplished geoscientists eager to extend the company's offshore dominance.[33]

In the mid-1960s, in preparation for the next general OCS lease sale, Shell greatly upgraded its marine seismic technology thanks to the digital computing revolution. The company introduced a new fleet of large, purpose-built offshore seismic vessels, the first to employ digital recording equipment. The *Phaedra* was launched in 1965, followed by the *Artemis* and *Niobe* in the next two years. One observer likened the electronics room on the *Niobe* to a set on the *Star Trek* science fiction television series. Shooting and recording took place on the same ship, each capable of recording large numbers of seismic traces.[34]

Digital technology had enabled the industry to solve many of the problems that plagued early reflection operations offshore. New recording capabilities allowed dynamite charges to be replaced with air-gun devices such as the Vibroseis, developed by Conoco, which emitted a pulse as the sound signal, saving both money and aquatic life. Deconvolution, a sophisticated method of inverse filtering made possible by digital technology, solved the aggravating problem of water reverberation in which seismic signals reflected back and forth between the top and bottom surfaces of the water created a "ringing" noise that could mask deep reflections. With digital processing and recording, geophysicists analyzing offshore prospects could measure the relative wave amplitudes between seismic traces for the first time. True amplitude recovery (TAR), as Shell called this kind of measurement, gave geophysicists much higher quality seismic data than they had ever known. Digital filtering techniques could even be applied to data recorded in the days before such techniques had been developed.[35]

Between 1962 and 1967, there was a lull in general lease sales in the Gulf of Mexico, which gave companies time to develop their large lease inventories and improve on technologies. A small federal-state auction of drainage tracts in April 1964, however, drew over $60 million in bonus payments, indicating the industry's growing interest in offshore tracts, especially those offsetting proven production. In the 1964 drainage sale, Shell bought a lease at South Timbalier Block 26 next to Chevron's major production on the south flank of the Bay Marchand field, one of the largest in south Louisiana. Based on work done earlier by Joe Broussard under Ronnie Knecht, Shell's Bay Marchand lease yielded a nice discovery. "We thought we had gas," remembered Billy Flowers, "but it turned out to be oil, which was a great advantage as far as economics were concerned." Another advantage was that Shell Oil paid only $1.27 million for the 625-acre tract, compared to $13.1 million paid by Gulf and Phillips for nearby 1,250-acre Block 28 with similar production.[36]

The next big lease action came at the July 1967 federal sale offshore Louisiana. Armed with improved geology and geophysics, Flowers organized a special team led by Claude McMichael to prepare for it. Like Jerry O'Brien's group prior to the 1962 sale, the team included geologists and geophysicists who studied in detail the geologic characteristics of existing oil and gas fields. The McMichael study resulted in an enormous data base that served as a valuable reference work for this and later sales. The group also developed an equation for predicting volumes, but the equation was not used because of some critical flaws. "You weighted the source rock, you weighted this, you weighted that, and so forth and so on, and you got a number out on the prospect," explained Miner Long, who was then offshore division geologist. As the chief geophysicist in New Orleans, Glen Robertson, pointed out at the time, if any one of the weighted parameters went to zero, then there was no oilfield, even though that would not be accurately reflected in the sum of the weighted numbers. More useful was a new bidding strategy developed by a team of mathematicians employing statistical analyses and game theory. In the past, bidding strategies had not been that sophisticated. "It was kind of like, 'this is what we think it is going to take to win,'" explained Don Russell, who coordinated the study of bid strategy.[37] Russell's group looked at old files and determined, tract by tract, what Shell had thought prospects were worth in past sales and what the industry had bid on them. Analyzing statistically the relationship between Shell's value and the industry's bids, they developed a mathematical model to guide the bidding and predict how much money the company should spend in the upcoming sale to maximize its profitability.

Shell laid out big money in the 1967 sale. The industry spent $510 million for nearly 750,000 acres in waters ranging out to three hundred feet, another spending record for offshore leases that astonished even veteran observers. "To a great extent," the *Oil and Gas Journal* concluded, "the bids show the respect which the industry has for offshore geophysical work in areas untested by the drill." Shell accounted for a large portion of the total winning bids, $101 million for sixteen tracts. "Shell put up more blue chips and took the most pots June 13 in the offshore gamble to find and produce gas and oil from that magic Mecca known as 'Offshore Louisiana,'" wrote *Ocean Industry*.[38]

Shell acquired some good tracts in the sale, including South Pass 70 and South Pass 65, both of which turned out to contain major oilfields. South Pass 70 was a pricey tract in three hundred feet of water, costing $21.3 million, whereas South Pass 65 was a bargain at $775,000. The latter field was on the east side of the Delta and obscured by gas-filled mud, which caused a problem for the seismic data. Jim Wilson and Billy Flowers nevertheless thought they had enough data to decipher the structure and the trap. They argued for the prospect over some objections and were right. To the relief of top managers in the company after recent bad news from the West Coast, the Gulf of Mexico was still yielding riches. But not everyone profited. A review of the 1967 sale five years later found that only Shell and a few others (Kerr-McGee, Chevron-Mobil, Offshore Operators, and the Signal/LL&E/America/Marathon group) ended up in the "plus column." [39]

Shell's relief at the time, however, was also mixed with disappointment. No matter how good or careful the geophysical work and mathematical analysis, geology could still fool even the best companies in areas "untested by the drill." The New Orleans office found this out the hard way on a tract it purchased in the Main Pass Area East Addition. Prospect 370, as they called it, looked like a beauty. It was a big faulted anticline structure with simple closure and what seemed to be good sands. Shell bid $33 million for three adjoining blocks, Main Pass 236, 206, and 121. This was the highest figure the company had ever paid for a single prospect. Glenn Robertson, who had just purchased a new Ford Mustang automobile, pointed out that this amount was enough to buy ten thousand Mustangs. But exploration manager Jim Hohler was so confident the prospect would be a winner he assured naysayers that it was a "lead pipe cinch." [40]

When drilled, however, the prospect contained nothing commercial. The lead pipe cinch was a $33 million dud, not even counting the cost of drilling it. This traumatic news provoked serious soul searching in New Orleans. "It was very bad," remembered Miner Long. Understanding why they failed became exceedingly important for the exploration team as Shell geared up for the next big sale off Texas in 1968. The problem, as it became apparent to Billy Flowers, Miner Long, and others, including Jim Wilson, who had expressed reservations about the prospect before the sale, was that they did not understand the role faulting had played at Prospect 370. [41]

Shell was determined to learn from its mistake. The Texas area up for sale in 1968 was a very sandy section with many faulted anticlinal structures, much like the Main Pass area that had attracted the big bucks in 1967. This meant Shell geoscientists would have to rethink their whole understanding of the factors controlling hydrocarbon traps in the Gulf. The Shell head office sent a team that included geologists Miner Long and Urban Allen, geophysicist Mike Forrest from New Orleans, and geochemist J. T. Smith from the head office to examine the problem and help the offshore division in Houston prepare for the sale. Allen developed a model for entrapment in the Gulf that assumed that all faults will leak oil or gas unless impervious shale is across the fault to act as a seal. This, Allen argued, had been the flaw with Prospect 370. Tested extensively, the model also explained how hydrocarbons could migrate from great depths to younger rocks. Meanwhile, the model for hydrocarbon generation developed by BRC under J. T. Smith demonstrated that the source rocks offshore Texas, based on their temperature histories, would have generated mostly natural gas, which was worth much less than oil. "You could draw a big oil field on those anticlines," said Long. "But if you looked at the sandy section with all those faults, all of a sudden you got very concerned about the probability of large accumulations."[42]

As a result of this work, it became clear that the traps would probably be small and that the most likely hydrocarbon would be natural gas. Accordingly, the Houston division cautiously bid low in the sale. To its competitors' surprise, Shell won only one tract for $1.5 million. By contrast, Humble and Texaco, bidding individually and jointly, together spent an exorbitant $375 million for about thirty tracts. These companies were clearly forecasting large reservoirs of oil. But at the end of the day, in view of Shell's restraint in the sale, Humble and Texaco officials must have felt a little nervous about the expensive tracts they had just purchased. Indeed, those tracts proved to be virtually worthless. Drilling confirmed that these structures contained only small gas accumulations. Shell was able to turn misfortune at Prospect 370 to its advantage, while Humble and Texaco committed a colossal mistake, as expensive as any made offshore to that point.[43]

A landmark development in geophysics also emerged from Shell Oil's work at the 1967 Louisiana and 1968 Texas sales. In mapping the subsurface structure on Prospect 370 using some of the best seismic data available at that time, Mike Forrest, a staff geophysicist in New Orleans, observed what he called a "strong seismic reflector," or an echo, on the top of the structures, where production would most likely be found. He also observed strong seismic reflectors on the crests of structures in the Plio-Pleistocene trend in both offshore Texas and Louisiana. These events continued to puzzle him until one day, in the spring of 1968, he came across some old Russian geophysical abstracts translated into English that described how Russians had mapped oil and gas pays using seismic data. The abstracts led Forrest to wonder: "Could this be what I am seeing in the seismic lines?" Was it possible his interpretations revealed the exciting possibility of directly detecting hydrocarbons from seismic amplitude readings? The overall theory behind the direct detection of hydrocarbons had been around for years. The idea was that the acoustic impedance of a loosely cemented rock filled with hydrocarbons was different from that of a similar water-filled rock and that with sophisticated enough technology, this difference could possibly be detected. In 1960, Carl Savit of Western Geophysical had described the phenomenon and suggested that it might be useful in discovering stratigraphic traps. But most geophysicists had generally ignored it because seismic data had not been good enough to be diagnostic. Perceived wisdom held that seismic data only enabled the understanding of structure. One still had to take the risk of drilling to find oil and gas. But if hydrocarbons could actually be seen on the seismic record, and if this could be factored into bidding for leases offshore, then it would remove a large element of risk. It would change the nature of the game.[44]

With the "strong events" he saw using true amplitude recovery of digital seismic data, Forrest thought he might be onto something. "Mike was not the first to predict it, and maybe not the first to notice it," said Billy Flowers, "but he was the first guy to say we have to do something about it." Nobody at first took his observations seriously. In 1968–69, Forrest continued to observe high amplitude reflections on the south flank of the Bay Marchand salt dome and in the Plio-Pleistocene trend. Mapping prospects for the next big federal sale offshore Louisiana (scheduled in 1969 but postponed until late 1970), he again noticed that these reflections occurred where structural traps would be expected. This could not be just coincidence. Yet he could not provide a technical explanation for it. "I was the joke of the office," Forrest said. Other geophysicists in the marine division started calling his bold reflections "bright spots," since they appeared to light up on the seismic section. A common coffee room joke at the time referred to them as "B.S.," which is what some geophysicists thought of bright spots. At holiday time, one colleague strung a line of Christmas lights across a seismic section taped to his office wall, pointing out all the bright spots to passers-by. "It was too simple, too easy, too good to be true," explained Flowers.[45]

In the spring of 1969, at the urging of Glen Robertson, Forrest decided to document his bright spots with well data. "I'm a data hound," he admitted. By calculating the acoustic velocities of rocks from well log data, it was possible to estimate what the seismic traces should look like at a particular well site. If the "syn-

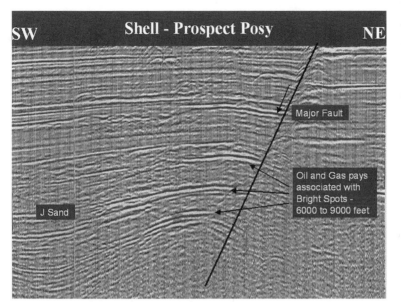

17
The seismic section of Shell Oil's "Posy" prospect showing a bright spot amplitude. *Photo courtesy Mike Forrest.*

thetic seismic trace" in the producing zone of a well matched the bright spot amplitudes on the actual seismic trace, then Forrest could support his hypothesis. So he gathered data on six fields where he knew there was shallow gas. He spent about two weeks putting together a study that correlated the seismic data with the well logs. Lo and behold, they showed a relationship between bright spot reflections and the gas pays and made a compelling case for his theory. In June, Forrest submitted his study to Billy Flowers, now chief geophysicist in New Orleans, and John Bookout, who had just become New Orleans Area vice president, and then left on a two-week summer vacation.[46]

When he returned, the office was abuzz. "You missed it," they told him. "Missed what?" he replied. "We showed your data package to McAdams and he ran with it." Within days, McAdams ordered the creation of a special Amplitude Analysis Project team in the Basic Measurements and Theory Section at the Bellaire lab to test Forrest's empirical observations. This section was supervised by Aaron Seriff, who was already looking at ways to understand seismic amplitudes better, and the project leader was Emmanuel "Manny" Baskir. Using Forrest's well data, the Bellaire team recognized the physical principles that caused the bright spot phenomenon, made new theoretical calculations, and in a matter of days verified them in the lab. In September 1969, Seriff and Baskir published within Shell Development Company a seminal Technical Progress Report, "Reflection Coefficients of Hydrocarbon-Filled Reservoirs," on quantitative predictions and assessments of seismic amplitudes. By providing the basis in physics for Forrest's correlations, Bellaire's research team gave Shell management the confidence to apply this wonderful new interpretive facility. The only question was how. Said Billy Flowers: "The issue was, do we sit on this so tight that only a few of us use it or do we spread it through the company and let everybody use it? How fast would the news get out? Our feeling was that it wouldn't take long."[47]

The exploration managers decided to spread it and use it wherever possible while the company still enjoyed a competitive advantage. But during late 1969 and early 1970, Shell experienced mixed success. The porous Gulf Coast deltaic rocks were amenable to the technique but not heavily cemented rocks in most onshore provinces. Surface noises on land also tended to distort the amplitudes. One exception was in the Sacramento Basin of California, where gas-filled tertiary sandstones infilling the two-thousand-foot-thick Markley Gorge could be identified with bright spots. In 1969–70, the team, led by Elwood Hardeman and Bob Weiner, made six discoveries out of seven wells, but the total gas reserves were only 100–150 billion cubic feet (bcf) total. Even in the Gulf of Mexico, which was more amenable to bright spots than elsewhere, geophysicists could not always tell how many valid bright spots they had in their seismic readings. The first one they saw might be valid, but the second and third might have tailed from the first. "As with any new idea, any new technology," said Forrest, "there were peaks and valleys."[48]

In 1970, under geologic lease sale coordinator Leighton Steward, Shell began to apply bright spot evaluation in its offshore bidding. The first opportunity was in January, at a small drainage sale. The New Orleans staff recognized a bright spot on a low-relief structure at Main Pass 18 east of the Mississippi River. But they were outbid by Mobil. Ray Thomasson, Shell's offshore exploration manager, believed that Mobil's high bid indicated it was onto the phenomenon and that perhaps the competition knew more than Shell realized. So the New Orleans office geared itself up for its first big chance, the December 1970 federal lease sale offshore Louisiana—the "Posy" sale, as Shell called it. The company had started assigning code names instead of numbers to prospects, for simplicity's sake as well as security. It did not mean much to an eavesdropper if the conversation was about posies, orchids, and tulips. Aaron Seriff's team at Bellaire, including Manny Baskir, Bobby Long, David DeMartini, and Jim Robinson, continued to work on various aspects of the quantitative description problem to help New Orleans analyze prospects for the sale. In that sale, Shell bought six prospects, all of them decent fields, and two of them were big, especially Prospect Posy (Eugene Island 331), which was interpreted by Chuck Roripaugh. However, Forrest and others felt that the company did not push bright spots hard enough due to lingering insecurity about the validity of the tool. Posy, or the Eugene Island 330 field, turned out to be the largest field in federal waters on the Louisiana-Texas shelf, ultimately containing over 700 million barrels and covering several blocks in water depths out to one thousand feet. In hindsight, Shell missed a chance to buy the best block on the crest of the structure (Eugene Island 330), purchased by Pennzoil Oil and Gas Operators (POGO), and two other high-volume Posy blocks where geophysicists had mapped multiple bright spots. Shell Oil's original volume in the Eugene 330 field was 143 million barrels of oil equivalent (BOE) whereas Pennzoil's Eugene Island 330 tract alone contained 251 million BOE.[49] "We were very proud of what we accomplished, but we could have done better," recalled Forrest. Shell had not pressed its advantage as an early adopter of the technology, and competitors were not far behind. Manny Baskir and David DeMartini conducted a post-sale study for Shell Development of the bidding patterns of Shell's largest competitors in the sale and concluded that Humble and possibly Mobil and POGO had used amplitude anomalies, or bright spots, in their bidding.[50]

After the sale, the men responsible for offshore research and operations set out to allay top management's insecurity about bright spots. Under the leadership of Billy Flowers, research and operations used the data from the 1970 sale to improve the application of the technology. In 1969 and 1970, many technical people believed that bright spots were useful only to detect natural gas. But the study of petrophysical data demonstrated that they could also indicate oil, which was much more valuable, but that the so-called amplitude change was less than for gas. The main question going into the next major sales, in September and December 1972, was whether Shell could identify oil versus gas before bidding on the leases. Brilliant research performed at Bellaire gave managers greater confidence than ever that they could do this. With contributions by Bob Chuoke, Al Bourgeois, and Harry Hasenpflug in a

Head Office Task Force led by Paul Terrasson, Bellaire researchers improved deconvolution and made correlations with velocity and density logs from wells. Bill Scaife, Gene McMahan, and Aubrey Bassett in the offshore division devised programs to measure and interpret the mass of data from all the pertinent reflections on all the seismic lines. "That was a big advance over just saying there is hydrocarbon there," said Billy Flowers. "We could start to say there was this much hydrocarbon there, and we feel the sand is of this quality."[51]

This work was timely, for other oil companies had caught on to the idea of bright spots. In the September and December 1972 sales, Shell employed bright spots more aggressively than in 1970. Although it won only one tract in the September sale in the eastern Gulf, it led all plungers in the December sale of tracts in the western Gulf, which was the big one. The industry exposed a breathtaking $6.19 billion in total bid money and paid a record $1.67 billion in total high bids. Because of the tremendous sums at play, many companies took on bidding partners, including Shell Oil, which normally bid alone. In partnership with Transco, the company spent $221.4 million for thirteen tracts. Analysts at the time pointed to the prospect of higher natural gas prices and creative methods of financing by groups such as Pennzoil's POGO as the key factors driving up bids. But bright spots, or "hydrocarbon indicators" as Mobil called them, were clearly a significant factor unrecognized by outside observers.[52]

Using probability analysis, the Shell geological and geophysical team led by Dick Grolla and Ed Maunder rated each bright spot with a likelihood of oil, gas, or water and then made a histogram that displayed the reserve estimate distribution plotted against the probability of success for each prospective sale block. This methodology produced success indeed in the December sale when Shell Oil acquired another major oil field, Prospect Pine (South Marsh Island 130 and 131, containing a discovery volume of 174 million BOE), for $76.7 million. On the other hand, this success was tempered by the realization that Shell had found "phony" bright spots on the west part of the Pine Prospect. Shell geophysicists discovered that they could be fooled by sands with small amounts of gas looking like oil. They could also be fooled by coal beds and marine shale. Other companies who had discovered bright spots, but who had not done as much technical work as Shell, subsequently pushed it even harder, spending big money on leases with invalid bright spots. In the June 1973 offshore Texas lease sale,

Shell won only one tract. The company bid on a total of seventeen and finished second thirteen times. The high bids, however, were up to three times greater than what Shell bid. "Mobil and POGO went berserk on these things," remembered Charlie Blackburn, vice president of what had become the Southern E&P region by this time. "It turned out most of these things were spurious and we were just fortunate that we didn't buy them."[53]

Over the next several years, however, bright spot technology improved considerably and became a very reliable tool for exploration in all geologic provinces and at greater formation depths than initially thought. By reducing the potential for drilling a dry hole and modifying the weighting of risk, it allowed Shell to put more money into its lease bids and more than make up for it in decreased drilling costs. From mid-1969 forward, Shell shot more seismic data than anyone else and scanned every prospect for bright spots. Shell's seismic crews shot half-mile grids while other companies used one- or two-mile grids. Forrest remembered meeting with some outsiders who looked at one of Shell's maps and exclaimed with astonishment, "Are those all seismic lines?" In 1974, the *Wall Street Journal* reported that most other oil companies were following Shell's lead in working on their own direct detection methods and that "Bright Spot is suddenly the hottest game in town."[54]

Offshore lease sale evaluation had long been the company's strength. A paper presented by a group of academic researchers to the Society of Petroleum Engineers in 1978 found that from the beginning of federal leasing in 1954 through 1976, Shell Oil had far outpaced all major rivals in the ratio of revenue earned to bonus bid (2.34 for Shell to 1.55 for its closest major rival, Gulf Oil).[55] But this did not even include most of the revenue that would be earned from the highly productive leases obtained largely by bright spots during 1970–74. During this period, Shell Oil and the industry found more reserves offshore in the Gulf of Mexico than during any other comparable time period prior to the deepwater era of the 1990s. In a 1987 "lookback" on offshore lease sales in the Gulf of Mexico for Shell Offshore Inc., Doug Holmes noted that this "was the period of large oily prospects, like Posy, Pine, Bourbon, and Cognac, as well as some big gas prospects, like Papaya, Albatross, Felix, Cypress, Mako, and Gypsum, that SOI discovered." The industry's discovery volume for this period was nearly 8.8 billion BOE (more than half of the 15.5 billion BOE discovered during 1970–86). Shell Oil's discovery volume of 782 million BOE (and ultimately over 1 billion BOE) across the four play areas (Plio-Pleistocene, East Louisiana Miocene, West Louisiana Miocene, and Texas Miocene) was second only to the 851 million BOE discovered by Mobil, Shell's chief rival in the bright spot game. However, in terms of bonus bid per BOE discovered—a key indicator of exploration success—Shell Oil demolished all competitors at $0.62 per BOE. Holmes's lookback study maintained that the company's greatest edge had been "expertise in offshore lease sale evaluation," that is, in recognizing productive tracts, evaluating the probability of economic production, and estimating volumes.[56]

Just as important, this exploration expertise kept Shell out of some areas that did not show bright spot amplitudes, which saved the company tens of millions of dollars it might otherwise have spent on leases and drilling. Indeed, Shell Oil's exploration strategy in the early 1970s was relatively conservative, almost too conservative in hindsight. As Mike Forrest claimed, the company did not push bright spots as hard as it could have, and Holmes's lookback study argued that the company's federal offshore bonus bids "were too low for the volume potential." Buoyed by success and the soaring price of crude, however, the company took greater risks after 1974. Although Shell E&P deemphasized offshore Gulf of Mexico during the mid-1970s in favor of other offshore regions, especially Alaska, the great strides made by Shell Oil geologists and geophysicists in the 1960s would nevertheless help the company find more than two billion BOE offshore Louisiana and Texas by 1986.[57]

Shell Oil E&P met the challenges of exploration in the 1960s head on by continuing to emphasize and trust its technological capabilities. The setbacks in Alaska and off the West Coast did not demoralize the organization but provided new incentive to build on its strengths. Geophysical and geological success in Michigan and the Gulf of Mexico gave Shell managers added confidence in their technical staff, investment strategies, and operating efficiency. This confidence turned out to be absolutely crucial in meeting the new challenges of the 1970s caused by great social, political, and economic transformations in American society that profoundly altered the environment in which Shell Oil did business.

The End of Business as Usual

By the early 1970s, a broad front of social movements in the United States—civil rights, women's rights, environmentalism, labor activism, anti–Vietnam War—challenged politics and business as usual. The OPEC embargo of 1973 and subsequent nationalizations, which restructured the international oil market, magnified the challenge. In the United States, political fallout from the embargo and soaring gasoline prices resulted in government price controls on oil, the loss of the depletion allowance, and threatened divestiture of integrated companies. Meanwhile, the unprecedented combination of a severe recession and spiraling inflation, caused by heavy spending on the war, produced the strange new phenomenon of "stagflation." In the midst of it all, the Watergate scandal engulfed the Nixon administration and ultimately led to the president's resignation.

One of the most serious challenges to corporate America was the rising concern about environment, health, and safety issues, touched off by the publication of Rachel Carson's 1962 exposé of the potential damages of pesticides, *Silent Spring,* and the 1969 blowout at a Union Oil offshore platform in the Santa Barbara channel. As society demanded much higher standards of performance from industry, Shell and its competitors came under intense public scrutiny. A 1970 blowout and fire at a Shell platform in the Bay Marchand offshore field was a watershed event for the company in forcing the E&P organization to refocus on safeguarding the environment and workers.

Shell Oil was forced to change in other ways. Social and political pressures to hire more women and integrate its segregated industrial plants led to the gradual adoption of new personnel policies to diversify the workforce. Labor strikes at Shell refineries in 1969 and 1973 encouraged management to seek a better relationship with its employees. Less newsworthy at the time but of great long-term importance was the mounting difficulty of managing the flow of information and products within a large and expanding industrial organization. This drove efforts to centralize companywide information and computer systems and eventually led to the relocation of the head office to Houston, a dynamic city that provided a better environment than New York for this kind of centralization and closer to the company's increasingly vital operations in the Gulf of Mexico.

By the early 1970s, Shell Oil was remaking its organization and establishing a new corporate culture. None too soon. After 1973, the U.S. oil industry would never be the same. The era of price stability and domestic abundance in a protected market was over. The new order would be characterized by price fluctuations in an international market largely controlled by OPEC producers. On top of the assortment of pressures and challenges faced by Shell Oil was the enduring problem of diminishing domestic oil reserves. M. King Hubbert's warning about peak production for U.S. oil came true in 1970.

Shell Oil's domestic reserves were dwindling at an alarming rate, and access to low-cost foreign crude had become an increasingly vital concern. Shell's expensive offshore wells, which provided more than 40 percent of its crude oil output, had nearly reached peak production, and it was too soon to tell what might be found in deeper federal waters. The company had no position on Alaska's North Slope, and other onshore basins in the United States offered little promise for big finds. Furthermore, Shell's joint program of exploration with Shell Canada had proved to be a disappointment. The deepening energy crisis of the early 1970s forced the company to consider all possible ways to meet the country's exploding energy demands. It invested in alternative resources such as shale, tar sands, geothermal, and solar, and became a major U.S. coal producer. Under Pres. Harry Bridges, Shell Oil muscled its way into the international oil game, creating some awkward tension with its majority shareholder, Royal Dutch/Shell. It also negotiated a deal with Saudi Arabia for access to long-term crude supplies in exchange for Shell's investment in a petrochemical complex. However, Shell could not stake the future of the company on any of these areas of investment. Its main business was finding and producing hydrocarbons in the United States. John Bookout, who took over as president in 1976, refrained from diversifying further afield into unrelated industries, unlike some other major oil companies. Instead, the company plowed its profits back into exploration and production, pushing even harder to find new petroleum resources, especially offshore in the Gulf of Mexico.

Centralization

By the 1960s, Shell Oil had grown into a huge and complex organization. Across the country, the company and its subsidiaries employed thirty-five thousand people, some eight thousand of whom were involved full time in handling data and information. And the numbers were growing. Armies of secretaries and stenographers, almost all of them women, produced and reproduced the myriad forms, memoranda, and correspondence needed to run the business and keep people in communication with each other. It was a manual operation drowning in paperwork.

Another problem was the lack of uniform data throughout the company. Communications for operational purposes were generally quite good and record-keeping thorough. But operating departments more often than not developed their own record-keeping procedures. E&P, for example, often kept two sets of records—one for reporting engineering analysis and oil and gas reserves and another that met the requirements of the financial department for accounting, corporate consolidation, and tax determination. As John Redmond pointed out, "there was not a common definition for a barrel of oil, a cubic foot of gas, or a dollar throughout Shell."[1]

The electronic computer seemed to offer salvation. The company installed its first stored program computer at the Bellaire lab in 1954, and for the rest of the decade the use of computers at Shell was mainly limited to scientific applications. But the success in these applications encouraged the company to consider using computers in commercial operations to help control the rising tide of data and information. In 1958, Shell created a Data Processing Department to coordinate the introduction of computers in the company. Nevertheless, computer operations sprang up in a fragmented and decentralized way, according to the varied needs of different departments.

The need for greater control and coordination seemed clear. In 1959, John Redmond spearheaded the move to consolidate Shell Oil's data systems across functional lines, which proved to be a monumental enterprise that took almost a decade. By 1969, Shell's Information & Computer Services (I&CS) organization had set up regional data service centers, running so-called "third generation" computers, such as the IBM System/360 series. In addition, at the Houston Data Service Center, I&CS installed a large

UNIVAC 1108 scientific computer. It was accessible by remote control and could read cards and print information from twenty terminals at Shell locations around the country. Although primitive compared to recent advances in computing power and Internet connectivity, this system saved valuable time for large computing jobs and eliminated much of the pre-processing work and manual card and paper handling required by computers without remote-access capability.[2]

The amalgamation and upgrading of Shell's data systems set in motion a larger trend toward centralization in the company. In 1968, Shell Oil E&P embarked on a major consolidation, reducing the number of operating areas to two: the Western Region headquartered in Houston and the Southern Region based in New Orleans. Houston took over Midland, Denver, the West Coast, and Alaska, while New Orleans handled offshore Louisiana and activities in the eastern half of the country. In 1971, E&P closed its Los Angeles office and moved West Coast management to Houston. The consolidation enabled better communications and cut out redundant overhead during a period of tight budgets, but it was traumatic. Shell Oil's heritage ran deep on the West Coast. E&P operations in California were among the company's oldest, dating back to the 1920s. Deeply entrenched were the employees and their families, who were reluctant to be transferred east, trading the temperate West Coast for the swampy Gulf Coast. The closing of the Los Angeles office was not a decision taken lightly or easily. Said John Redmond, who returned to Shell Oil from Shell Canada in 1971 as executive vice president, "I have often thought of the move as one that could be compared with the closing of one of the older dioceses of the Roman Catholic or Episcopal churches. This is not said, or meant, in jest."[3]

The expansion of the Houston E&P office bolstered the Bayou City's emergence as a center of gravity for Shell. Houston had been a Shell town for many decades. By the 1960s, it had become the booming hub of the U.S. petroleum industry as well. More than that, it was one of the fastest growing and most affordable cities in the United States. Shell already employed more than five thousand people in the area, the largest single concentration of Shell employees in the country. During the early 1960s, the company started to outgrow its offices in the Shell Building, the Fannin Building, and the Prudential Building. In 1966, Gene Bankston, Shell's senior officer in Houston, arranged to install most of Shell's

Houston employees in a new skyscraper to be built by Gerald D. Hines.

Hines was a relatively small-time Houston developer. His tallest building up to that point had been a sixteen-story apartment project. Nobody on his staff had any skyscraper experience. But he brought in deep-pocketed investors, such as John and Charles Duncan, as limited partners, and he produced an innovative design prepared by architect Bruce Graham of the famous Chicago architectural firm, Skidmore, Owings, and Merrill. Most of all, Hines had charisma and the talent for selling a deal. He burst into one key meeting with Shell executives wielding a set of bronze door handles and promising that "this is the kind of quality I'm going to put in your building." Most importantly, he gave Shell a very favorable lease on half of the building. He was obligated to find tenants for the other half and personally guaranteed 100 percent of the forty million dollars in construction costs. As it turned out, Hines was able to fill the offices not occupied by Shell at above-market rates.[4]

With Shell as the primary tenant, Hines agreed to name his building One Shell Plaza. Over the next three years, Houstonians watched a modern-style tower rise up 650 feet to dominate the downtown skyline. Clad in gleaming-white, Italian Travertine marble quarried from the same region as the marble in Rome's Coliseum, the fifty-story One Shell Plaza laid claim to being the tallest building west of the Mississippi and the highest reinforced concrete structure in the world.[5] One Shell Plaza put Gerald Hines on the map as a big-time Texas developer. He would go on to build Houston's famous Galleria shopping mall and architecturally innovative and modern Houston skyscrapers such as the Texas Commerce Tower, designed by I. M. Pei, and the Transco Tower and Pennzoil Place, designed by Philip Johnson, in addition to other famous structures far beyond Houston.

In 1967, as the groundwork was laid for One Shell, the lease on Shell Oil's offices in New York City's RCA building came up for renewal. Manhattan's rising costs and deteriorating working conditions prompted Pres. Dick McCurdy to pose the question: "Do we really want to be in New York?" Increasingly, Shell employees were turning down offers to be transferred to New York. They would say: 'I've got three kids. I can't afford it.'"[6] Jack Horner, vice president of Finance, headed up a team to study possible cities to where the company could relocate its operating headquarters.

During eighteen months of study aided by the Stanford Research Institute, Horner's team collected data on a half-dozen cities but eventually narrowed down the list to two, Dallas and Houston. Both cities had low costs of living, room for growth, and a Sun-Belt ambience. And they were located in the Central Time Zone, making intracompany communications easier.

Despite being a strong candidate from the outset, Houston was not a predetermined choice. Some argued against putting the headquarters in the same city, cheek by jowl, with so many line operations. Ultimately, the availability of office space gave Houston the edge. Gerald Hines was planning to build a high-rise garage across the street from One Shell Plaza and agreed to turn the building into a twenty-eight-story combination garage-office that could provide the extra space needed. Shell leased out half of the new building, which became Two Shell Plaza, and in August 1969 announced it was headed to Houston. Executive offices, however, were to remain behind in New York City, near the financial community and in closer liaison to Royal Dutch/Shell.

"Shell to Move 1,000 Workers Here" read the banner headline across the front page of the *Houston Chronicle* the day after the announcement. The final number was closer to 1,400. The Texas-sized move took nearly a year beginning in November 1969. Some 700 employees who could not imagine leaving New York for Houston refused the offer to relocate. For many of those who chose to transplant themselves, it took time to adjust to their new surroundings, especially to the air-conditioned hibernation that Houston's stifling heat and humidity forces upon inhabitants during the summer months. "I was in a state of shock for hours," said one spouse after hearing of her husband's transfer.[7]

Many others welcomed the move and adapted quickly. In Houston, where the cost of living was far less than in New York, people were able to buy homes often twice the size of the ones they left behind in the suburbs of Connecticut or Long Island. The move eliminated the long commute people had been taking into Manhattan and gave the company at least an hour more work per day out of each individual. Shell quickly recouped the thirty-five million dollars it spent to relocate its offices and people in the form of lower administrative and operating costs and greater productivity.

If moving Shell's headquarters to Houston did not provide enough excitement, the succession struggle at the top of the company did. Having reached the mandatory retirement age of 60, Dick McCurdy retired at the end of 1969 and handed off the presidency to Denis Kemball-Cook. At age 59, however, Kemball-Cook's tenure was brief. During the holding operation in 1970, the three executive vice presidents—Bob Hart, in charge of MTM (manufacturing-transportation-marketing), Jack St. Clair, in charge of chemicals, and Ed Christianson, head of E&P— competed openly to become chief executive. Each was intelligent and accomplished. But Shell's board elected Harry Bridges, the president of Shell Canada, to succeed Kimball-Cook.

All three men were disappointed, Christianson so much so that he left the company (to be replaced by John Redmond). St. Clair stayed to lead Shell Chemical through a prosperous period during the 1970s, and Hart ascended the management ranks of the Group, eventually rising to managing director. The new president, Bridges, was an Englishman who had spent more than thirty years with the Group in more than a dozen countries. He was a tough administrator with extensive knowledge of how oil moved around the world. These characteristics served the company well during the turbulent period of the early 1970s, when dependence on foreign sources of oil became critical to the U.S. oil industry.

Assuming the presidency in July 1971, Bridges pulled up all of Shell's stakes in New York City and transferred the executive offices to Houston. The original move had reduced the head office in New York from seven floors of the RCA to two. "But even then, it was costly," recalled Bridges.[8] Furthermore, the umbilical cord between New York and Houston proved difficult to maintain. So in 1972, Shell's entire head office took up residence in One Shell Plaza, shortening the lines of communication. By moving the head office right in the middle of operating land, Shell brought everything a lot closer.

The savings and efficiencies achieved by the move opened up the possibilities of consolidating other activities. The ongoing advances in computer technology allowed Shell to carry out the centralization of data processing one step further. In 1972–73, I&CS closed the regional data centers, with the exception of the credit card center in Tulsa, and moved all data processing into a new building on the south side of Houston near the Astrodome. The ten-story computer center was part of a five-hundred-acre commercial and residential real estate project developed by Shell, called Plaza del Oro. Once finished and occupied, Shell's computer center had more computing power than any place in the Houston area, with the exception of the NASA Space Center in Clear Lake. "We had all kinds of security worries because the nerve center of the company was there, and we had to protect it," said Don Russell, who took over I&CS in 1972. "It was kind of interesting!"[9]

The final act in the centralization of Shell was the closing of Emeryville and the construction of a sprawling new research complex at on the west side of Houston. The Westhollow Research Center, completed in 1975, housed all of Shell Development's research laboratories and offices, except those for exploration and production. The move eliminated duplication of certain research activities. In addition to the cost savings of the consolidation, the move increased research effectiveness by placing researchers in more intimate contact with the people who ran the company's businesses.

The centralization of the Shell organization in Houston capped a decade of modernization. Since 1961, an aggressive marketing drive had taken the company to second place in national gasoline sales, behind only Texaco. The expansion and extension of Shell's refineries and transportation systems provided crucial new capacity to supply a nation thirsty for oil products. The emphasis on petrochemicals gave the company a leg up in a market poised to grow tremendously in the 1970s. Big bets on E&P technology developed the capabilities needed to replace dwindling crude reserves. By 1973, Shell managers had grown more confident in their technological sophistication, investment strategies, and operating efficiency. This was absolutely crucial to meeting the two big challenges of the 1970s: environmentalism and domestic oil shortages.

Offshore Environmental and Workplace Issues

In the late 1960s, Shell Oil's offshore operations came under new scrutiny. Ongoing trouble with mobile drilling worldwide and several major disasters on production platforms in the United States began to draw heightened media attention and widespread public indignation. In December 1967, the tragic loss of the *Sea Gem* jack-up drilling vessel in the British sector of the North Sea, which killed thirteen people, spurred the first real government interest in the safety of offshore units. Although this interest was mainly in Europe, the *Sea Gem* incident set in motion a general reevaluation of mobile drilling that would eventually affect operations in the United States. The big wake-up call for the U.S. offshore industry was the January 1969 blowout at Union Oil's Platform A-21 in the Santa Barbara Channel, which spilled fifty to seventy thousand barrels of oil. Although the blowout happened off California, the fallout reverberated nationally. Santa Barbara catalyzed the national environmental movement and set the stage for the passage of the National Environmental Policy Act (NEPA). Gulf Coast operators, whose practices in the past had rarely been examined or challenged, suddenly faced a potentially hostile political and regulatory climate.

As the industry protested and resisted a stringent new set of OCS regulations handed down by Interior Secretary Walter Hickel in August 1969, calamities in the Gulf undermined their case. In February 1970, Chevron's Platform C in Main Pass block 41 blew out and caught fire. Oil pollution from that blowout postponed a federal lease sale, damaged wildlife, and drew a $31.5 million suit against the company by Louisiana oyster fisherman and a $70 million suit from the shrimpers. A U.S. District Court also fined Chevron $1 million for failing to maintain storm chokes and other required safety equipment, the first prosecution under the 1953 Outer Continental Shelf Lands Act. The Justice Department obtained judgments against other major oil and gas companies, including Shell Oil, for similar violations.[10]

Then, on December 1, 1970, Platform B in the Bay March-and block being developed by Shell burst into flames. As drilling contractors worked to complete twenty-two production wells,

with a combined capacity of fifteen thousand barrels of crude oil
a day, an accident in one of the wells sparked a fire that spread
to other wells, engulfing the platform. Oil from well B-21, which
ruptured twelve feet above the water line, fed the blaze. Flames
from the platform soared four hundred feet into the air, creating
a giant plume of smoke, and burning oil encircled the platform.
Some of the Shell officials who took charge of the effort were informed of the disaster as they took part in the December federal
lease sale in New Orleans. Arriving at Bay Marchand, they came
upon a frightening scene that represented one of the worst night-
mares of offshore developers. Sixty men had been at work on the
platform at the time of the blowout. Many had no option but to
leap from great height into the water to escape. Four men employed by the drilling company died at the scene; another died
later. Thirty-seven men suffered serious burns. A week after the
blowout, ten other wells were burning too. One of the company's
early reports on the Bay Marchand disaster concluded that "nothing remotely like this emergency has happened to Shell in the
twenty years it has been working in the Gulf." Indeed, only a few
similar accidents had occurred in the history of offshore develop-

ment; the company found itself in frontier territory as it sought to bring the well under control.[11]

In the past, the company might have decided to plug each well individually, despite the fact that this approach would have spilled large quantities of oil into the Gulf. But the heightened concern for pollution control in this era dictated a different approach. Shell asserted that "concern for environmental protection has been the key to our activities from the moment the Platform B fire roared into life." To minimize pollution, Shell "followed a plan of keeping it burning and eating up as much escaping oil as possible until relief drilling could kill the wild wells." An array of barges and buoys collected the oil that did escape into the water. The company pursued this plan relentlessly for the 155 days it took to put out the fires and control the wells.[12]

Controlling the fire itself was dangerous work. Shifting winds, fog, rough seas, and the intense heat from the blaze hampered the efforts of fire fighters to reduce the flames and then extinguish the fires below the surface with thousands of barrels of drilling mud. Two firefighting firms, Jet Barges Jaraffe and Red Adair, sprayed the platform with fourteen thousand barrels of water per minute to try to reduce the heat and preserve as much as possible the structural integrity of the legs of the platform. Chemical dispersants were applied around the platform and along the shoreline from east of Timbalier Island to west of Belle Pass. Shell also employed skimmers, straw, and booms to contain and collect the spreading slick, which nevertheless entered Timbalier Bay in late December. Well B-21, where the fire had started, continued to burn so ferociously that firefighters named it the "the wicked witch." Relief drilling killed the fire in the wicked witch on December 30, a month after it had begun. Despite cries of "the wicked witch is dead," workers knew that they still faced hard challenges in killing the remaining ten wells that had caught fire from B-21.[13]

The trick was to continue a controlled burn of oil flowing to the platform until relief drilling had directed the flow away from the fire. To do so, workers had to maintain a constant vigil, monitoring conditions on the burning platform from vessels hundreds of feet away. When tubing of the wild wells succumbed to the heat and looped back into the bed of the fire, action was needed to be certain that the oil flow continued to burn instead of spilling into the Gulf. One innovative solution was the use of a sharpshooter—Shell production supervisor Ken Ring, a former army rifle instructor—to shoot holes in the tubing with armor-piercing bullets. When oil spurted from the bullet holes and caught fire, workers knew that the oil was still flowing and that they could continue their work in killing the well. Similar creative solutions were required when tubing from other wells melted and looped into the water, preventing some escaping oil from burning. In these cases, Shell engineers rigged a makeshift sandblaster on a 150-foot boom and "sawed" the ends off the tubing to assure a fuller burn. Such ingenuity kept the work moving toward its final goal, killing the wild wells.[14]

The second front, drilling relief wells, went forward smoothly. Five drilling rigs used directional drilling to reach eleven different wells via ten different relief wells. Shell observers first had to make educated guesses based on observations of the burning platform as to which reservoirs were on fire. Then the drillers had to find creative ways to reach them as quickly as possible. In a co-ordinated effort that cost more than thirty million dollars, these relief wells succeeded during more than five months in reducing the flow of oil to the burning platform so that fire fighters there could finish their work. By mid-April, oil released from the platform had been reduced to twenty barrels per day.[15]

Parallel with the completion of the technical work to kill the wells was a systematic public relations effort aimed at preventing Shell itself from being killed by criticism. Acknowledging its responsibility for the disaster, the company sought to keep the public informed about the on-going effort to bring it under control. Open cooperation with the media and regulators paid off. The company made the best of a bad situation and drew praise from environmental regulators for its commitment to reduce the oil spill and minimize shoreline pollution. Shell also admitted to a bit of luck—the oil from Bay Marchand was light oil and thus had less visible environmental impact than heavier crudes. Moreover, Shell benefited from the fact that its handling of the Bay Marchand blowout compared favorably to Chevron's handling of its Platform C fire on Main Pass block 41.

The offshore disasters of 1969–70 forced Shell and other companies to get religion on the issues of environment protection and worker safety offshore. Until then, it was generally assumed that the polluting and hazardous nature of offshore operations was inherent in taking land-based technologies, equipment, and practices into the marine environment or, as in the case of diving, by inventing technologies using humans as laboratory subjects. In that adjustment, almost everything had to be redesigned and rethought. Marine engineering and construction advanced by improvisation and trial-and-error, and error could be harmful to the environment and debilitating or fatal to workers.

Environmental protection and worker safety had not been high priorities. Not that Shell Oil and other operators had been unconcerned with such issues, but they subordinated them to the imperatives of growth and profitability. "For years and years, we discharged everything overboard," remembered Lucius Trosclair, longtime Shell Oil employee and manager. "We could throw food overboard. The water that came off your platform, it could go overboard. Your bathrooms went overboard." "You always thought the Gulf was big enough to absorb all the trash and debris," added Alden Vining, a former roustabout and supervisor of production maintenance who started with Shell in 1957. "When they cut up the old grating and stuff, and pieces of pipe around these places to repair them, they just threw them in the water. We figured, well, it is plenty deep, nobody is going to come around here and see this, or nobody cares."[16]

Other kinds of environmental destruction could not be perceived at the time, such as the effects of thousands of miles of canals dredged and laid with pipelines in a spaghetti-like maze through the marshes and swamps of south Louisiana. The canals broke up natural barriers and provided easy conduits for saltwater intrusion and tidal scouring, which led to massive erosion and drowning of the marshes. In recent years, Louisiana has lost annually twenty-five to thirty-five square miles of coast, a land area larger than Manhattan. This land loss endangers wetlands petroleum operations and pipelines not designed for open waters. It also destroys the ecosystem that supports the nation's largest commercial fishing industry. Indeed, the receding coast threatens whole communities as well as the survival of Cajun culture.[17] The greatest factor in this tragedy is the containment of the Mississippi River by levees, which prevents soil replenishment by periodic flooding and the spreading of estuaries. But scientists believe that canals are responsible for no less than one-third of the total coastal-zone degradation. Furthermore, some geologists are convinced that the industry's removal of billions of barrels of oil and saline formation water, along with trillions of cubic feet of natural gas, has caused serious subsidence in the wetlands and offshore, and thus is also responsible for the receding coastline and the heightened vulnerability of the region to hurricane devastation, which was put on shocking display during Hurricanes Katrina and Rita in September 2005.[18]

Shell and other companies viewed death and injury as the unfortunate prices to pay for technological progress. In any business or occupation that involves people working with extremely heavy equipment, such as drill pipe, highly pressurized and combustible materials, such as oil and natural gas, and imperfect technology, there are going to be accidents and, unfortunately, fatalities. Accidents happened when workers fell from heights, had equipment fall on them, or were injured when ropes and cables broke. In addition to the awful physics of the machinery itself, workers had to contend with bad weather, dangerous boarding procedures, drilling mishaps, fires, and explosions far from land.

Safety was compromised not only by the sheer technological challenges but by the necessity to complete work as quickly as possible. Offshore installations were expensive to build and operate. The sooner production could be brought online, the more profitable the project. Drilling vessels contracted on day-rates, which increased time-cost pressures. Production processes were highly interdependent, and delay in one section could cause delays elsewhere. And delays cost money. So there was incredible time pressure to drill the wells, install the platforms, and get the oil and gas flowing. Operators, and especially contractors, like the owners of mobile drilling vessels, did not overly concern themselves with safety. At times, they even cut corners. The major accident rate (loss exceeding one million dollars) for mobile drilling vessels worldwide soared from 1 to 2 percent of the total vessels operating during 1958–64 to 7 percent in 1965, although still much lower than the 14 percent of 1955. This caused Lloyd's of London to raise insurance rates on most types of vessels to almost 10 percent of equipment value per year (not including liability coverages and costs of uninsured exposures such as downtime and lost production). The major accident rate declined back to 2 percent in the late 1960s but remained unacceptably high, especially for jack-ups.[19] Safe processes and designs on platforms as well as mobile drilling vessels either did not exist or remained untested ideas in the minds of researchers. Facilities engineering on production platforms was a novel concept. Platforms were often stick-built with equipment squeezed or slapped together on the deck with little concern or foresight for worker safety.[20]

Although Shell trained and drilled its workers about safe operating practices, the company really did not force them to manage safety or manage change in safe ways. "We talked safety but we were really not that much into safety," said Lou Trosclair, "Of course, we tried not to hurt anybody, don't get me wrong—I mean, we were not barbaric or anything—but it seemed like the work had to be done and you did it as fast as you could." Shell offshore veteran Ken Viater further explained that Shell was not "as conscious of safety at the beginning. I can remember working around noisy machinery with no ear plugs . . . the older guys could tell you 'watch this' or 'watch that.' But there was nobody watching you and telling you 'you had *better* do this' or 'you had *better* do that.'" But, Viater added, "the last ten to fifteen years I worked with Shell it got to be more and more to where everything had to be checked." Cliff Hernandez had a similar assessment, though expressed more frankly: "When I first started working, they did not care whether they killed you or not! In other words, 'we are going to get it done regardless.' There was no suing like people are suing now. Back then, if you got hurt, they just pushed you to the side and put somebody else in. I mean, a lot of people got hurt and did not get paid for it. Crippled. But nowadays, they are trying to look after you different. It is a better deal."[21]

Before about 1970, few pressures from outside individual companies insured greater protection of workers. There were no union pressures, since upstream oil operations were not unionized. OCS orders were worded very generally and thus did not encourage the standardization of safe procedures. The U.S. Geological Survey (USGS), which was responsible for the regulation of offshore operations on federal leases, did not inspect installations on a regular basis. State and federal regulatory bodies were underfunded and understaffed. Some supervisors were political appointees, and even those with the appropriate training and competence often did not have the requisite experience in the oil business to grasp its changing technological capabilities.[22]

After 1970, as Hernandez pointed out, Shell Oil indeed strove to give its workers a "better deal." The explosion of personal injury lawsuits in the 1960s provided one impetus. Initially, the Longshoreman and Harbor Workers' Compensation Act (LHWC) of 1927 covered most offshore workers. This act was designed to fill a gap between the Jones Act (1920), which protects seaman, and state workers' compensation, which covers injuries incurred in a particular state. LHWC provides medical and disability benefits, rehabilitation services, and wrongful death benefits to survivors for injuries, illness, or death sustained during maritime employment on navigable waters of the United States. Maritime employment includes loading/unloading, building, and repairing vessels and offshore structures. In 1959, however, the U.S. Fifth Circuit Court of Appeals ruled (*Offshore Co. v. Robison,* 266 F.2d) that workers regularly assigned to "special purpose vessels" such as mobile offshore drilling units could be treated as seamen under the Jones Act. The significance of this decision was that the Jones Act not only entitled seamen to "transportation, wages, maintenance and cure," which was equivalent to workers' compensation for seamen, but also allowed injured seamen to obtain damages for pain and suffering from their employers if it could be determined that the injuries resulted from negligence by the shipowner, captain, or crew. After the Robison decision, a steady stream of personal injury lawsuits hit offshore operators, drillers, and construction companies. In earlier years, remembered Lou Trosclair, "we could hurt somebody and the poor guy would be hurt and that would be it. But then, the lawsuits started and paperwork really started because you had to document everything." Many companies also found themselves paying workers' compensation and getting sued as well.[23]

After the Bay Marchand disaster, Shell Oil engineers revamped operating procedures offshore and, as a lead member of the Offshore Operators Committee and the API Offshore Safety and Anti-Pollution Equipment Committee, worked with the USGS to strengthen federal safety and environmental regulations. New orders required additional safety features on platforms and pipelines, including the first-time requirement that subsurface safety valves be installed on all producing wells (OCS Order No. 5–3). Although subsurface valves had been commercially introduced in 1954, problems and costs associated with them prevented their universal application. The new order issued in 1973, however, led to the rapid improvement and refinement of the technology. Other orders mandated the testing of safety devices prior to and when in use; more careful control of drilling and casing operations; prior approval of plans and equipment for exploration and development drilling; and new practices and procedures for installing and operating platforms. To enforce the new regulations, the USGS tripled its inspectors and engineers, introduced a more systematic inspection program, and stopped using industry-furnished transportation for inspection purposes.[24]

In addition to their work in revising OCS orders, Shell facilities engineers contributed to the drafting, in a short period of about six months, of a new set of API "recommended practice documents" for the selection, installation, and testing of various kinds of safety devices and piping systems, as well as for platform design. As Ken Arnold, a former Shell facilities engineer who was a key figure in drafting Recommended Practice 14C, covering basic surface safety systems for offshore platforms, explained: "We changed in a period of very few years and got most people to buy into it. Then, the operating people had to maintain them, which was important, and the suppliers started to supply better gizmos, better three-way valves, better sensors, and we learned how to incorporate these sensors into designs in ways that they actually worked and did not give false signals."[25] At the same time, Shell published and disseminated for the first time a "Safe Operating Practices" manual for offshore workers. In addition, Shell and other offshore operators revamped personnel training for offshore operations with the aid of the API, universities, and suppliers. Shell Oil found that its offshore division had been experiencing a high attrition rate as its offshore workforce requirements expanded during 1970–71, which "contributed substantially to the failure of the apprenticeship method to provide training, especially basic training." The new training program increased worker retention, boosted morale, and contributed to an improved safety record.[26]

The new attention to worker safety was part of larger changes in the nature and organization of work at Shell Oil. One catalyst for change in the Shell workplace was the sharp business downturn at the end of the 1960s. During 1969–70, Shell's per-share earnings fell from $4.32 to $3.52. The balance between crude production, refinery output, and product sales at Shell had become too heavily weighted on the downstream end. Behind Texaco, Shell was the second largest marketer in the country, with an 8.25 percent national market share. But it did not have the upstream supply to support this position. "Short of a miracle," wrote *Forbes* magazine in 1972, "Shell doesn't stand a prayer now of ever achieving the kind of efficient balance that enables Texaco and one or two other big international outfits to lead the profitability parade."[27]

Shell had to squeeze more efficiency out of its downstream operations while it intensified the search for petroleum. For the first time since 1961–62, the company reduced its overall workforce. Much of this reduction came through the elimination of several chemical lines and the relocation of the head office to Houston. But managers made cuts elsewhere in the organization. From 1969 to 1973, the total number of Shell Oil employees declined from 39,000 to 32,000. On the E&P side, however, the problem was preventing attrition and retaining and hiring skilled personnel, especially as it expanded offshore. To prevent large gaps in technical experience, the production department convinced management to recruit engineers at a constant level, regardless of economic conditions. In the late 1960s, Keith Doig, head office manager for mechanical engineering, developed a "dual ladder of promotion" for technical specialists so they could achieve jobs at higher pay levels beyond that of "staff engineer" without having to switch over to the management track. According to Steve Siebenhausen, this was "a big factor in enabling Shell to hold onto many of its talented and experienced specialists in the face of continual efforts by contractors, competitors, and consulting groups to hire them."[28]

Probably the most significant change in the nature of the Shell Oil workforce offshore and throughout the company was the new mandate to increase the percentage of racial and ethnic minorities and women among its workers. Up to that point, like the majority of American corporations, Shell Oil employees from top to bottom were largely white males. In 1965, women accounted for 14.5 percent of Shell's total employees; the large majority of those held clerical and secretarial positions. Minorities comprised only 3.8 percent of the total; most worked in lower-rung, labor-gang jobs at the refineries.[29] In the 1960s, however, pervasive racial segregation and gender discrimination in American society began to break down. The civil rights and women's liberation movements demanded that the nation broaden its definition of citizenship and open new educational and employment avenues for women and minorities. The dismantling of formal segregation in the South, the movement of middle-class white women out of the home and into the workforce, and the growing resolve of the federal government to assure equal employment opportunities forced corporate America, including Shell Oil, to pay more attention to diversity and equality in personnel decisions.

Beginning in the mid-1960s, Shell stepped up its efforts to hire and promote racial minorities. In July 1964, Shell signed a joint statement with Pres. Lyndon Johnson concerning a cooperative program called "Plans for Progress" aimed at promoting equal employment. Johnson went on to issue executive orders in 1965 and 1967 that prohibited discrimination on the basis of sex or race in any projects involving a government contract. This meant that Shell and other oil companies on the Gulf Coast had to search for ways to give minorities equal employment opportunities. It was not an easy process, however. Gulf Coast Texas, Louisiana, Mississippi, and Alabama were some of the most rigidly segregated and racist places in the nations. Shell settled many racial discrimination cases along the way. Logan Fromenthal remembered the first African American worker being hired offshore around 1967. Before then, he did not recall working with any African Americans or women offshore Gulf of Mexico or onshore Louisiana. "It was men. White men. Maybe some Hispanics." Over the next ten years, the percentage of minority employees at Shell rose from 3.8 percent to 12.5 percent, while those in higher-paying managerial and professional positions increased from 1.5 percent to 4.8 percent.[30]

Women also made strides at Shell. Although the total number of female employees only rose from 14.5 percent to 17.2 percent between 1965 and 1975, over 5 percent were holding professional or managerial positions and increasing numbers of women were moving into work traditionally done by men. For the first time since the manpower shortage of World War II, women were hired into operating and craft positions in Shell refineries. The few women who began to graduate from universities with science and engineering degrees could also find jobs at Shell. In 1972, the company hired its first female geophysicist, Marisol Garcia. However, older-generation refinery workers, roughnecks, scientists, and managers did not readily embrace the idea of minorities, and especially women, working beside them in what they considered to be male-centered, white-privilege occupations. The macho culture and extraordinary physical and time demands of offshore work made it difficult for women to break into those occupations, even more difficult than for African American men.[31]

The Energy Crisis

On October 6, 1973, just as the Hebrew Yom Kippur celebrations began, Egyptian jets attacked the Sinai Peninsula, where Israel had occupied strategic territory along the Suez Canal since 1954. At the same time, Syrian troops moved into borderlands also claimed by Israel in the Golan Heights. The ensuing Yom Kippur War (also known as the October War) shook the international petroleum industry to its core. Eleven days after the attack, the Nixon administration announced that the United States would supply arms to Israel to counter support given to Arab forces by the Soviet Union. In response, King Faisal of Saudi Arabia, a strong foe of Israel and an ally of Egypt, cut his kingdom's oil output and ordered an embargo against the United States and all countries supporting Israel. Other Arab members of the Organization of Petroleum Exporting Countries (OPEC) joined the embargo. The next week, OPEC raised the posted price of oil from $3 to $5.11 a barrel. Two months later, the cartel raised it again to $11.65. The new control asserted by producing countries over the pricing of oil, observed U.S. Secretary of State Henry Kissinger, "altered irrevocably the world as it had grown up in the postwar period." [32]

The Yom Kippur War triggered a crisis that had been building for a long time. During the 1960s, oil consumption by the world's industrial economies had exploded, yet oil prices had stubbornly remained under $3 a barrel. By 1973, easily accessible oil and gas in the United States was becoming difficult to find. On the eve of the war, spare oil production capacity had dropped to virtually nothing. The world was not running out of oil, but the geography of supply was shifting away from the United States to the Middle East. Americans grew quietly but significantly more dependent on that region's oil; Saudi Arabia replaced Texas as the swing producer for the whole world. "If you're wondering how this could have happened in the land of the free and the home of plenty," wrote *Shell News* in September 1973, "look around and count your 'blessings'—your big, powerful car, your air-conditioning, your house full of labor-saving devices, your ever-improving standard of living. The U.S. has been called a 'high-energy civilization,' which is a prettied-up way of saying that, when it comes to energy, we have been acting as if there were no tomorrow." [33] With the embargo, tomorrow had arrived.

Panic at the pump and growing skepticism of big business battered the public image of the oil industry. Newspaper columnists and congressional committees vilified oil companies as dangerous conglomerates who had conspired to raise the price of oil and who were now enjoying "excess profits." Although the profitability picture was much more complicated than this, the crisis was severe enough to spur the federal government into action. In an effort to assure adequate supplies, Nixon imposed mandatory oil price controls and allocations. And to satisfy the critics of the oil industry, Congress reduced the depletion allowance and threatened to break up integrated oil companies.

As early as 1949, M. King Hubbert, world-famous geologist and associate director of exploration at Shell's Bellaire laboratory, had

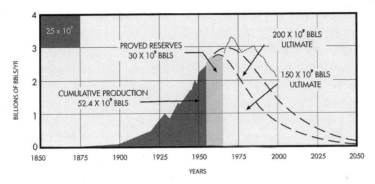

BILLIONS OF BBLS/YR

PROVED RESERVES
30 X 10⁹ BBLS

25 x 10⁹

200 X 10⁹ BBLS
ULTIMATE

150 X 10⁹ BBLS
ULTIMATE

CUMULATIVE PRODUCTION
52.4 X 10⁹ BBLS

1850 1875 1900 1925 1950 1975 2000 2025 2050

YEARS

20
Modified version of King Hubbert's original 1956 graph. The lower dashed line was Hubbert's estimate of U.S. oil production if the ultimate discoverable oil was 150 billion barrels. The upper dashed line, for the ultimate discoverable oil of 200 billion barrels, was his famous prediction that U.S. oil production would peak around 1970. The dotted line is actual U.S. production between 1956 and 2000, which was more than Hubbert predicted, largely due to Prudhoe Bay and offshore Gulf of Mexico. Modified from Hubbert's original 1956 graph.

warned of an eventual oil and gas shortage in the United States. His figures indicated that oil was being pumped out of the ground at a rate that would double every few years. Using widely accepted estimates that the United States had 150 billion to 200 billion barrels of recoverable oil underground, Hubbert predicted that these reserves would vanish much faster than suggested by the previous one hundred years' experience. Borrowing a nineteenth-century Belgian biologist's formula for calculating the growth of animal populations, he developed a production curve, which some called "Hubbert's Pimple" or the more palatable term, "Hubbert's Peak," showing that U.S. oil production would peak sometime around 1970 and then drop as fast as it had risen, as producers drilled for shrinking supplies that were harder to find. "Growth, growth, growth, that's all we've known," Hubbert later explained. "World automobile production is doubling every 10 years, civilian air travel doubles every 10 years, human population growth is like nothing that has happened in all of geologic history. The world will only tolerate so many doublings of anything, whether it's power plants or grasshoppers."[34]

Presenting his findings formally for the first time in 1956 before a regional meeting of the American Petroleum Institute, Hubbert provoked heated debates over the amount of recoverable oil in the United States. His prediction was not taken seriously, even though it was based on mostly undisputed statistics and the simple assumption that producers naturally go after the best rocks and the easiest-to-acquire oil, driving production to a peak before it inevitably slides down the other side of the curve. In the midst of the great postwar boom in U.S. oil production, nobody was prepared for this pessimistic scenario. "It jolted the hell out of the petroleum industry, including my own company," confessed Hubbert. "They were genuinely, honestly shocked. Ten minutes before

I was to speak, the phone rang and it was the New York office. Couldn't I tone it down a bit? Take out the 'sensational' parts?"[35]

By 1971, Hubbert's prognostications no longer seemed sensational. U.S. crude oil production had closely followed Hubbert's Peak. The reserve-to-production ratios for both oil and gas in the United States had declined steadily from their peaks in 1960. In other words, the industry was not finding enough oil and gas to

meet soaring demand. Shell Oil's gross crude oil and natural gas production grew from 444,000 barrels per day in 1966 to 629,000 barrels per day in 1971, while the company's estimated proved reserves declined from 2.9 to 2.5 billion barrels over the same period. Spare or "shut in" capacity in the oil fields disappeared. Environmental regulations restricted high-sulfur coal and nuclear-powered electric generation, thus increasing the strain on oil and gas supplies. The winter of 1969–70, the coldest in thirty years, witnessed alarming fuel shortages. The following summer, capacity constraints on utilities caused brownouts in cities all along the Atlantic Coast. In 1971, David Barran, chairman of Shell Transport and Trading, declared: "The buyer's market for oil is over."[36] Confirmation of this came in 1972, when the Texas Railroad Commission ended production rationing in Texas.

Meanwhile, price controls on oil, imposed in 1971 as part of Nixon's anti-inflation program, simultaneously discouraged domestic oil production and stimulated consumption. Natural gas supplies became tight, largely because of a regulatory system that could not keep up with changes in markets. As the cost of foreign oil delivered to the United States caught up with the domestic wellhead price, imports began to surge, aided by loopholes and exceptions to the oil quota system. The phrase "energy crisis" crept into national political discussion. In April 1973, James Akins, the chief oil expert at the State Department, officially sounded the alarm about the loss of America's energy independence in a widely read *Foreign Affairs* article entitled, "The Oil Crisis: This Time the Wolf Is Here."

That same month, with summer gasoline shortages looming, Richard Nixon delivered a major address on energy, announcing that he was abolishing import quotas. Domestic production, even under quota protection, could not satisfy America's thirst for oil. Under mounting political pressure from Capitol Hill, Nixon then introduced a "voluntary" allocation system, intended to assure supplies to independent refiners and marketers. "Those two acts, coming one on top of the other," writes Daniel Yergin, "perfectly symbolized how circumstances had changed: Quotas were meant to manage and limit supplies in a world of surplus, while allocations were aimed at distributing whatever supplies were available in a world of shortage."[37]

In 1973, the Nixon administration adopted mandatory allocation and price controls on domestic oil, which blossomed into a bewildering and ever-changing system of regulations, administered by the Federal Energy Agency (FEA, later renamed the Federal Energy Administration).[38] Controls and allocations brought about unnecessary product dislocations and artificial shortages, making life especially difficult for jobbers and gasoline dealers, some of whom were forced out of business. Consumers hoarded gasoline and frantically topped off their tanks. Angry drivers queued up at service stations, overwhelming and frustrating the dealers.

Anger intensified as oil companies reported record profits from the surge in oil prices. Shell Oil's income nearly doubled to $621 million. After dropping back in 1975, oil company profits rose again during the next several years but only enough to keep up with inflation. Although huge in absolute terms, the rates of return for the oil industry—except for the banner year of 1974—were just below the average rate for all American industry. The profits registered during the embargo nevertheless lit up on the nation's political radar screen. The large oil companies were denounced as traitors, conspiring with foreign powers at the expense of the United States. The revelations of the Watergate scandal had already fueled public outrage; the Senate investigation into Nixon's secret campaign contributions found that Ashland Oil and Gulf Oil had violated the law. As it had in the Teapot Dome scandal of the early 1920s, oil came to symbolize the corrupt influence by big business on politics, a view encapsulated in the slogan that used a play on "Exxon" as a symbol for the oil industry, "Impeach Nixxon." In early 1974, with tensions running high, senators and congressmen rushed to call hearings on the role of the oil companies in the energy crisis.[39]

The most dramatic were those held by the Senate Permanent Subcommittee on Investigations, chaired by Sen. Henry "Scoop" Jackson of Washington. Over a three-day period beginning on January 21, Jackson's committee interrogated top executives from the seven major oil companies. The hearings took place in the baroque Senate caucus room, the very same room where the misdeeds of Watergate had recently been exposed. The seven representatives appeared together before a packed crowd and under glaring television lights, lined up like a murderer's row at one long table. Standing to take the oath together in front of Jackson, they

nevertheless provided a captivating front-page photo: the "Seven Sisters" personified.

Jackson opened the hearings by ridiculing a Shell newspaper advertisement explaining its recent rise in profits, which he held up for the cameras. Then he launched into a series of hostile questions, relentlessly castigating the oil companies for "blatant corporate disloyalty" to the United States. He charged them with making "obscene profits," a catchphrase repeated in the national media. "The American people want to know," Jackson declared, with a piercing stare and grave tone, "if this so-called energy crisis is only a pretext; a cover to eliminate the major source of competition—the independents, to raise prices, to repeal environmental laws, and to force adoption of new tax subsidies." If the companies did not confess, then Jackson promised to uncover the truth himself.[40]

With the 1976 presidential election on the horizon, the senator was seeking publicity for his run at the Democratic nomination. Once a champion of environmental legislation from his position as chairman of the Senate Interior Committee, he had disappointed liberals with his support for the Trans-Alaska Pipeline, which transported oil that would be eventually unloaded in his state of Washington. Furthermore, the Senate Watergate Committee had recently revealed his receipt of a ten-thousand-dollar secret campaign contribution from Gulf Oil. Here was an opportunity to redeem himself by attacking the industry.

The hearings became mostly an exercise in demagoguery. Investigative journalist Robert Sherrill, who was no friend of the oil industry, called them "the dumbest hearings on a major topic to be heard or seen on Capitol Hill in a decade." The interrogating senators betrayed a surprising ignorance of the oil industry, asking trivial questions and acting flustered when the oilmen responded with piles of industry statistics. "Oddly, the general press wasn't critical," Sherrill wrote. "Odder still was that the most intelligent criticism came from one of the oilmen, Harold Bridges, president of Shell Oil." Conducting the hearings with representatives of seven competing companies appearing as a panel of witnesses, rather than individually, was not the best way to proceed, Bridges noted. Each company had its own unique operations, problems, and views. Time did not permit each member of the subcommittee to ask each witness more than two or three questions. And if Gulf Oil gave an answer that did not apply to Shell, that left Shell "with the choice of either interrupting the questioner and infringing upon his limited time, or seeking in vain (as I found myself doing on several occasions during the hearings) to catch the attention of the interrogator by the raising of the hand. Such would not be the case if witnesses appeared individually."[41]

Putting the seven oilmen on the same stage made for better political theater. They came across as defensive and surly in the face of Jackson's humiliating onslaught. The forum prevented them from making their case aggressively. It did not help that on the last day Exxon released its 1973 earnings, up 59 percent over 1972. The hearings abruptly ended without offering much enlightenment on the issues.[42]

As retailers struggled to satisfy motorists on the front lines and oil company executives went on trial before the public, Shell Oil's transportation and supply organization scrambled to acquire new sources of crude oil for the refineries, which were running flat out at 99 percent of capacity. In fact, well before the embargo, Shell had begun to focus on improving its access to international supplies. Still resisting Hubbert's prediction, many industry forecasts in 1970 pointed to 1975 as the magic year when U.S. production capacity would peak and large imports would begin filling the gap. The fact that the peak came the same year the forecasts were made did not come as a surprise to some people in Shell Oil. One of these was Stan Stiles, who became head of Transportation & Supply in 1970. A World War II veteran, Stiles had joined the company in 1947 as a mechanical engineer in Los Angeles E&P.

He spent the next two decades in different assignments across the country for Shell and working for the Group in such places as the Netherlands, Trinidad, Indonesia, and London. When Stiles first came to Transportation & Supply as a general manager in 1967, he discovered that two years earlier, Tony Dempster Jr. had begun making a long-term crude oil forecast (LTF) for the supply organization of T&S. Dempster's LTF showed that U.S. production would peak by the mid-1970s and then gradually decline. The gap between demand and production would then double every three to five years. The study also revealed that spare shut-in capacity was much less than what was being reported, which placed the day of reckoning much sooner than anyone anticipated.

From its beginnings in the United States, Shell Oil had supplied itself largely with crude oil produced domestically and to a lesser extent in Canada. But the company could no longer enjoy the domestic self-sufficiency that it had in the past. Now, teams of Shell negotiators traveled around the world to purchase crude on both the spot market and by long-term contract. In stiff competition with other oil companies and representatives from Europe and Japan, Shell's procurement efforts raised the company's crude oil imports from 30,000 barrels a day in 1972 to over 200,000 barrels a day in 1974. "The only way to get ahead in this business of transportation and supplies is to learn Arabic," Stiles used to tell his people.[43]

Fortunately, Shell already had made arrangements for transporting this big flow of foreign oil to its refineries. It negotiated a one-third stock ownership in a multimillion-dollar transshipment terminal in Curaçao and signed a contract with a foreign shipping company to move crude for Shell from the Persian Gulf to Gulf of Mexico ports via the Grand Bahama terminal. "We made the commitments for the moving of oil before we had the oil—but that's a risk we had to take," said Stiles. It turned out to be a smart bet when supertankers started heading toward the United States. These terminals were designed to transfer cargoes from very large crude carriers (VLCCs)—a new class of deep-draft tankers capable of transporting 300,000 dead-weight tons of oil—to smaller tankers that could be accommodated in U.S. ports.[44]

This was a good interim solution, but Gulf Coast refineries immediately realized that it would be easier and cheaper to offload these supertankers directly. So Shell joined a five-company consortium to build the country's first superport capable of handling VLCCs, the Louisiana Offshore Oil Port (LOOP) off Lafourche

Parish, Louisiana. Environmental opposition and stiff regulatory requirements, however, delayed completion of the project. In May 1981, ten years after it was first proposed, the $700 million LOOP finally opened for business. Linking more than a quarter of the nation's refining capacity to supertankers calling from overseas, LOOP symbolized the growing dependence of the United States on foreign oil and solidified M. King Hubbert's reputation as a national prophet.

Going International

As they walked down New York City's West 50th Street on an October day in 1970, Harry Bridges—just named executive vice president of Shell Oil—had a brief but fateful conversation with Sir David Barran, chairman of both Shell T&T and Shell Oil.

"David, we are getting short of crude oil," said Bridges. "Would the Group agree to supply our shortfall at a competitive price?"

"No, we can't," replied Barran, a monocled British aristocrat who enjoyed immense stature in the world of oil. At that moment, he was leading the major oil companies in a fight against escalating demands by the OPEC cartel and was unsure about the Group's ability to supply its own refineries. "We are short of crude ourselves," Barran explained.

"In that case, Shell Oil *has* to go overseas," Bridges informed his chairman.[45] Barran conceded the point. Within weeks, Shell Oil had exploration teams traveling to countries outside of North America for the first time in its history, and Bridges soon took over as president of Shell Oil.

He was wading into a delicate situation. His discussion with Barran was the climax of years of discussion over allowing Shell Oil to explore abroad on its own. There had been open competition for his job among some of the vice presidents, and Bridges wondered how he would be received, especially as a British citizen following strong presidents of U.S. nationality. With his extensive experience in international exploration, however, Harry Bridges was eminently qualified to lead the company overseas. His career with the Group had virtually spanned the globe. The son of a coal mining official from Sheffield, England, the young Bridges distinguished himself in math and physics. Hoping to save enough money to obtain an advanced degree in aeronautical engineering, he took a job with Royal Dutch/Shell, heading up a geophysical crew in the humid swamps of Papua New Guinea, in a remote corner of the world where money could be earned but not spent. After serving as a navigational instructor for the Royal Australian Air Force during the war, he returned to exploring for oil in the jungles, this time in Ecuador and Columbia. "I'd be one of the pioneers going in to start up another exploration venture. The first 10 years of my career, I suppose you would say I lived mostly in a tent."[46] He moved on to various managerial positions in India, Qatar, Indonesia, Nigeria, and Venezuela before returning to the Group as a coordinator in 1959.

During the 1960s, Bridges assumed increasing responsibility for the Group's interests in the Western Hemisphere, first as regional coordinator for the Caribbean, Central America, and South America and then as president of Shell Canada. While in Canada, he was given the choice to return to the Group as a managing director or take over the reigns of Shell Oil. "It took me three seconds to make up my mind," remembered Bridges. "It was the difference between running such a big company, which was probably one-third or more of the entire Group, or being one of a committee of seven. I was very glad to go down to New York. And I never regretted that decision."[47]

The question of whether or not Shell Oil should "go international" had long been under consideration. In 1966, Monroe Spaght, in his new position as Royal Dutch/Shell managing director, had asked Shell Oil Executive Vice President Ned Clark to put together a presentation assessing the company's capability to explore and develop oil and gas resources outside of the United States. Some people on both sides of the Atlantic believed Shell might have difficulty producing enough oil to supply the huge downstream investments it was making in its drive for greater market share. Bridges, who at the time was the Group's manager of international supplies, projected a coming supply crunch for the Group's refineries around the world. The Group had not been as successful in exploration as some of its international competitors, especially in the Middle East. In case of a desperate shortage, Shell Oil could not count on back-up supplies from the Royal Dutch family. Feeling a special obligation to Shell Oil's public shareholders, Spaght believed that the company ought to be given chance to go abroad where prospects looked appealing.[48]

In the early 1950s, interestingly, the Group had invited Shell to participate in some of its overseas ventures. At the time, John Loudon told Max Burns: "We've got our hands full everyplace. If you want to come and take any part of this, please talk to us about where you want to go and let's see if we can't make a partnership deal." But after serious study and soul-searching, Burns told Group directors Shell Oil had enough business to keep it occupied in the United States. The company just had too many technical commitments to places like offshore Louisiana and the Williston Basin, which were beginning to heat up. Thus, the opportunity passed until Spaght brought up the question again in the mid-1960s. "If Shell had gone overseas early, they might have been able somehow to do better than the Group, who had the world open to all this," he later reflected. "But of course you're guessing that one member of the Shell family would have been smarter than the rest."[49]

In 1966, Clark instructed John Bookout, as manager of the E&P Economics Department, to lead a Shell delegation across the Atlantic to discuss the issue. At a pivotal meeting in London with the Group's senior management and E&P coordinator, Bookout presented an analysis of various international plays where he thought Shell Oil would do well. Group officials, however, questioned Shell's ability as a high-tech (and thus relatively high-cost) company to adapt to the very different kind of environment outside the United States. Shell's average finding costs in the United States were 50 cents per barrel, whereas finding costs overseas were more on the order of 10 cents. Bookout responded to this concern by pointing out that Shell had competed very successfully with American international oil firms in the United States; since those firms were successful outside the country, there was no reason why Shell could not be, too.[50]

No immediate decision came out of the London meeting, and the issue lay dormant for several years until a group of Shell's minority shareholders revived it. In June 1969, on behalf of Shell Oil, Robert Halpern and other plaintiffs entered in the Supreme Court of New York and the Court of Chancery in Delaware parallel shareholder derivative actions against the Royal Dutch/Shell shareholders in Shell Oil. The Halpern plaintiffs alleged that between 1959 and 1969 the Group "violated its fiduciary duties" in transactions with Shell Oil by requiring the American affiliate to purchase substantial amounts of crude oil and other hydrocarbons from Group companies at prices substantially exceeding those prevailing in the free market.[51] Beginning in the late 1950s, the international glut of crude oil had produced an increasing disparity between the "posted" or official price, which was held constant, and the actual market or "arms-length" price at which crude was sold, which was dropping. In other words, companies offered bigger and bigger price discounts for market oil, while the posted price could not be lowered, largely for the political reason that it served as the basis for the revenues of the producing countries. That affiliates could be charged higher prices for crude oil than those paid by independent refiners was so widely acknowledged, it gave rise to the expression "only fools or affiliates pay posted prices."[52]

Shell Oil shareholders were not fools, but neither was Shell an ordinary affiliate. The Royal Dutch/Shell defendants denied the allegations and moved to strike the complaint as a sham. The main evidence for the charges cited by the plaintiffs were speeches and articles published between 1962 and 1964 by oil analysts unconnected to the Group. The articles only referred to crude oil transactions and not natural gas, petroleum products, and other services mentioned in the complaint. In May 1970, the Delaware judge found that the "articles do not provide good ground for charges of self-dealing, mismanagement and the like during a ten-year period, half of it after the last of the articles was published, in transactions involving many subjects other than crude oil."[53] Yet the plaintiffs kept the case alive with an amended complaint. A decision by the Delaware court in November 1973 granted the motion of the defendants to dismiss claims accrued prior to June 1966, on the grounds that the statute of limitations barred such action, but ordered further discovery of information for the period January 1966 to June 1969. This case and two other minority shareholder suits submitted by the same plaintiffs in March 1974 dragged on for another seven years before a settlement was finally reached (see below).

The implicit allegation of the Halpern suit was that Royal Dutch/Shell had prevented Shell Oil from operating in more profitable oil-producing areas abroad and kept it dependent on Group companies for supplies. Although the Group continued to dispute the charges, the minority shareholder pressure was too strong to ignore. "This antitrust action was always a worry to the directors in London and The Hague," recalled Harry Bridges. "Much less of a worry to people in the United States because we realized it was a part of living."[54] Antitrust was indeed a fact of life for most major corporations doing business in the United States. Whether or not the Halpern case had merit, it pressured Shell Oil to consider the possibility of venturing outside the United States. U.S. reserves were declining, oil prices were rising, and all of Shell Oil's major competitors had their own overseas programs. Minority shareholders wanted to know: Why didn't Shell Oil?

With the Halpern case simmering in the background, Group directors suggested to Bridges, who was president of Shell Canada at the time, that his company take in Shell Oil as an exploration partner. Shell Canada had some large leases in frontier basins, but it did not have the capital or technology to pursue them. A joint venture seemed like a good fit for both companies. In early 1970, Bridges negotiated an agreement with Shell Oil in which two wholly owned subsidiaries of the American company (Shell Explorer and Shell Prospector) would join Shell Canada in exploring off the east and west coasts of Canada as well as in the Mackenzie River delta of the Northwest Territories. The joint venture also would examine ways of extracting heavy oil from Alberta tar sands (Peace River and Athabasca), something Bellaire researchers had studied in the late 1950s under a previous joint venture, which Shell Oil had pulled out of in 1962.

The most exciting area was offshore Nova Scotia, where Shell Canada's leases covered thirty-four million acres all along the coast of the province and miles out to sea. The geology showed large structures and salt domes. The seismic data looked like another Gulf Coast or Nigeria. In 1970–71, the joint venture drilled twenty-four wildcat wells, but it found little commercial oil and gas. Although the structures were impressive, the sand layers were so large that shifting along a fault did not normally create a petroleum trap like it did in the Gulf. "It turned out to be a complete flop, the biggest disappointment of my life," admitted Bridges, who had moved to Shell Oil shortly after negotiating the joint venture of Shell Oil and Shell Canada. On the West Coast, drilling off Vancouver Island and in Queen Charlotte Sound proved equally disappointing. The same went for the MacKenzie Delta. Upon hearing the early reports of dry holes in the fall of 1970, Bridges decided to approach David Barran with his request to take Shell Oil international.[55]

The Sign of the Pecten

After receiving encouragement from Barran, Bridges wasted no time in establishing a new International Ventures Organization under E&P. In 1971, Gerry Burton, vice president for E&P on the Pacific Coast, was appointed to head it. Burton had spent many years with Shell as an exploration manager (he joined the company in 1938). But as exploration opportunities on the Pacific Coast shrank, he found his responsibilities reduced. International Ventures offered a new area where he could apply his talents.

Burton was good at assessing risk, and the risks of exploring overseas were much greater than in the United States. In 1949, he had been the first manager of the Marine Division in New Orleans, one of Shell Oil's most successful risks. Bridges hoped that International Ventures would have a similarly auspicious beginning under Burton's guidance.

With a small staff and little overseas experience or data, however, International Ventures began as a kind of hole-in-the-wall operation. "When we first got started," explained Franklin "Hoppy" Conger, chief geologist in the new organization, "we knew we were novices in this area."[56] But Burton and Conger diligently built up an expanding international presence. In early 1971, Shell Oil established its first exploration venture outside of North America by forming a subsidiary in Colombia. The company took a 25 percent stake in this subsidiary, a partnership with Conoco and Ecopetrol, the Colombian national oil company. Others followed in Peru, Honduras, Malaysia, New Zealand, Indonesia, and Vietnam. Each company was named "Pecten" (Pecten Colombia, Pecten Vietnam, etc.), the scientific name for the scallop, whose shell was the Shell Oil trademark. Pecten was adopted as the corporate name in the international field, since there, "Shell" was the exclusive property of Royal Dutch/Shell.

While the Pecten companies' explorations in Latin America yielded little, Southeast Asia offered greater promise. In 1973, near the end of the Vietnam War, Pecten Vietnam leased offshore tracts in the South China Sea from the tottering Saigon regime. The Nixon administration viewed the participation of American oil companies in the search for oil offshore Vietnam as a way to buttress its plan for the "Vietnamization" of the war (removing American troops to allow "Asians to fight Asians"). Oil investment would replace American forces and aid as a crucial prop to the South Vietnamese economy. In late 1974, Pecten Vietnam made two minor wildcat discoveries, heartening both the Nixon administration and the Saigon regime.

These and other strikes in the general vicinity made by Mobil, however, backfired on the Americans and South Vietnamese. Ecstatic over the finds, Saigon played up their significance to strengthen support for its regime, which distressed the North Vietnamese and possibly triggered the North's final offensive. As communist forces overthrew Saigon, Pecten plugged its offshore wells and abandoned the area. The U.S. trade embargo on the new regime prevented Pecten Vietnam and the other American oil companies from returning. As a newcomer to the international oil game, Shell Oil learned a quick lesson in political risk.[57]

International Ventures enjoyed better results in deals made with the Group, which had strong positions in many countries. In 1973, Shell Oil took a one-sixteenth interest in a consortium led by Shell New Zealand and British Petroleum to develop the Maui gas field offshore New Zealand. When brought on-stream in 1978, the field ranked as one of the world's biggest. At the same time, a Royal Dutch/Shell subsidiary, Sabah Shell Petroleum, invited Shell Oil (Pecten Malaysia) to participate on a 50–50 basis, without prior expenditures, in exploring and developing a 2.88-million-acre area offshore the Malaysian state of Sabah, formerly North Borneo. The Group had recently discovered the Samarang field but was having problems interpreting the geology and developing it properly.[58] Pecten Malaysia provided valuable technical input and by 1976 had helped Sabah Shell develop production of 50,000 barrels per day, which was shared with Petronas, the Malaysian government-owned oil company. In January 1976, the first barrel of crude oil ever produced from an International Ventures' well arrived at Shell Oil's Anacortes Refinery from the Samarang field.[59]

The next showcase operation was offshore Cameroon, the one-time French-English colony on the hinge of West Africa. In early 1976, Shell established its Pecten Cameroon subsidiary. But other operators, including a Group company called Shell Camrex (Societe Shell Camerounaise de Recherches et d'Exploitation), already held all the exploration permits. These operators knew there was oil in the offshore extension of the Niger Delta, but they had only discovered pockets of oil and gas. So they all were sitting on their leases. After culling through some of Shell Camrex's data, Hoppy Conger became convinced that Shell's U.S. Gulf Coast technicians could make the area pay off. The Niger Delta is geologically similar to the Mississippi Delta offshore Louisiana. Shell could use its bright spot seismic knowledge, which was unavailable to the other operators, as effectively offshore Cameroon as it had in the Gulf.[60]

Under exploration vice president Jack Threet, Pecten Cameroon began modestly, taking a farmout from the Group, and then biting off larger and larger pieces until it had built up a 49 percent interest in the most prolific part of a delta field, in partnership with the French firm, Elf-Serepca.[61] Then Pecten started striking oil. "We drilled about 20 wildcat discoveries," remembered Marlan Downey, who replaced Threet. The fields were indeed small. But Shell's bright spot technology (which by agreement Shell did not divulge to Elf) assured that the drilling rigs did not miss many of them. "Our French partners and the government of Cameroon couldn't figure out what in the hell was going on," Downey said. By 1979, Pecten Cameroon's production was 60,000 barrels per day and rising. To work the production, the company imported Cajun roughnecks and engineers, who felt right at home among the mangrove swamps, sweltering humidity, and French-influenced culture and cuisine.[62]

Unfortunately, the government of Cameroon felt Pecten was getting too comfortable for its own good. In 1980, it passed a new petroleum law giving it the right to acquire a 50 percent equity interest in Pecten Cameroon, which dampened some of Shell's enthusiasm over its recent discoveries. Remarked Downey: "Ever try arguing with a sovereign government?" Even so, the company still had a good contract, which allowed it to recoup its costs plus 13 percent of the remaining oil as profit. Pecten also received consolation later in the year when the Group settled its long-standing U.S. minority shareholder suits out of court by turning over all its remaining interests in Cameroon to Pecten. In return, the plaintiffs withdrew all their complaints, ending a bitter chapter in shareholder politics at Shell. Pecten Cameroon went on to become the most successful and profitable operation for Shell Oil overseas. By 1984, it had a share in production of 160,000 barrels per day from Cameroon, which accounted for the majority of Shell Oil's net foreign production of 33,000 barrels per day (the amount left over after the shares for partners and host countries were distributed).[63]

Although modest compared to Shell's domestic total of 464,000 barrels per day, foreign production gave the crude-short company an important safety valve and reinforced its proud independence from the Group. Considering the relatively brief time Shell had been in business overseas, the worldwide Pecten organization was an impressive achievement. Pecten companies had explored or were exploring in twenty-five countries and had a permanent staff of four hundred, plus three hundred foreign nationals on the payroll. In 1982, International Ventures was transformed into Pecten International, a holding company for Shell's E&P operations outside the United States, with Marlan Downey as its first president. By this time, Pecten had become one of the most active international explorers, with interests in about 125 million acres (about seven times the land Shell held in the United States) around the world. Pecten's arrival as an international oil company in its own right was symbolized in 1984 by its ability to borrow $350 million in international markets, without recourse to Shell Oil, for further development work in Cameroon.[64]

Still, the international expansion of Pecten had limitations and constraints. For one, it created growing tension with Royal Dutch/ Shell. Because of the minority shareholder issue, Pecten could not discuss its business—such as bidding, concessions, and financing—with the Group. Some Group managers did not appreciate competition from another member of the Royal Dutch/Shell family. Mike Forrest, who succeeded Downey as president of Pecten, remembered one conversation with a Group general manager:

"Mike, tell me how many years you guys have been in the oil business in Pecten?" said the gentleman.

"Pecten has been around for twenty years," replied Forrest.

"That's my point. We have been doing this for one hundred years. You guys don't know anything about this international oil business!"[65]

Indeed, Pecten may have been a first-rate E&P organization, but it did not have the experience and influence often needed to navigate through difficult political terrain. It was better at managing technical risk than political risk. Part of the problem was timing. Shell was trying to establish overseas positions during a period of international crisis, intense Third World nationalism, and anti-Americanism. Host governments commonly kept large shares of the profits from foreign oil investments, up to 90 percent in some cases. The best prospects open to Pecten were often in volatile and corrupt countries. Said Don Russell, who preceded Downey as president of International Ventures, "It seemed like I spent all my time trying to make deals with the dictators of the world! Marcos in the Philippines. Assad in Syria. Stroessner in Paraguay." As a straight-shooting company, however, Shell refused to pay bribes or under-the-table payments. "They would try to shake you down as much as they could, and you couldn't allow it," said Russell.[66]

When Pecten eventually did gain a foothold in Syria, international politics temporarily tripped it up. In 1986, the threat of sanctions by the U.S. government against any U.S. company operating in Syria, which had been identified as a terrorist country, forced Pecten to transfer its interests to Royal Dutch/Shell under a standstill agreement. At the end of 1987, however, Pecten acquired its non-operator interest back from Royal Dutch/Shell (under the terms of the standstill agreement) and continued to participate in the project. Gross production increased to 400,000 barrels per day by the end of 1991—the most productive international project Pecten ever undertook. In the mid-1990s, as Shell Oil lost its managerial autonomy (see chapter 7), Pecten eventually traded its Syria interests to the Group.

Bookout Takes Over

During the late 1970s, the drive to diversify took many U.S. oil firms into completely different businesses. Threatened by proposed divestiture and sobered by declining U.S. oil reserves, companies hedged their bets outside the energy sector. Although such investments were relatively small in the overall portfolios of diversifying firms, they were substantial nonetheless. Mobil bought the Montgomery Ward department store chain, Exxon went into office equipment, and ARCO and Socal invested in copper. Gulf Oil even bought the Ringling Brothers and Barnum & Bailey Circus. "That, more than anything else," writes Daniel Yergin, "did seem to prove that the clamorous new era—of OPEC's imperium and high oil prices, of confusion, bitter debates, and energy wars in Washington—really was a circus."[67] When oil companies bought nonrelated properties, it played into the hands of the industry's harshest critics who made headlines with claims that Big Oil was using its "immoral profits" to beget even greater profits instead of spending the money to ease the energy crisis.

Prevailing opinion in the business press held that oil had an excellent short-term outlook but an uncertain long-term future in a conservation-conscious world. Why shouldn't the industry use its brainpower and cash flow to get into other businesses? The first question John Bookout confronted when he became president and chief executive in May 1976 to replace the retiring Harry Bridges was: "Should Shell Oil diversify?" For much of his first year as president, Bookout holed up in his forty-fifth-floor office in One Shell Plaza developing a strategic plan for the company. He created a planning department and numerous groups to study investment opportunities inside and outside the oil business, from forest products to biotechnology. "At the end of a year and a half," he said, "I was in a position to say we shouldn't diversify. It had to be almost egotistical to think that Shell could pay a premium to take over a company in an industry we knew nothing about and cause it to perform 2 to 2 1/2 times better than it had been, which would have been necessary to get the return on investment we needed."[68] Although oil and gas was becoming harder to find in the United States, Bookout believed that the company could continue to expand profitably its domestic exploration and production base. During his twelve years of leadership (1976–88), Shell remained true to the faith that its greatest successes would be in areas where it had the opportunity to develop and apply its technology.

A tall, lanky man with a soft Gulf Coast drawl, John Frank Bookout Jr. was dedicated to the oil business. The son of a Shreveport, Louisiana, railroad engineer, he had grown up around oil wells and drilling rigs. In high school, he used to delight in driving to the Caddo gas field at night and reading his textbooks by the light of the gas flares. He chose geology as his college major "because I didn't want to be cooped up in an office, and the concept of a geologist in those days was someone who spent a lot of time in the oil fields, setting wells, catching samples, and so forth."[69] As he was about to enroll at the University of Texas, however, World War II intervened. He joined the Army Air Force and flew twenty-three missions as a B-17 bomber pilot in the European theater. In 1950, he completed his master's degree in geology at Texas and went to work for Shell as a geologist in the Tulsa Area office.

Bookout rose steadily up the exploration ranks and gathered wide-ranging experience in the company's most active geographic areas. In 1961, after stints in Amarillo and Wichita Falls, Texas, he became exploration manager of the Denver E&P Area. He then took a six-month assignment with Royal Dutch/Shell, visiting all the Group operations in the Persian Gulf. This helped him evaluate Shell Oil's capability to go international when he was asked in 1965 to make such a presentation to the Group as manager of E&P Economics. In 1967, Bookout became vice president for the New Orleans E&P Area, where he helped steer the company into the deeper waters of the Gulf of Mexico. In 1970, he was the obvious candidate to replace Harry Bridges as president of Shell Canada, which was just preparing for a major play offshore Nova Scotia. "When I was asked to pick my successor in Shell Canada, I toured all the E&P Areas and met John Bookout," Bridges recalled. "He was clearly the outstanding man. I went back to Dick McCurdy and Denis Kemball-Cook and said: 'I want John Bookout.' They replied: 'We thought you would!'"[70] The Shell Canada job gave

Bookout valuable experience running an integrated company about one-fifth the size of Shell Oil and served as a stepping stone to the position of executive vice president of Shell Oil in 1974, replacing the retiring John Redmond.

Bookout impressed everyone with his intellectual dynamism and intensity. "He was the most intense person I ever met," said Bob Nanz. "Boy, he just ate and slept this stuff." But Bookout also could disarm people with a cool, easy-smiling style. "He's a little like the country boy who plays it low but comes out 10 miles ahead of you," one long-time oil executive described him.[71] Charlie Blackburn, who was general manager of the onshore division and later executive vice president for E&P under Bookout, testified to this. He remembered making a presentation with Don Russell, who was regional production manager in New Orleans, about whether or not a construction project should be done union or open shop. He and Russell recommended nonunion. "John sat right there and talked us out of it, both of us," said Blackburn. When they agreed to change their minds, Bookout contradicted them again. They were right the first time, he told them. "What a lesson!" recalled Blackburn. "He raised so many objections that he finally convinced us that we were wrong when, all along, he agreed with us. He just wanted to make sure we had thought it through."[72]

Although he did not seek or receive the publicity that some other American CEOs did, Bookout was one of the hardest-working men in the business. He scheduled little time for vacations, recreation, or even a game of golf now and then. "I guess I never felt I could take five to six hours off to play," Bookout explained. He enjoyed hunting and collecting paintings of birds, but he devoted most of his time to Shell. "John Bookout was the only guy I ever knew who wore a watch and never looked at it," said Jack Little, who became Shell Oil president in 1998. "I don't know why he wore it. It didn't make any difference what time it was or what his schedule was, John could keep people waiting hours because he wanted to talk about something that was then on his mind." Bookout convened meetings on the weekends and phoned people in the middle of the night; he expected everyone to be on the job all the time and performing. Nothing bothered him more than company lethargy—moments "when people haven't done the best they are capable of doing."[73] Although Bookout's demands could be trying, many of Shell's top managers felt invigorated by the challenge and developed tremendous respect for him.

Bookout was a superb industrial politician. He was not afraid to butt heads in order to get his way. "We watched his progress through the ranks with fascination," said a former colleague. "Every time he got promoted, there would be a confrontation between him and someone else. John always won. He settled authority issues quickly." His leadership and resolve vaulted him to the pinnacle of the company at the relatively young age of 53. "Take a decision, take it firmly and don't look back on it," he often preached. When he assumed the presidency of Shell Oil, Bookout faced a series of major decisions about the allocation of resources within the company. The big run-up in oil prices and leap in revenues had raised questions about where Shell should invest its capital, not only outside its traditional lines of business but within them.[74]

As had been the case off and on since the mid-1960s, the main struggles occurred between E&P and chemicals. In 1975, Shell Chemical had started up the Norco Gasoline Olefins Plant (GO-I) and was planning to build two more world-scale olefins plants, costing more than $500 million each, that would rely on heavy oil as a feedstock and sell a large part of their output (mainly ethylene) to the merchant market. One would be at Deer Park in

Houston (OP-III) and the other at Norco outside of New Orleans (OP-V). Connected to each other by a pipeline, they were going to be mammoth processing units, each producing 1.5 billion pounds per year of ethylene. OP-V also would produce 1 billion pounds per year of propylene and 500 million pounds per year of butadiene. Petrochemicals had rebounded from the slump of the early 1970s, and Shell Chemical was determined to lead the industry in supplying critical shortages in ethylene.[75]

A dyed-in-the-wool E&P man, Bookout opposed such a giant expansion in olefins, insisting that money would be better spent looking for oil and supporting the traditional downstream businesses of refining and marketing. Jack St. Clair, the president of Shell Chemical, argued that petrochemicals were the wave of the future. Chemical plants could upgrade hydrocarbons into a wider and more profitable range of derivatives. Compared to refining, St. Clair emphasized, the raw material costs of a chemical plant were a small proportion of the total sales price of upgraded derivatives. This still did not reassure Bookout of the wisdom in building capacity in a commodity product so far in excess of Shell's own internal needs (Shell had not built up a strong position in the derivative businesses). He believed that only one plant was necessary. "I argued with Jack until I was blue in the face," Bookout remembered. "We can supply the market with one facility out of either Norco or Deer Park."[76] St. Clair stood firm: "We are going to have both of these plants."

Bookout and St. Clair were cordial and respectful toward each other, but they had not seen eye-to-eye on many issues in general executive committee meetings prior to Bookout becoming president. Bookout was skeptical of the crude supply deals being negotiated in the Middle East, though he eventually came around to supporting Shell Oil's big petrochemical investment in Saudi Arabia.[77] He also was uncomfortable with the management-by-committee matrix form of organization that St. Clair had introduced into Shell's downstream businesses. The disagreement over the olefins plants brought the brewing conflict between the two men out in the open.

It finally came to a head at an October 1976 board meeting, the first of Bookout's presidency. But instead of presenting a divided opinion to the board on the olefins investment strategy and forcing the directors into an awkward position of taking sides, Bookout proposed an alternative approach that would allow the projects to go forward while at the same time spreading the risk.

He agreed to support the building of two plants but only if the second plant, OP-V at Norco, took on a partner. This compromise was acceptable to everyone, especially in view of the massive construction and start-up costs and the anticipated effects they would have on the profit performance of Shell Chemical. So Shell brought in Dupont as a 50 percent partner in the plant. "That was the best thing that ever happened to us," said Bookout. "We lost our ass on that plant for years." Already by 1978, several years before Shell's olefins plants were completed, oil companies found themselves "sloshing in ethylene" due to an overexpansion of capacity. Everyone had rushed to expand at the same time, but nobody had committed as heavily to ethylene as Shell.[78]

In the wake of the conflict over expanding olefin capacity, as financial prospects for the new olefins plants quickly dimmed, John Bookout took charge of the show. He gained the board's approval for a new delegation of authorities that clearly established one line to the president. Financial and investment decision-making thereafter became highly centralized. In 1979, the retirement of Jack St. Clair removed a strong, independent voice from Shell Chemical and enabled Bookout to consolidate power further. Learning from the mistake the company had made in betting that the rise in demand for petrochemicals would double the rate of growth in gross national product (GNP), Bookout forced the company to sharpen its planning and forecasting. "Planning methodology was from an earlier era, a surprise-free era," he explained. Plans were done in the functions, and functional plans were combined in an annual meeting called the "Strategy Planning Conference." "There was little strategic about it. I could accept the plan. Rejection was not really an option. It might better have been called the 'Medical Planning Conference.' The corporate patient got minor surgery or a money transfusion."[79]

Bookout demanded greater say over functional planning. He designated himself as the chief strategic planner and drew all the functional planning into an overall Corporate Planning Department (previously called planning and economics). All planners then worked with the same analytic tools, which made it easier to compare investment options and gave the functions greater insight into how they were impacted by each other. Under this system, officers below the CEO concentrated on preparing risk-discounting scenarios for an investment decision rather than making that decision themselves. Risk-discounting meant determining how likely a given project, be it service stations, drilling programs, or chemical plants, would be to fail under different conditions. Once a project was underway, it continued to be analyzed on a "lookback" basis, its results constantly compared with the results forecast for it. Future projections were then revised accordingly.

John Bookout's strategic plan, formulated in 1978, called for reducing capital spending in the chemical sector and plowing investment into the company's conventional oil and gas business. The plan intended to reverse declines in domestic oil production and increase, not just replace, the company's reserve base. In retrospect, this objective should not have seemed controversial. But in the climate of the late 1970s—highly charged by projections of dire oil shortages, pressures for diversification, and fears about divestiture and restricted access to federal lands—it was. Bookout sounded almost defensive in describing Shell's strategy to outsiders. "I don't want you to think that we are overly complacent, or that we don't recognize vulnerabilities," he told the Los Angeles Society of Financial Analysts in October 1978. "Obviously, oil and gas are limited resources; and eventually, the oil industry we know today will undergo change. However, we do not agree with those who say it will be over soon." [80]

International exploration and production, coal mining, alternative energies, and international crude supply agreements all helped Shell expand the quest for energy resources. Still, Bookout made sure the company played to its greatest strength: the application of technology to the development of domestic oil and gas reserves. A winning strategy for over seventy years, it was too soon to abandon it now.

Improving oil recovery from existing fields, a special realm of Shell expertise, offered great promise under the new regime of high prices. Using sophisticated steam and carbon dioxide flooding techniques, Shell gave old reservoirs a new lease on life and helped the company bring in new volumes of heavy oil. In 1979, Shell Oil purchased the Belridge Oil Company for $3.65 billion, the largest corporate acquisition in history up to that point. Belridge possessed several hundred million barrels of heavy, molasses-like oil, and Shell was one of the few companies with the technology to extract it profitably. In the first five years of development after the acquisition, Shell tripled production from the Belridge properties from 42,000 barrels per day to 125,000 barrels per day. In 1986, the South Belridge field produced 60 million barrels of crude, making it the most prolific field then producing in the forty-eight contiguous states, trailing only the two huge Alaska reservoirs among active U.S. fields. Belridge augmented Shell Oil's recoverable reserves by 35 percent. Observers who at first criticized Shell's Belridge acquisition soon changed their opinion. "That's the production story of the decade," said Bill Rintoul, a Bakersfield-based oil historian and journalist.[81]

Bookout and Shell, however, were not content merely to increase production from existing fields. They had faith that there was still a lot more oil to be found in the United States. If it could be done so profitably, domestic oil was always more attractive than foreign oil. It did not have to transported overseas by tanker, and the producer was not at the mercy of some international cartel or foreign government. The most attractive and promising place for Shell Oil to find new domestic oil reserves, once again, was offshore in the Gulf of Mexico.

The Offshore Imperative

"There's a romance about big, offshore structures," said Pat Dunn, Shell's manager of civil engineering, back in 1989. "There's something about seeing them out there on the frontier."[1] Since the time Bouwe Dykstra teamed up with Doc Laborde to build the *Mr. Charlie* submersible drilling vessel, Shell Oil had carried on a passionate affair with these structures. But as they rapidly evolved, words could hardly describe their mind-boggling size and complexity. In the 1970s, Shell led the industry in dramatically extending the depth threshold for fixed platforms from 350 feet to over 1,000 feet. In the 1980s, Shell's geoscientists and engineers continued to push the offshore frontier in both exploration and production, moving the industry to take another quantum leap, this time off the edge of the continental shelf into the ultra-deepwater of 3,000 feet and beyond.

Offshore was more than just a romance for Shell Oil E&P. It was its heart and soul, a symbol of longstanding technological leadership, and a main source of income for the entire company. Along with investments in the enhanced recovery of heavy oil in California, offshore development was the key component of Shell's multifaceted strategy in the 1970s to expand the quest for energy resources. And the Gulf of Mexico remained the hotbed of activity. There, feverish exploration and platform installation followed the OPEC embargo, with Shell's Cognac platform in 1,025 feet of water establishing a benchmark that redefined the concept of deepwater. Yet, in the midst of the boom, many in the industry believed that the Gulf had begun to play out. Overall production declined, and ultra-deepwater seemed technologically and economically unfathomable.

In the mid-1970s, Shell and other companies began to shift their long-range sights to other unexplored U.S. offshore provinces, such as the Atlantic basin, California, and Alaska. But political controversies, environmental opposition, and dry holes delayed or limited drilling in most of these areas. Desperate for new reserves, Shell once again staked its future on the Gulf. Summoning the courage to probe deeper into uncharted territory, it embraced advanced seismic technologies, gambled on deepwater leases, and developed new deepwater platform and subsea systems that enabled production beyond the continental shelf. The deepwater play of the 1980s was a tough sell to some of Shell's directors, who were understandably concerned about taking such giant, costly, and speculative steps into the virtual unknown. This might have been the greatest risk Shell Oil had ever faced. The big question, asked in the early days of offshore, was revisited: Even if the technology could be developed, would deepwater ever pay? But strong E&P leadership, driven by confidence in Shell's marine engineering capabilities and a courageous belief in the potential of the Gulf to yield large new fields, persuaded the company to take the risk.

Confidence and Courage

The willingness to take on massive technological challenges against conventional wisdom defined the corporate culture of Shell E&P. Research was closely integrated with operations and engineering, and personnel moved fluidly back and forth between Bellaire and the area offices. Long before the "team concept" or "matrix form" of organization came into vogue, Shell's E&P management had encouraged the formation of task forces—collections of line, staff, and research people with different skills working on a problem together. This mutual support system emboldened managers to go into big projects with a certain level of technical understanding, confident they would come out with more knowledge than they started with. "We have a very unique Shell culture in E&P," explained Charlie Blackburn. "It is composed of a lot of different people with many different ideas which they can advance without being afraid of being stepped on just because the idea may be a novel one. We try to encourage innovative thinking, and we think we have a system that allows us to do that."[2]

The top two men who led Shell's initial thrust into deeper offshore terrain in the 1970s, Pres. Harry Bridges and his executive vice president, John Redmond, were technically oriented managers who drew faithfully on the talents of the organization. However, they did not have any substantial experience with Shell Oil's E&P operations for a period prior to 1971, particularly offshore. Bridges had not had a real opportunity to become acquainted with the E&P staff, and although Redmond knew many of the staff quite well and had experience with offshore, he had not been associated with Shell Oil for at least ten years. But these two men respected the fact that Shell's exploration and production staff were proven leaders in the offshore industry. The confidence this staff had in what they were doing was Shell Oil's great strength. The challenge for Bridges and Redmond was to sustain the offshore momentum, assure the staff of their commitment and cooperation, and prepare for major decisions to come.

John Bookout, who first replaced the retiring John Redmond as leader of E&P and then Harry Bridges at the top of the company, further sharpened Shell Oil's focus on the offshore frontier. Bookout believed in the offshore and was fully conversant with Shell's evolving capabilities in this area. In fact, he was the first Shell president ever to have real hands-on experience with offshore oil and gas. In the late 1960s, as vice president for the New Orleans Area office, he had overseen the development of bright spot seismic technology. Upon his return from Canada to become executive vice president of E&P and then Shell's CEO, Bookout emphasized offshore development and campaigned hard to have the U.S. federal government open up the nation's continental shelves to oil exploration. His counterparts in the industry regarded him as one of the brightest and best-informed men among them. In 1981, they elected him chairman of the National Petroleum Council, the industry organization that compiles reports for the U.S. secretary of energy. In 1984, he became chairman of the American Petroleum Institute, the first Shell Oil president since Max Burns to head the industry's chief trade association. Bookout's exceptional strength as a leader and the respect he commanded both within and outside the company were instrumental in convincing Shell's board to continue moving deeper offshore.

Top E&P management under Bridges and Bookout consisted of people who had distinguished themselves technically during a period when offshore had taken center stage in E&P. Bookout's executive vice president for E&P, Charlie Blackburn, was a petrophysicist and protégé of Gus Archie, the man who invented the field. Graduating from the University of Oklahoma with a degree in engineering physics, Blackburn joined Shell in 1952 and eventually held key assignments as area petrophysicist in New Orleans, chief petroleum engineer in Houston, and the head office E&P budget coordinator. Blackburn had been on the *Blue Water 1* when it drilled its first well in 1962, a dramatic event that convinced him of the potential of offshore. Although a production man, he was a quick study in the science and art of exploration, and he had a gambler's feel for the game of offshore bidding. As vice president of the southern E&P region, he ran the bidding in the important federal offshore lease sales in the Gulf of Mexico in 1970 and 1972, when Shell first deployed its bright spot technology, and he managed Shell's deft handling of the Bay Marchand blowout. Known for his sly humor and keen intelligence, Blackburn became another towering figure in the company. Bookout's alter ego and second-in-command for ten years, he also served on the boards of the API and the National Ocean Industries Association, as well as on the energy committee of the U.S. Chamber of Commerce. As executive vice president, Blackburn felt it essential that Shell "exercise its technical muscle to the fullest degree" to stretch deeper offshore.[3]

Bookout's E&P vice presidents, all technically accomplished, contributed in their own way to Shell's offshore vision. On the exploration side, geophysicist Billy Flowers was a prime mover in getting Shell to apply state-of-the-art geophysics offshore. Geologist Bob Nanz, a pillar in Shell Oil's E&P research organization for many years, orchestrated Shell's crusade for greater access to federal offshore lands. In 1970, he succeeded the retiring McAdams as vice president of exploration. The emphasis on technology, so touted by McAdams, continued as Nanz focused the exploration effort in the following years on the Gulf of Mexico, Alaska offshore areas, and selected onshore areas such as Michigan. Jack Threet, a veteran of exploration in many frontier basins with both Shell Oil and the Royal Dutch/Shell Group, helped push the company into the ultra-deepwaters of the Gulf of Mexico. So did Tom Hart, one of the most memorable characters of his generation in the company. During the late 1960s, pivotal years for offshore exploration at Shell, Hart had been exploration general manager for the Southeastern Region in New Orleans. Named vice president for exploration in 1975, he went on to spend three years as a planning coordinator for the Group before returning as senior vice president of exploration in 1981. Six-feet five-inches tall and 250 pounds, Hart never hesitated to speak his mind. Brash and colorful, he was an engaging storyteller, master performer, and all-around bon vivant. A Harvard graduate with a master's degree from Louisiana State University, he could establish an intimate rapport with any audience, from ivory-tower scientists to oil-patch roughnecks. He was also a practical joker. Production manager Sam Paine never forgot the time he loaned Hart a set of vintage books on the Civil War, and when they were returned, found that each was inscribed with the words: "To my good friend

Sam Paine," signed "Tom Hart." More than anything else, however, Hart was a first-rate explorationist who was comfortable working in a variety of geographic, geologic, and political settings. "Tom could distill complicated details down to general concepts better than anyone I've ever seen," remembered Blackburn.[4]

Managers on the production side also had demonstrated exceptional technical abilities and offshore experience. For most of Bookout's presidency, Gene Bankston and Don Russell headed the production organization. In the late 1950s, Bankston (vice president for production, 1972–79) had contributed to developing the "big picture" policies on how management decisions should be made in Shell E&P and the economic model that supported the first push into what then was considered deepwater (past 200 feet). Russell (vice president for production, 1980–86) had been a star researcher at Bellaire in the area of reservoir engineering and had helped develop more rigorous quantitative methods for evaluating offshore leases. In the late 1960s, he also had been regional production manager in New Orleans. Under Bankston and Russell, Shell E&P refined and improved its sophisticated methods for preparing economic scenarios for given offshore prospects, using statistical projections of volumes, prices, profitability, drilling costs, and success ratios. As offshore development approached deeper water, and as competition for leases intensified, production economics became ever more important to formulating bids.

While the 1960s marked the great leap forward in exploration technology, the 1970s witnessed similar progress in offshore production technology. In this area, Shell was well prepared to take the lead. First of all, it was committed to the Gulf of Mexico, which at the beginning of the decade accounted for over 50 percent of the company's domestic crude oil and natural gas liquids production. Shell knew offshore Gulf of Mexico as well as any company in the business. It was the only area in the Shell E&P organization that had kept engineers and technical teams in place continuously over the post–World War II period. In 1971, the marine division in New Orleans was split, renamed the Offshore East Division and Offshore West Division, and then in 1974 combined into one large Offshore Production Division, the company's largest. In 1972, Shell confirmed its commanding presence in Louisiana by moving fifteen hundred employees into a giant, new, fifty-one-story skyscraper in downtown New Orleans, called One Shell Square. Built in white marble by Gerald D. Hines, One Shell Square was twin to One Shell Plaza in Houston. Housing Shell's

Southern Exploration and Production region, it towered above the New Orleans skyline and laid claim to the distinction of the tallest building in the Deep South.[5]

A Steep Learning Curve

During this period, Shell Oil concentrated on expanding its deep-water production capability. In 1972, activity in the Gulf began to taper off. There was still a lot of development in so-called shallow water (out to 300-foot depths), but the industry was not really expanding into deeper water. Lease sales had been postponed due to rising environmental concerns over offshore development and the fallout over several platform disasters. Yet there was another factor that contributed to the lull. The industry was still trying to figure out how to operate at greater depths in the proven oil province of the Gulf of Mexico. Fixed platforms had become standard for waters extending out to 350 feet, but moving deeper—toward 600 feet and beyond—introduced fundamentally new problems. Jackets would be more slender and therefore more susceptible to stresses caused by wave dynamics and metal fatigue, which could be safely ignored in shallow water. So Shell continued to explore alternatives for producing at these depths. By the early 1970s, the company had elite engineering groups working on a range of different technologies, including subsea wells, fixed platforms, and tension-leg platform designs. Some professional competition existed between the various groups, but it was congenial, as everyone realized that they were striving toward a common goal.

In September 1972, just as Shell was moving its Southern Region employees into One Shell Square, Shell's subsea engineers reached a milestone by completing a seafloor well in dry atmospheric conditions at a record depth of 375 feet in the Main Pass area. Developed for Shell by Lockheed Petroleum Services, the system involved the use of a permanent one-atmosphere chamber to house wellhead equipment. A diving bell mated and sealed with a collar on the well chamber, permitting entrance for a three-man maintenance crew to work on the wellhead in a shirt-sleeve environment.

One-atmosphere chambers designed to house oil field equipment first had been developed in the 1950s. They had provided personnel access through a removable caisson. But water depth limitations of the caisson caused this chamber concept to be abandoned in favor of "wet" systems, operated and maintained by remote, diverless devices. The Shell-Lockheed system revived the chamber concept using a diving bell that, in theory, had virtually unlimited depth capability. It promised to revolutionize subsea operations by extending them to water depths of 3,000 feet or deeper. A year later, the well was reentered for minor inspection, and by January 1975, the system, operating on a dually completed well, had produced more than half a million barrels of oil.[6]

The system was aimed at the deepwater of the future, however, rather than the present. The one-atmosphere chamber was merely the first phase of a three-phase project. It led to the development of an ocean-floor gathering and collecting system and adaptations to allow equipment to withstand pressures at 3,000 feet. As of 1972, such advances were still years away. The test was a success, but many kinks and technical problems still had to be worked out with subsea wellheads, as well as with the long flow lines and production risers linking them to the surface. The big constraint, however, was cost. Wellheads required high-priced hardware and complex controls on the seafloor. Each well had to be drilled and completed by an expensive drilling vessel, a semi-submersible or drillship. "If you have a lot of wells to drill in close proximity," explained Carl Wickizer, project manager for Shell's subsea system development program at BRC, "then that group of wells will cost a lot more if they are subsea wells than if they are wells drilled from a platform."[7]

As the 1970s progressed, subsea systems found their greatest application in the North Sea. After major discoveries earlier in the decade, this offshore province boomed. The North Sea's deep, tempestuous waters and high-volume wells made subsea systems more attractive for some fields than platforms; on-bottom equipment escaped the constant battering of giant waves. Oil firms could afford to spend more money for subsea systems because the platform alternatives were so expensive in such an environment. Many Shell Oil engineers working on subsea technology were reassigned to the North Sea to work for Shell U.K. Exploration & Production (Shell Expro), the Group's offshore joint venture with Esso. In the late 1970s, Brazil's state oil company, Petrobrás, forged even more boldly ahead with subsea technology in its Campos Basin development, eventually becoming the leading innovator in this area.

Back in the Gulf, meanwhile, the name of the game remained platforms, almost exclusively. "We poured one heck of a lot of money into subsea, and there were fits and starts for 25 years," said Pat Dunn, who took over the Head Office Central Engineering Group in 1969. "Simultaneously, the fixed platform bunch was merrily going on and on and on."[8] First organized in 1965 and headed by Bob Bea, the Central Engineering Group assumed the task of designing and overseeing fabrication and installation of all of Shell Oil's offshore structures. Previously, the operating divisions (e.g. New Orleans, Houston) had performed the engineering. But with the increasing challenges of offshore engineering and Shell Oil's expanding portfolio of leases, a more specialized and concentrated effort was needed. During its first year, the Central Engineering Group designed and managed the construction of thirty-three platforms in water depths ranging from 30 to 300 feet offshore Louisiana and Texas. This was the most ever designed and constructed in a single year than any other period in Shell Oil's history. This group kept finding ways to extend the depth capability of platforms on a more cost-effective basis than what could be done with subsea systems. Dunn declared that "the period 1964–1972 was, in my opinion, the most active in terms of platform technological development in the whole history of the offshore."[9]

Shell stayed at the forefront of innovation, but advances in deepwater production resulted from the acquisition of knowledge

24

Shell Oil's first Central Offshore Engineering Group, 1965, with a scale model of a typical mid-1960s offshore production platform. Bob Bea, chief engineer-manager is standing on the right side. Lead engineers were Gene Strohbeck (standing, back left), Jimmy Mayfield (standing, back left next to Strohbeck), Arv Fisher (seated on left in front of Mayfield), and Peter Marshall (seated on right in front of Bea). *Courtesy Bob Bea.*

and skills by the industry as a whole. The introduction of the digital computer in the early 1960s revolutionized design techniques, and construction companies built bigger launch barges to assist the installation of increasingly ponderous platforms. Onshore support industries and communities had sprung up all along the Gulf Coast, in places like Morgan City and Lafayette, Louisiana, helping to spread and standardize skills among offshore operators. But the industry had more to learn about production before tackling the 600- to 1,000-foot depths that had become routine for exploratory drilling. In 1969, several events with lasting implications accelerated the industry's learning process and opened up a new era of deepwater development.

One of these events happened in May 1969, when five thousand people, including most of the top officials in the industry, gathered together in Houston for the first annual Offshore Technology Conference (OTC). Nearly 380 companies displayed the latest products and equipment, and attendees obtained a close look at the progress of the industry through more than one hundred technical papers. It was only appropriate that Dick McCurdy, president of Shell Oil, the undisputed leader in the Gulf of Mexico, gave the keynote address to the convention. McCurdy's speech highlighted the need for better technology, better economic incentives, and better public relations as prerequisites for advancing into deeper water.[10] The OTC, which has met annually in Houston ever since, became a key forum for publishing and exchanging technical information, recognizing individual achievements, and rallying the industry behind a common sense of technological purpose. "It loosened the secrecy surrounding companies' research efforts," explained Dunn, "making it much easier to release important technical results."[11]

The year 1969 also witnessed the publication of the first API Recommended Practice (RP) document for offshore. The RP and its successors established rational guidelines and criteria for designing, constructing, and inspecting platforms. They were written by the most knowledgeable engineers, including many from Shell, led by Pat Dunn, who integrated their experience and research efforts into consensus documents for building safe, reliable, cost-effective platforms. The standards writing process dovetailed with the technical knowledge coming out of the OTC and fostered further cooperation and sharing of information. Much of this work was published in order to receive an open, academic-type review to help gain acceptance. Of utmost significance, the RP guidelines were later incorporated into the federal government's regulatory program. The Bureau of Land Management (and later the Minerals Management Service and Coast Guard) actively participated in creating the guidelines and resolved disagreements before the documents were issued. Moreover, noted Dunn, "regulators and industry personnel got to know each other and were able to understand each other's positions better. This had intangible benefits over and above the standards writing process."[12] In 1993, Dunn received the OTC Distinguished Achievement Award, in part for his influence on innovations in the design and construction of offshore structures, and in part for his role in championing government-industry efforts in the development of API recommended practices.

The third major event of 1969 was Hurricane Camille, the third and most devastating "100-year storm" to hit the Gulf during the decade. In fact, Camille was so powerful that it was classified as a "400-year storm." In August of that year, it crashed through the offshore alley east of the Mississippi Delta, inflicting widespread carnage and passing over South Pass Block 70 where Shell had only five months earlier installed two large platforms in record depths of 300-plus feet. Right after the storm, Maurice Patterson, a Shell oceanographer who had installed instrumentation programs on these platforms, flew out in a helicopter to retrieve his data tapes. Arriving on the scene, he noticed that one platform had been heavily damaged and the other one was missing! Search parties subsequently found the platform lying sideways on the bottom about 100 feet to the southeast, in the downslope direction and against the waves. While the other platform still stood, it had been displaced by 4 feet in the same direction. Investigations subsequently determined that soil movements undoubtedly caused these two failures. All told, the damages to Shell's offshore structures caused by Camille amounted to $11 million. Other companies reported similar damages to production platforms—at least several were swept away in addition to Shell's—costing the industry over $100 million total.[13]

Despite destroying offshore structures, Camille proved a blessing in disguise for the industry. Across the Gulf of Mexico in the deepest water, Patterson and other scientists had outfitted platforms with instruments that held up through the storm. "It was a really marvelous achievement for them," remembered Skip Ward, an oceanographer with Shell Development at the time. "And so, we were presented not only with the challenge of Camille but also an awful lot of data with which to try to quantify and better understand the hurricane." During the storm, these instrumentation programs recorded a measured wave height of 75 feet, which captured the attention of every operator in the Gulf. Before then, most companies had abided by a design wave height of 58 feet (the distance between wave trough and crest) for a 100-year storm in water past 150-foot depths, which had been the recommendation of hydrodynamics experts. R&D groups in the major companies, including Shell, Chevron, and Exxon, had been trying to upgrade this wave height but faced resistance from management uneasy about the escalation in costs. Chevron, for one, had increased their design wave height to almost 85 feet. Shell's conservative design philosophy had protected their other deepwater platforms exposed to Camille's fury. After the storm, Shell and most other companies pushed their design wave height to around 72 feet.[14]

Thus, Camille became an important calibrator for understanding the dynamics of wave forces and verifying the strength and size of platforms, as well as for generating "hindcasts" of the effects of past storms in order to predict future extremes. Moreover, soil movements generated by Camille's high waves had destroyed Shell's two South Pass platforms. In fact, Bob Bea, chief engineer-manager of the Central Engineering Group, had actually predicted that a 100-year storm would cause a subsea soil slide at the very location where the platforms were lost. However, the Central Engineering Group had calculated that the chance of losing a platform due to soil movement was less than losing it by wave and wind forces and that the oil field had about a 30-year life. So the platforms were installed. But Camille—the granddaddy of all hurricanes—drove home the threat of sea-floor slides to offshore structures and prompted more intense study by oil companies, consultants, and universities. As a result, engineers learned how to determine safer locations for platform placement and how to develop slide-resistant platforms for unstable areas.[15]

The data gathered from Camille and shared within the industry helped build technical consensus around all kinds of design criteria. This consensus eventually was reflected in the API Recommended Practice documents and federal regulations. "Everybody ended up having the same view on extremes of wind, waves and currents due to hurricanes in the Gulf of Mexico," said Ward. "Many of the same players then used those tools and methodology to develop criteria all over the world."[16]

Other developments during this period helped the Gulf of Mexico offshore industry to prepare for new challenges. As mentioned in chapter 5, dramatic blowouts in 1969–70 spurred oil companies to improve operational procedures and revise safety guidelines, again under the aegis of the API. These incidents also taught the industry valuable lessons about the importance of maintaining public support for the whole enterprise of offshore development. Lessons learned in the tough North Sea environment, furthermore, helped improve practices in the Gulf. In the early 1970s, several North Sea platforms installed in 500 feet of water, under the most inhospitable conditions, provided invaluable knowledge about wave dynamics and metal fatigue.

The North Sea also provided an example of how quickly costs rose with increasing depth. Offshore leaders such as Shell Oil knew that safe, reliable platforms could be built in much deeper water—but at a steep price. The big question was: Could they afford it?

Project Cognac

The OPEC oil embargo of 1973 provided the answer. The skyrocketing price of crude oil and an aggressive federal leasing system gave new impetus to offshore expansion. With prices at ten dollars per barrel instead of three dollars per barrel, companies found they could justify much more expensive offshore drilling and development. And the federal government eagerly encouraged them. Under the mandate of Project Independence, the Nixon administration increased the pace of leasing in the Gulf of Mexico and resumed Outer Continental Shelf sales off the Atlantic, Pacific, and Alaskan coasts, all of which had been closed to drilling after the Santa Barbara blowout. In 1973, even before the embargo, the government held sales in the central Gulf, offshore Texas, and in the so-called MAFLA region—offshore Mississippi, Alabama, and Florida. After the embargo, Interior Secretary Rogers Morton announced that the government aimed in 1975 to lease *ten million* acres of offshore property to oil companies, as much as had been handed out in the entire twenty-year history of Outer Continental Shelf leasing. Most people outside the government regarded this goal as totally unrealistic. It also raised the hackles of environmentalists, who geared up for confrontation in California and along the Atlantic coast. Nevertheless, the announcement accelerated plans to offer "deepwater" tracts in the Gulf, where environmental opposition hardly registered.

In anticipation of deepwater sales in the Gulf, Shell exploration's seismic surveys located a number of attractive features for testing, and some confirmation drilling was completed. One of the most attractive prospects, code-named "Cognac" by Shell, was on tracts in the Mobile South area. The amplitude reflections, or bright spots, on the seismic records gave a high probability of finding several oil and gas plays on the structure. "The prospect was full of bright spots," said Mike Forrest, geophysical project leader for the Offshore Division at the time. Shell's technical analysts estimated that the field might contain 150 million barrels. Although not large by Middle Eastern standards, this was potentially a major field by Gulf of Mexico standards.[17]

Cognac, which was located in 1,000-foot water depths, would establish a new offshore frontier, one far deeper than that which had so concerned Bouwe Dykstra in the late 1950s. It would be another giant and risky step, with very expensive drilling and facility costs. Shell managers also expected very high bids for the tracts, as much as $100 million for a single, five-thousand-acre tract. This worried president Harry Bridges. He felt that the Board of Directors needed to be more fully apprised of the methods used for evaluating the bidding before they could consider and approve the company's move into such deep waters. Heretofore, the Board had been advised but not in great detail. It was a security problem. Bids reached into the millions of dollars, and the bidding was very competitive. But because of the escalating cost and added risk, Bridges decided to give the Board a more detailed strategy presentation.[18]

As the price and competition for offshore leases increased significantly, the process of deciding which tracts to bid on in a lease sale and how much to bid had become an increasingly lengthy and secretive process involving the work of hundreds of people over a period of several years. After the Department of Interior called for nominations on tracts, Shell and other companies would submit a list of tracts to the government, based on the ongoing collection of seismic and geological data and information from previous lease sales. As the sale approached, Shell would undertake intensive seismic work, with a geophysicist and geologist assigned to each prospect—the geophysicist analyzing the seismic data and mapping out the subsurface structure and the geologist acting as the evaluator, determining which possible oil or gas reservoirs should be pursued and estimating the volumes of oil or gas on each tract.[19] Beginning in the late 1960s, the technical team also worked with other parts of E&P—production, economics, platform design, and drilling—to establish a most probable tract value within a range of values, discounting for operational and geologic risks. The list of proposed tracts was then culled out through a District Review and then a Division Review. A month before the sale, a final Division Review was held and general bids were attached to the tracts. The head office then reviewed the bids with all the E&P vice presidents and finally with the executive vice president and president, who placed the final numbers on the sealed bids. The few people involved at this stage, the better, since a competitor would only need to bid one penny higher to take a given tract away from Shell.

The meeting with the Board on Cognac took place several months before the scheduled March 1974 lease sale for the Gulf. John Redmond and Bob Nanz made the presentation, outlining Shell's detailed methods of evaluating prospects. They also reviewed the application of Shell Oil's bright spot technology for the very first time. The main objective of the presentation was to show the Board that Shell's bids were based on the value of a particular tract to Shell and never at an amount just to be higher than a competitor. Following the strategy presentation, Redmond and Nanz discussed the specific tracts they were targeting and the bidding levels. As usual, they did not name the individual tracts nor specify their location. But Cognac was one of the prospects reviewed, and the price per tract was around $108 million. The Board approved the recommendations and, as Redmond remembered, "we were sure that the strategy presentation had helped in gaining their confidence and support." [20]

Shell Oil E&P had enough confidence and support from the Board to bid alone in the approaching March sale. But in this case, Shell Oil took on bidding partners. Traditionally, Shell had not bid with partners, preferring to go it alone and protect its technology. As the prices for leases soared, however, Shell decided to reduce its front-end financial risk by taking on partners. [21] This decision also was forced by smaller operators, who were priced out of the picture and who had been pressuring the majors to give them some representation in the bidding. But, in Shell's case, these smaller partners had little input. They did not know what kind of technical work was involved, nor did they even know where the lease was or how much was being spent. Their participation was essentially an investment in Shell's proven record offshore. [22]

Shell Oil bid strong in the March 1974 sale and got most of what it wanted. The U.S. government opened up over two hundred tracts (940,000 acres) in the central Gulf, including forty-two deepwater tracts (199,000 acres) on the continental slope offered for the first time. Oil companies spent a record $2.16 billion in bonuses in the landmark sale. Only thirteen of the deepwater tracts were bid on, but the eleven bids that were accepted pulled in an impressive $321 million. While Exxon was the top spender in the sale ($245 million for six tracts), the Shell Oil group made the biggest bet on deepwater—spending $214.3 million in bonuses for three adjoining blocks on the Cognac prospect (Mississippi Canyon 151, 194, and 195, originally called Mobile South No. 2) and paying over $112 million for the most prized of the three blocks. However, Shell did not win all of the blocks on the prospect. A group led by Amoco won the fourth block (Mississippi Canyon 150) with a bid of $81 million, beating out Shell by about $10 million. [23]

The Cognac prospect was so far beyond working depths that Shell Oil did not even have a vessel that could drill it at the time of the lease sale. "But we had a lot of guys who thought we could figure out how to do it," said Lloyd Otteman, division production manager for the Offshore Division. They soon found a semi-submersible, the *Pacesetter II*, and reequipped it with added mooring, larger conductors, and other modifications. Not until June 1975 did Shell have the unit ready to drill on the prospect in 1,000 feet of water, more than twice the depth in the Gulf previously plumbed for commercial production. In July, the first exploratory well for Shell by the *Pacesetter II* struck oil. "The discovery well was logged in the middle of a Friday night," recalled Mike Forrest. "Leighton Steward, division exploration manager, supplied the champagne and led a 6 AM toast." While on vacation in Michigan during the drilling phase of the project, Lloyd Otteman received a call from Larry Smith, division engineering manager, reporting that the well log showed 140 feet of pay, more than enough to go forward with a platform. "That's great news!" replied Otteman. But on the two-day drive back to Houston from Michigan, all he could think of was "now we've got to produce! We've got to build a platform."[24]

During the next year, Shell drilled eleven more tests on the four blocks—eight discovered oil and gas. As it turned out, Amoco had obtained the best acreage (Shell Oil's discovery volume was 196 million BOE). From Shell's perspective, the logical way to develop the field was to unitize the operations of the two groups of partners. But Amoco used its reserve estimates as a strong bargaining chip. After two years of difficult negotiations, Shell and Amoco formed a joint venture with Shell as operator (Shell 42.8 percent, Amoco 21 percent). By 1977, when the agreement was signed, Shell was already building the jacket, strengthening the company's bargaining leverage over Amoco, which had not yet figured out how to develop its interests. "They didn't have a design, and we were already building the bottom section of the platform," explained Sam Paine. "So we knew we had them. We traded hard. We didn't back off. And they finally agreed with us."[25]

When Shell bought the leases at the March 1974 sale, however, its engineers had not yet come up with a design concept for producing in 1,000 feet of water—thus the source of Otteman's concern. At the actual time of bidding, according to John Redmond, "if we had had to commit to a completion method, we probably would have gone the subsea route . . . although the idea of placing a man in a chamber at 900 feet-plus water depth was not all that appealing—to me, not at all." A year earlier, Pat Dunn's civil engineering group had begun to analyze the problems of designing and installing a fixed platform for such a water depth. To withstand the day-to-day waves of deep ocean water as well as the extreme winds and waves of hurricanes, it would have to be mammoth-sized and heavily reinforced, dwarfing anything ever built. The base of the structure also had to be very sturdy to withstand tremendous forces from mud slides. Design, however, was the easy part. Finding a way to install it was the main challenge. Along the Gulf Coast, there were no construction yards and launch barges even remotely big enough, or tow-out water depths deep enough, to handle a one-piece, 1,040-foot-tall steel jacket. Still, Shell engineers knew they could find a way. After visiting the platform design team in New Orleans on the eve of the lease sale, John Redmond asked Sam Paine, general manager of offshore, if his boys could do it—Paine replied emphatically, "Yes sir. We can!" [26]

The only conceivable solution at the time was to build and install the jacket in sections small enough to be floated and lifted by available equipment and then mate the sections in the water. Such a project would be incredibly complex, requiring untried procedures. At the same time, Exxon was working on installing the "Hondo" platform in 850 feet of water in the Santa Barbara Channel. Exxon engineers decided to launch the jacket in two pieces and then mate them horizontally in protected water. The mated jacket would then be uprighted on the bottom. Shell engineers considered a Hondo-style mating operation but ruled it out due to the risks of this time-consuming procedure in the hurricane-prone Gulf. Instead, they settled on a unique and innovative concept: building and launching the jacket in *three* pieces, mated *vertically*, or stacked, under water. Launching each section would be a separate and relatively quick operation. And the mating would take place deep enough to be protected from strong wave action.

This was easier said than done. "It took many agonizing hours of planning, thinking, re-thinking the problems," said Gordon Sterling, who supervised the detailed engineering design of the structure. "How would the base section be connected safely and securely to the middle and top sections? How would the base be leveled on the Gulf floor? There were many other questions. But the answers started coming." Peter Casbarian, Shell's project manager for Cognac, assigned an elite team of specialists to engineer various parts: Gene Strohbeck was responsible for engineering and pile hammer design; Bobby Cox for launching the base section; Wayne Simpson for the electronic instrumentation package for guiding and monitoring each step of the installation; and Rick Smith for the hydraulically actuated mud mats for leveling the base section, as well as on ballasting, pile cementing, and flooding systems for installation. Said Dan Godfrey, the engineer in charge of fabricating the base: "Whether a man was working on the base, the middle or the top section of this construction effort, no one—even old timers in the yard—had ever been a part of something so special."[27]

Shell awarded the contract for building and installing the Cognac structure to J. Ray McDermott, one of the leading offshore construction firms in the Gulf. In April 1975, even before exploration drilling had started, steel was ordered, and in December, fabrication of the base section began at McDermott's Bayou Boeuf yard in Morgan City. Slowly, tons and tons of steel filtered into the yard. Joint by joint, brace by brace, the base began to take form like a giant jigsaw puzzle. It grew even larger than initially planned. The discovery of huge reserves and the exploding U.S. demand for oil in the mid-1970s compelled Shell to speed up development. Consequently, the original jacket, designed for a forty-well, single-rig platform, was enlarged to accommodate two drilling rigs and sixty-two wells. In turn, the total weight of the jacket increased from nineteen thousand tons to forty-nine-thousand tons, with an attendant escalation in costs.

On a fast-track schedule, a project of such unprecedented magnitude and complexity was bound to run into problems and setbacks. "It seemed (to one who was involved) that there was at least one crisis a day during design and construction," remembered Pat Dunn. The general design for the jacket as a single piece was essentially a deepwater application of the basic API drilling-production platform used in shallower waters. Yet, translating the design into metal called for exceptional accuracy in fabrication to ensure that the pieces would fit when mated. All measurements had to be temperature calibrated to take into account the expansion and contraction of the steel, from the hot Louisiana sun to the cold depths of the Gulf of Mexico. Braces even millimeters out of alignment, for example, had to be replaced. "In terms of construction tolerances, there's absolutely no comparison with any other job," Dan Godfrey noted. "If the sections don't mate, you can write off the whole thing."[28]

Shell and McDermott employed space-age technology to ensure that the sections would fit. A survey by Boeing Aerospace predicted how well the sections would match under varying offshore conditions. An on-site construction survey used infrared devices to check Boeing's figures. The fit of jacket members was checked with photogrammetry, a computerized method of correlating photographed targets developed by the U.S. Army for high-altitude mapping. Even the most sophisticated measuring techniques, however, could not completely eliminate uncertainty. The

25
Towout of the
Cognac platform
base section, July
1977. *Photo courtesy
Peter Marshall.*

confident assurances Sam Paine gave John Redmond about the
Cognac platform hid a degree of personal doubt about it. Notwith-
standing top management's confidence, Paine recalled the sleep-
less nights he spent fretting about the construction of the jacket.
On visits to McDermott's yard, he would repeatedly ask Pete Casbarian, with some trepidation: "Is this thing really going to fit?"
Invariably, Casbarian's reply was: "Don't worry, Sam. It is going
to fit!"[29]

These doubts would never be completely laid to rest until the
sections were mated in the water. Installation was even trickier
and more worrisome than fabrication. There was no margin for
error. Over half of the three hundred man-years of engineering
logged on the entire project dealt with installation procedures.[30]
In July 1977, the massive base section, the size of the Houston
Astrodome, left the yard and was towed ever so carefully out to
location. "I guess we were like a bunch of expectant fathers waiting on the new arrival," said Norris Dodge, in charge of the base
section installation. "Except that all the fathers knew that what
was coming out of the Morgan City delivery room was something
more on the order of the Jolly Green Giant."[31]

On July 25, after all systems were go, the base section was
launched from the barge. "This sight is one we've all awaited for
a long time," exclaimed Casbarian, as he watched. Two derrick
barges, stabilized by a special twelve-point mooring system and
winching out four lines apiece, began guiding it gently down-

ward. Precise monitoring took place through a system of underwater television cameras and Honeywell-designed acoustic telemetry devices, whose transponder signals were processed by computers to pinpoint locations. After two days, the jacket base came to rest within 20 feet of the target on the seafloor. There, hydraulic mud mats jacked the structure to the level position. To nail the base to the bottom, McDermott introduced a new commercial technology—underwater pile drivers. These 40-foot-long hammers were housed inside air-filled chambers, which looked like upside-down quart jars with hoses hooked to the top, and guided into the pile sleeves by Honeywell's acoustic positioning system. Four of the twenty-four "skirt" piles were driven before Hurricanes Anita and Babe halted operations in late summer. The base section remained pinned to the bottom, unaffected by the storms. When calm weather resumed, the hammers completed driving the 625-foot-long, 60-inch-diameter piles deep into the seabed. Teams from Taylor Diving supported installation and set endurance records for divers working in depths of 850 to 1,025 feet. "Cognac is taking such a large step beyond anything that has gone before," wrote *Offshore Engineer* in 1978, "that it may be a long time, if ever, before those divers go as deep again at a fixed platform." [32]

In July 1978, ten months after the base section was set, the middle section arrived. Despite the size and precise positioning of the base, returning crews could not find it! They searched for a day and a half. "I don't think word of that ever got back to New Orleans," said Bruce Collipp, who was on the scene to supervise the mating of the sections. "We were busy doing this, we were busy doing that. . . . it's got to be down there!" Once the structure was finally located, the middle section was then launched and maneuvered over the base. Mating the two sections was the most critical aspect of the installation operation, the moment of truth for the entire project. The giant, twisting, floating mass of steel bracing that was the middle section had to be steered down to the right place in the right orientation, within just a few feet of the docking cones on the corners of the base section. Information on the jacket position, heading, attitude, winch tension, amount of water in the legs, currents, docking pole penetration—all gathered by Honeywell's network of cables and transponders—was transmitted to a computerized control room on board one of McDermott's derrick barges and processed into steering instructions. [33]

Miraculously, the middle section approached the base in al-

26
Workmen lower a 625-foot piling into the Gulf of Mexico to nail down the Cognac base section. *Photo courtesy the Associated Press.*

most perfect alignment. Crews quickly slid the docking poles into the cones, and the mating sequence came off without a hitch. Immediately after launching the mid-section, the launch barge returned with the top section and the sequence was repeated. The connection between the middle and top pieces was a little "tight," remembered Sam Paine. But the pieces eventually fastened together after a tug boat tied onto the top section and "jiggled it a little bit." For the numerous engineers who had labored for years over the installation, the successful mating of the sections brought relief and jubilation. During the next few months, construction barges installed the well conductors and platform decks. In September 1979, more than five years after the leases were bought, the Cognac platform began producing oil and gas. By the summer of 1981, all the wells had been drilled, permanent production facilities had been installed, and the world's deepest platform-to-shore oil pipeline had been laid. At the end of 1982, Cognac was producing 72,000 barrels of oil and 100 million cubic feet of gas per day.[34]

Cognac was the most sophisticated fixed platform installation ever completed; at $240 million, the platform was also the most costly. From start to finish, the overall project cost Shell and its co-owners nearly $800 million. Other companies built subsequent platforms in similar depths with less steel and launched them in one piece from larger barges for much less money. Union Oil named its two 1,000-foot platforms, installed in 1980 and 1981, "Cerveza" and "Cerveza Light" ("beer" and "light beer") to emphasize their cost-savings compared to Cognac. But these beer-budget projects could not have happened without the deepwater precedent established by Cognac. It marked an unparalleled advance in the technology of offshore structures and set records for the deepest water, largest number of wells, and heaviest steel platform, among numerous other innovations. In 1980, the American Society of Civil Engineers (ASCE) honored Cognac with the annual award for the "Outstanding Civil Engineering Achievement," the first ever received by an oil company. Along with Exxon's Hondo and developments in the North Sea, Cognac opened a new era for truly enormous, offshore engineering-construction projects. It introduced the "team" or "project line" concept to the industry, marrying disciplines such as naval architecture, structural engineering, and mechanical engineering. Company engineers also worked more closely with the fabrication and installation contractors than ever before, taking project management to a whole new level. "It was a giant step," said Sam Paine. "I mean, a *giant* step."[35]

Controversies, Delays, Disappointments

As the energy crisis of the mid-1970s intensified, and as onshore prospects in the United States dried up, U.S. oil companies looked increasingly offshore to expand domestic reserves. But even with Cognac paving the way deeper into the Gulf of Mexico, many oil-men, including Shell Oil's own, believed that after twenty-five years of development only lean prospects remained in the Gulf. The best hope for increasing national reserves, they insisted, was to open up the unexplored sedimentary basins off the East and West Coasts and off Alaska.[36]

The industry's drive to explore these areas, however, collided with opposition from environmentalists and coastal communities. Of the nineteen thousand wells drilled in U.S. waters up to 1975, only four had caused major oil spills. But those four had been relatively recent and spectacular. As the industry moved into deeper, rougher waters, environmentalists feared that the likelihood of spills increased—with potentially ruinous consequences for marine ecology and recreational beaches along places like Long Island and Southern California. New England fishermen, furthermore, did not want oil companies invading their territory. Governors and politicians from coastal states, unprepared to cope with the onshore impacts of an aggressive leasing program, objected to providing costly services and facilities for offshore development. They wanted to be consulted about the federal leasing program, which they increasingly argued would be inconsistent with the requirements of state coastal and marine management programs. "People seem to want new oil sources developed, but they don't want it where they live," complained John Bookout. "We have been far less willing to open up our continental shelves than most countries."[37]

Bookout emerged as a vocal and articulate spokesman for expanded access to frontier areas. "Offshore represents the major domestic potential yet to be explored," he repeatedly emphasized. Other Shell executives also spoke out. Already sensitized to environmental concerns and convinced of the need to establish a more open relationship with the public, Shell Oil sent its E&P managers out to plead the case to government officials and coastal communities. Exploration vice president Bob Nanz spearheaded the effort, organizing and presenting detailed information before numerous groups on what he called the "Offshore Imperative." Nanz and other Shell representatives participated in industry efforts organized by the API and coordinated with the National Ocean Industries Association (NOIA) to help overcome local and governmental resistance to offshore development. "We did a lot of work with fishing groups in different areas, because they were one of our primary opponents," remembered O. J. Shirley, Shell's Southern E&P Region safety and environmental conservation manager, who was active in these efforts. "We worked with the governor of Massachusetts in trying to get access to Georges Bank. We worked with New Jersey people for access to the Mid-Atlantic. It was easy to identify who our adversaries were, and we tried to get an opportunity to speak to them."[38]

It was a tough battle. The adversaries were not easily converted. Shirley recalled being physically threatened at a public hearing in Boston by a burly fisherman, who eventually had to be scolded back to his seat by a Department of Interior official, a young woman formerly of the Sierra Club. Oil company representatives struggled to convince people of the industry's renewed commitment to safety and environmental protection. Shirley had been a founder of the Clean Gulf Associates (CGA), an industry organization formed in 1972 to upgrade oil-spill handling capabilities in the Gulf. As lease sales were scheduled in the Mid-Atlantic, some of the same companies organized a new group, called the Clean Atlantic Associates (CAA), with Shirley as its first chairman. The CAA compiled an oil-spill contingency manual, identified areas of particular sensitivity to oil spills, and planned to stockpile oil-spill equipment for the North-Atlantic, Mid-Atlantic and South-Atlantic regions. The CAA sought to puncture the stereotype of offshore oilmen as insensitive to the environment and demonstrated the industry's willingness to abide by rigorous environmental protection standards. "Through strong personal contact, one-on-one discussions, and actual friendships, we formed relationships with the environmental community," said Shirley.[39]

These efforts helped break down public resistance, but obtaining leases and permits to drill still entailed protracted legal struggles. "It looked like, sometimes, that we were never going to get there," said Shirley, "but, looking back, we gained access to almost every area that we wanted to drill offshore." One promising area was the Baltimore Canyon Trench off the coasts of Delaware and New Jersey. In a 1976 federal sale, Shell and partners obtained twelve tracts in the relatively shallow waters of the Baltimore Canyon. The sale was contested in court, and not until March 1978, when the U.S. Supreme Court refused to hear an appeal of a lower court decision validating the sale, was drilling allowed to proceed.[40] A string of dry holes from the 1976 sale, however, including several by Shell Oil, dampened enthusiasm for a second sale held in 1979. Shell had been hoping for a bonanza, "one or more giant fields the size of Mexico's Golden Lane," said Jack Threet. There had been geologic reason to hope for such fields. "We knew we had reservoirs and we were almost certain we had traps," he explained. "But we think there was probably not enough oil generated in the Atlantic Basin to migrate into those traps."[41]

As companies began to write off the Baltimore Canyon, attention shifted to another promising area—the Georges Bank Trough southeast of Cape Cod, Massachusetts. But drilling there encountered even greater opposition. In 1976, the Conservation Law Foundation and the State of Massachusetts filed suit to block sales in Georges Bank. After two years of legal wrangling, the U.S. Supreme Court refused to grant a final request to cancel the Georges

Bank sale, which was finally held in December 1979. Shell and its bidding partners won three tracts for a price of $86 million. Obtaining permits to drill, however, dragged on for many months. In 1978, Congress passed the Outer Continental Shelf Land Act Amendments (OCSLAA), which opened up the offshore leasing process to wider public participation, involving more government agencies, with the intention of building public confidence in this activity. At least in New England, however, this act further delayed drilling. The permits issued by the U.S. Geological Survey and the Environmental Protection Agency and approved by state agencies in Connecticut, Massachusetts, Rhode Island, and Maine—pertaining to mud discharge, spill equipment, and protection of fisheries—were among the most stringent ever applied to offshore drilling.[42]

In 1981, once all the appropriate permits had been obtained, Shell finally drilled its first exploratory well in the Georges Bank. Alas, this and subsequent wells turned up dry. It was a good gamble against long odds, because even with high costs, the rewards looked rich enough to justify the search. But after years of fighting the modern-day "Battle of the Atlantic" for access to the East Coast's continental shelf, the industry gained only a better geologic understanding of this offshore basin and appreciation of the political dimensions of offshore development outside the Gulf of Mexico.

Californians put up even fiercer resistance to offshore drilling than Easterners. Offshore development was not new to California, but it had proceeded along a different and stranger trajectory than in the Gulf. Beginning in the 1930s, oil companies had erected drilling platforms built from piers from Santa Barbara down the coast to Long Beach. Because the ocean floor of the Pacific sloped off sharply from the shore, companies could not move deeper gradually as they could in the Gulf. Large structures that would have been placed far beyond view in the Gulf were clearly visible from California beaches. In the late 1950s, to appease residents who did not want their scenic ocean view spoiled by offshore platforms, Shell helped introduce artificial islands made of sand and rock to house and beautify them. In the 1960s, the THUMS Group—Texaco, Humble, Union, Mobil, and Shell—extended this artificial island concept by building four ten-acre islands off Long Beach. Each had elaborate façades to camouflage rigs and equipment and give the impression of real estate developments rather than offshore facilities.[43] Leasing off California came to a sudden halt, nevertheless, after the 1969 Santa Barbara oil spill, which galvanized local groups statewide to agitate for restrictions on offshore development.

Despite early setbacks, the movement to drill offshore gained political strength. In 1974, after the moratorium on drilling was lifted, the State of California unsuccessfully tried to block the first federal lease sale, maintaining that it did not meet the requirements of the National Environmental Policy Act. In the December 1975 sale, held in Los Angeles, Shell Oil and its partners spent $123 million, most of this for two 5,700-acre leases on a prospect called Beta, in water ranging from 220 to 1,000 feet in San Pedro Bay off Long Beach. The sale bolstered anti-industry forces, however, creating enough pressure to cancel the two federal sales proposed for 1976 and 1978. A suit brought by the County of Santa Barbara postponed the next sale, originally scheduled for 1977, until 1979. Meanwhile, the California Coastal Commission (CCC), backed by Gov. Edmund G. Brown, issued ever more restrictive requirements for federal leasing to ensure that it was consistent with the state's federally authorized coastal management program. Subsequent lease sales became so embroiled in lawsuits and subject to the withdrawal of the most attractive tracts due to environmental concerns that development offshore California screeched to a halt. Beginning in 1982, Congress inserted prohibitions into the Department of Interior's appropriation that effectively shut down leasing on the Outer Continental Shelf of both the East and West Coasts.[44]

Within this antagonistic political climate, Shell Oil pressed forward with the development of its Beta prospect. Of all the tracts leased in the 1975 sale, Beta yielded the only commercial discovery, in July 1976. Exploratory drilling revealed an estimated 150-million-barrel field, and Shell badly needed this oil to supply its West Coast refineries, which had been forced to purchase increasing amounts of crude from other companies. But bringing the field into production would prove to be neither simple nor inexpensive. Platform designs had to account for the possible impact of shock waves generated from earthquakes. Although the advent of powerful computers had improved the seismic analysis of offshore structures, knowledge of earthquake design was still not well developed, even by the early 1970s. Ensuring that a platform had enough structural resilience to absorb the energy of severe earth tremors, therefore, required conservative and thus costly designs. Development strategy also had to take into consideration the fact that the reservoir contained heavy oil and low natural pressures. Water injection and downhole electrical pumps would be needed to produce it. Using sophisticated computer simulation techniques to predict reservoir performance, Shell studied alternate plans, eventually deciding to build two offshore structures instead of a combined drilling-production platform. The two-platform complex allowed for the most efficient development of the Beta field and provided the large amounts of space needed to support the processing equipment.[45]

Political and regulatory obstacles, driven by growing opposition to offshore oil in California, hindered the project more than design considerations. But Shell was determined to see the project through by meeting or exceeding all state and federal safety requirements and environmental standards. Early on, Shell teams spelled out detailed development plans in face-to-face meetings with numerous community and civic groups, as well as with the appropriate local, state, and federal officials. They covered all the major impacts of the Beta project, including safety, air and water quality, marine traffic, oil spill prevention, and onshore activities. "The path that we adopted was to be completely open with them," said Phil Carroll, division production manager for Western E&P at the time. "No surprises or attempts to sneak something by. We did everything we could to accommodate them."[46]

Still, the permitting process dragged on for two years. Of the eleven different local, state, and federal agencies from which Shell had to obtain permits, the California Air Resources Board (CARB) threw up the most difficult roadblocks. Western E&P managers took a calculated risk, ordering fabrication of the components just as they began applying for permits. Brown & Root built the two platform jackets for Shell in Labuan, East Malaysia, while the deck sections, pilings, and conductors were made in Japan. "I was frequently asked," remembered Carroll, " 'My God, why don't you stop building those things until you are sure you can get the permits?' " But because the field required two major platforms in 260 feet of water, Shell compressed the construction schedule, contracting for components from multiple international contractors to speed up fabrication. In late 1979, the jacket for the drilling platform called Ellen was literally being towed by barge across the Pacific Ocean before Shell had obtained all the permits. Carroll planned to tow the jackets right out to location in San Pedro Bay and invite television crews out to see a major new source of energy desperately needed by the nation but that was being held up by regulatory red tape. Fortunately, the permits came through in time to avoid a showdown.[47]

Gaining permission to develop the Beta field was an impressive feat. Shell's frank and open discussions with government officials and community leaders cleared up many misconceptions about the impact of the project and paved the way through the permitting process. In early 1980, Shell installed the production platform Elly, linked by a 200-foot bridge to its sister drilling platform Ellen. Four years later, as the development drilling program on Ellen drew to a close, Shell installed a mammoth 700-foot drilling platform called Eureka to develop the much deeper southern portion of the field. Built by Kaiser Steel at Vallejo, near San Francisco, Eureka was the largest single-piece jacket installed up to that point on the West Coast and the sixth-largest overall in the world.[48]

The platform was not Shell's most technically difficult, but because the West Coast lacked the infrastructure that could be called on in the Gulf, the logistics of transportation created unique challenges. The jacket was so big that there were only four barges in the world that could carry it; the *Heerema H-109* came over from Japan to perform the job. In a dense fog, the jacket had to be towed carefully out of San Francisco Bay under the Golden Gate Bridge, while the deck, built by McDermott in Morgan City, had to be shipped through the Panama Canal. "It was like planning a military campaign," said Ralph Warrington, Eureka's project manager. All told, the Beta project cost $700 million and touched nearly every organization in the company over the course of a decade. By the late 1980s, Beta had hit peak production of about 20,000 barrels per day, the industry's only commercial success from the 1975 lease sale in Los Angeles.[49]

In Alaska, the last frontier area off U.S. coasts, Shell was not so fortunate. The first stumble came in the Cook Inlet, where Shell had enjoyed previous success in the Middle Ground Shoal Field. In a December 1973 state lease sale, the company tried to expand on that success by acquiring five tracts in the Kachemak Bay area of Cook Inlet. As Shell prepared to develop the leases, however, the coastal communities rose up against offshore operations in the Bay, a pristine, picturesque setting. In June 1976, after a protracted series of hearings, the state imposed a one-year moratorium on drilling in the Bay. A year later, state legislation authorizing condemnation of leases in the Kachemak Bay forced Shell to sell them back to the state.

Undaunted, Shell remained convinced of Alaska's oil potential and optimistic about the industry's chances at getting access to it. In the mid-1970s, Shell and other oil companies believed that federal territory in the Gulf of Alaska might have the same kind of big, concentrated oil deposits that were found at Prudhoe Bay. Sales of Gulf of Alaska leases by the federal government were supposed to follow the state sales at Prudhoe Bay, but the Santa Barbara blowout incurred the wrath of environmentalists and held up sales for years as research was done on the hazards of drilling there.[50] Finally, in April 1976, the federal government put the acreage up for lease, after failed attempts by the State of Alaska to block it. This sale, *Business Week* announced at the time, "may very well hold the last hope for an oilfield big enough to reverse the nation's four-year decline in oil production."[51] For Shell exploration managers, the sale also offered them a chance to redeem the company in Alaska after the failure at Prudhoe Bay.

The Gulf of Alaska was Shell's top candidate among the seventeen potential Outer Continental Shelf oil and gas provinces listed by the Bureau of Land Management in 1974. It was also a forbidding frontier region, one of the most hostile in the world. Its fierce, chilling winds drove waves cresting at one hundred feet. Fog often made helicopter transport impossible. Moreover, it was a seismically active area that would require earthquake-resistant platforms. "The Gulf of Alaska," said John Swearingen, chairman of Standard Oil of Indiana (Amoco), "will make the North Sea look like a kiddie pool."[52] Shell estimated that a production platform in three hundred feet of water in the Gulf of Alaska would cost as much as one in one thousand feet of water in the Gulf of Mexico. Unfortunately, the techniques Shell had laboriously developed for evaluating leases in the Gulf of Mexico were not applicable there. "There was no information other than seismic," remembered Marlan Downey, exploration manager for the Alaska division. "There wasn't a history of production. There wasn't anything that told you whether or not there would really be commercial oil there." Nonetheless, Shell was anxious to find out. In preparing for the sale, its Alaska division geophysicists identified several major structures. Although they did not find any verifiable bright spots on the seismic data, they saw hints of an unusual type of undersaturated oil that did not have gas. So they decided to bid aggressively, taking on ARCO as a partner, though, to spread the risk. The Shell-ARCO partners were the high bidders in the sale, together spending $276 million (Shell's share being

$148 million) out of an industry total of $572 million. They won twenty-nine tracts totaling 165,000 acres (nine of eleven prospects on which Shell bid).[53]

And they drilled nothing but dry holes. There was no source rock. It appeared that temperatures never got high enough in the formation to cook up the oil. "Everything looked good and the structures were there," said Bob Nanz. "Except oil was not generated in the particular ones we sampled." These dry holes were also expensive. Stormy weather and high formation pressures made drilling from semi-submersible vessels difficult, resulting in drilling costs from $10 million to $23 million per well. Shell's Gulf of Alaska venture was a complete failure, a miserable disappointment. When a second lease sale in the eastern Gulf came up a few years later, Nanz resisted any temptation to place another bet. "I feel like that monkey they put on the sled down there at NASA in the acceleration chamber," Nanz told his geologists. "He did not want to get back on that sled again and that is how I feel about this sale."[54]

Shell went to the sale but acquired only five tracts for $1.4 million. It was saving its money for other sales in Alaska's western and northern waters. Despite a string of controversies, delays, and failures in other frontier areas, Shell's exploration leaders still believed in the potential of offshore Alaska. In 1978, the company announced that it expected Alaska to provide 58 percent of the country's future crude and condensate discoveries. There were some very large structures offshore Alaska. If oil and gas had migrated out there, these structures could be "company makers." As Jack Little described Shell's thinking: "There was a huge, world-class field up there onshore. So there just had to be something, right, in the offshore?"[55]

New Horizons

During the 1970s, offshore oil in the United States became the subject of rising political controversy. Environmental opposition and the "Not-In-My-Backyard" (NIMBY) syndrome thwarted the industry's efforts to explore many frontier areas of the Outer Continental Shelf. As the oil industry also came under intense scrutiny for alleged profiteering after the OPEC embargo, questions about the competitiveness of offshore leasing increasingly entered into the discussion. Critics charged that the bidding system based on cash bonuses with fixed royalties did not always give the federal government a "fair value" on leases and that joint bidding by the major oil companies kept the smaller independents from operating in deeper waters.

Oilmen scoffed at the suggestion that lease sales were not competitive. They argued that even though smaller companies did not have the capital to develop leases on their own, many of them were often included in successful offers. Oil companies emphasized that the skyrocketing prices were ample evidence that the system was highly competitive. In 1977, Bob Nanz pointed out that of the average winning bids in the last twenty OCS sales, 45 percent of the bonus was "left on the table"—it was not needed to get the lease. "It's been more than competitive," he commented. "More like frantic." [56]

With the OCS Lands Act Amendments, the U.S. Congress attempted to reform the bidding process to make it even more competitive. The amendments required the Department of Interior, during a five-year experimental period beginning in September 1978, to try new bidding systems that reduced the amount of front-end money needed to obtain leases and thereby, in theory, enabled more companies to purchase leases. The traditional format consisted of a cash bonus bid for a given tract with a fixed percentage royalty on what was produced, whereas alternative systems included those that derived income for the federal government largely through variable royalties bids or net-profit sharing rather than through cash bonus bids. Shell, like other companies, did not like rising cash bonuses but still favored the traditional system over most of the alternatives, which company officials argued would only encourage speculation, impose new administrative burdens, and delay exploration. [57]

With the deepening energy crisis in the United States, the last thing the Department of the Interior wanted to do was delay or impede domestic exploration. The December 1978 overthrow of the Shah of Iran by Shiite Muslim revolutionaries cut off petroleum exports from Iran, lifting world crude oil prices from thirteen dollars per barrel to thirty-four dollars per barrel and precipitating a full-blown panic at the pump. In March 1979, U.S. Secretary of the Interior Cecil Andrus, as directed by the OCS Amendments, announced a five-year offshore leasing schedule aimed at expediting exploration and development. The program would average five sales a year with emphasis on the Gulf of Mexico and Alaska. Faced with new urgency to develop domestic oil deposits, the Department of the Interior continued to rely on the tried-and-true system of cash-bonus leasing and experimented with the different systems only in a limited way. After studying the comparisons, Interior found that these systems produced no statistically meaningful differences in industry competition, a view that the U.S. Supreme Court upheld in 1981.[58]

As Interior expanded its leasing program, Shell Oil geared up for the biggest push the company had ever made offshore. Company officials had often criticized the Department's leasing timetable in the past and thus were exhilarated by the promise of new areas being opened for exploration. Over the years, Shell had placed bigger and bigger bets on offshore development. Now, John Bookout and his lieutenants were prepared to stake the whole company's future on it. In their minds, there was really no alternative for a company whose central realm of business was in the United States. They could not see any more major finds onshore. Bob Nanz estimated that nearly 60 percent of the oil yet to be found in the United States was located offshore, most of it under federal control. The risks of pushing into the offshore frontier were staggering—huge bonuses, expensive drilling, and if all went well up to that point, the monumental costs of development. But they had to be taken for Shell to have a future as a major oil producer. The exploration department was looking for large-scope projects; these would involve higher risks, but if they came about would remake the company. "We worked so hard," remembered Mike Forrest. "Shell needed to find two hundred million barrels of oil a year just to stay even, to replace production."[59]

By the mid-1980s, roughly 60 percent of Shell's exploration dollars went to the offshore effort in the United States. "Exploration has been called a poker game," Jack Threet mused in 1984. "But there's more to it than that. In this game, we don't have chips or coins or dollar bills that can change hands over and over again. We're dealing with a declining resource base, and every barrel we find is never going to be found again."[60] Two places which Shell believed had oil to be found were Alaska and deepwater Gulf of Mexico. Environmental opposition had basically shut down leasing off California and Florida. Drilling in the Northern Atlantic and Eastern Gulf of Mexico (the MAFLA region) had found little. There were really no other virgin areas in the United States to explore for large oil accumulations. Shell believed that large oil fields would be discovered in Alaska and included the risked reserves in the company's ten-year long-term plan in the late 1970s and early 1980s. Mac McAdams was an early believer in moving out deeper in the Gulf of Mexico, and when he retired in 1970,

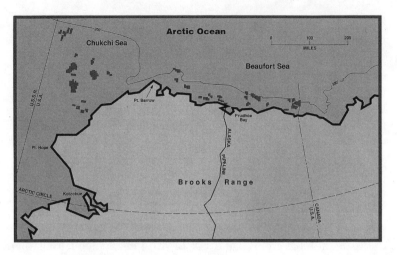

Billy Flowers and Bob Nanz took up the cause. However, the economics of so-called deepwater were still controversial (see below), so deepwater Gulf of Mexico did not really make it into Shell's long-term plan until the mid-1980s.

Despite the Gulf of Alaska bust, Interior and oil company officials considered other parts of offshore Alaska to have the highest resource potential of anywhere in the United States. It was big structure country. For years, Bob Nanz had led the charge in lobbying the Interior Department to accelerate leasing in Alaskan waters—particularly the Bering Sea and Beaufort Sea basins. After the second oil shock, his words finally appeared to carry more weight. In June 1979, Secretary Andrus revised the leasing schedule announced in March to give earlier consideration to the Alaskan sales. Although not entirely satisfied with the proposed pace of leasing, Nanz was encouraged by the announcement. He asserted that the technology was available for exploring most of the Alaska OCS basins. But extreme weather would make it difficult. Ice prevented seismic boats from even getting into Alaska's northern waters, except for maybe one year out of every five. Drilling crews would have to cope with minus-60 degree temperatures and twenty-four-hour darkness in the winter. Furthermore, there was no clear-cut method for producing oil from such an ice-ridden environment. Yet, Shell's credo held that if the fields could be found and the economic conditions were favorable, the technology would arrive to bring them into production. In 1979, with the price of crude soaring near forty dollars per barrel and the phasing out of price controls in the United States, almost any project seemed possible.

Shell Oil believed as fervently as anyone that Alaska might be the savior of the U.S. oil industry. Shell Western E&P performed exhaustive geophysical work on all of Alaska's offshore basins and, with Amoco as a key bidding partner, forked out millions of dollars in a succession of lease sales held between 1979 and 1985. In 1979, Shell spent $69 million in partnership with Amoco on leases in the first sale in the Diapir basin of the Beaufort Sea, north of Prudhoe Bay. In October 1982, the company joined Amoco, Union Oil, and Koch Oil in purchasing leases in another part of the Diapir basin, mostly on parcels that covered a huge structure called Mukluk. In April 1983, Shell Oil spent $78 million in a joint venture with Amoco and Marathon to acquire leases in the St. George and Norton basins of the Bering Sea.[61] A year later, a Shell-Amoco combine dominated a sale of tracts in the Navarin basin, with Shell putting up $175 million of the winning bids. The last major area was the Chukchi Sea, where, in lease sales held in 1985 and 1988, Shell outspent the competition for large tracts. In the final analysis, Shell spent more money and acquired more acreage than any other company in offshore Alaska lease sales.

All areas held tremendous promise. The Beaufort Sea possessed giant structures, Mukluk in particular. It looked very similar to the neighboring Prudhoe Bay field. It had the same reservoir rock, source rock, and geologic history. Even though Mukluk was only a one-to-two-billion-barrel prospect, about 10 percent the size of Prudhoe Bay, the industry—led by British Petroleum and its U.S. affiliate Sohio—had high hopes for it and spent nearly $1.5 *billion* on Mukluk leases.[62] Most of the tracts were in forty to one hundred feet of water covered with ice as thick as ten feet for eight months of the year. Shell and other companies turned to building artificial islands out of gravel to drill their exploration wells. Tragically, though, Mukluk turned out to be the most expensive dry hole in history.[63] Oil stains in the rocks indicated that it had once been a giant oil field. But sometime in geologic history the structure had been breeched, which allowed oil to leak to the surface. Or regional tilting had caused the oil to migrate elsewhere. "We drilled in the right place," said Richard Bray, the president of Sohio's production company. "We were simply 30 million years too late."[64]

Although Shell geologists had not assigned as high a probability of finding oil at Mukluk as some other companies, and thus did not bet as heavily on it (the company spent $162 million on leases), Shell Oil shared in the costly disappointment. Mukluk seemed to be nothing but bad luck. During a lunch break from a meeting in which Exploration Vice President Tom Hart was presenting the findings of the Mukluk dry hole to the general executive committee, a Shell Oil custodian accidentally shredded the Mukluk logs. This proved to be quite an embarrassment to Hart, who could never bring himself to tell Amoco, Shell's exploration partner at Mukluk, what happened to his copy of the logs.[65]

Shell and the industry did not fare any better in the other basins offshore Alaska. Either they found no source rocks or the deposits they did find were not large enough to be commercially viable. The company collected massive amounts of data on every prospect, drilled in every basin, and came up empty. The last gasp was in the remote, hostile waters of the Chukchi Sea. Shell had obtained acreage on several sizeable structures and, after struggling to satisfy environmental concerns to gain a federal drilling permit, discovered oil. The federal drilling permit was approved none too soon, on March 23, 1989, literally one day before the *Exxon Valdez* oil tanker rammed into a reef in Alaska's Prince William Sound, spilling 240,000 barrels of petroleum into those pristine waters. "We would probably still be up there trying to get a drilling permit if that had happened a day sooner," said Jack Little.[66] Even then, Shell had to jump through many hoops to prove it had the capability to drill in the tempestuous Arctic waters and built a $15 million oil-spill barge with state-of-the-art clean-up equipment.

The Chukchi deposits were too expensive to develop. The technological challenges were supreme, even for Shell. Because enormous sheets of floating ice would demolish conventional drilling and production platforms, the company looked to installing big, ice-breaker platforms and pipelines that could resist ice scouring. "It may have been a blessing in disguise that we didn't find commercial quantities," admitted Little. "We probably would have found the technological problems to be almost insurmountable." Even if the technology could have been found, however, the falling price of oil by the late 1980s made the development of the Chukchi deposits out of the question. During the 1980s, Shell spent an estimated $2 billion on leases and drilling offshore Alaska and came away with nothing to show for it.[67] So ended, for the time being, Shell's arduous, thirty-year quest to find bonanza reserves in Alaska.[68]

A New Deepwater Vista

As the failures followed one upon another in Alaska and other frontier areas, Shell started to shift the exploration spotlight back on the Gulf of Mexico, a proven oil province that in the late 1970s showed renewed signs of life with rising oil and gas prices. During 1975–77, Shell had actually deemphasized the "Cenozoic play" in the Gulf, in favor of exploration elsewhere. In 1970–74, Shell bid on 64 percent of the volumes discovered by the industry in the Gulf but on only 22 percent during the next three years. The company focused on geopressured natural gas prospects in the ultra-deep producing horizons of the Texas Miocene. Discoveries in 1975 at Prospects Manifold (Eugene Island 136) and Calcite (East Cameron 57) encouraged this search, and Shell subsequently dominated the Corsair trend with discoveries at Picaroon (Brazos A19, A20) and Doubloon (Brazos A23) in the June 1977 and May 1978 sales.[69]

The June 1977 Gulf of Mexico lease sale surprised industry observers by taking in $1.17 billion in high bids. Shell placed second in the bidding with $100 million in winning bonuses. "We were amazed at the high level and variety of interest," said Tom Hart at the time. "The competition on practically all of the tracts was tough."[70] Anticipated higher natural gas prices from the staged decontrol of gas, observers presumed, spurred on the bidding. Indeed, most of the discoveries on these leases—including significant ones by Shell on the Brazos and Matagorda Island tracts—were made in gas-rich areas stretching from the mouth of the Mississippi River westward to the Mustang Island area near Corpus Christi, Texas.[71] Over the next several years, lease acreage in the Gulf continued to draw spirited bidding. In the December 1978 sale, Shell outspent all others, laying out $184 million for ten tracts, again in natural-gas-producing areas. Two years later, Shell and its leasing partners announced a $1.2 billion program for developing fourteen central and western Gulf of Mexico gas fields discovered on these leases.[72]

Despite success discovering natural gas in the West Louisiana and Texas Miocene, Shell managers felt they could have done better. The company still led the industry in Gulf of Mexico discoveries in the mid-1970s with an ultimate volume (in 1987) of 349 million BEQ, compared to 229 million BEQ for its closest competitor, Gulf Oil. But these results did not meet the extremely high standards that Shell explorationists set for themselves in the Gulf. As the 1987 lookback study concluded, Shell had "lost a good opportunity to add volumes mostly by Bright Spot discoveries," and "a lot of smaller companies did well on the 78 percent of the volume SOI [Shell Offshore, Inc.] did not bid."[73]

Beginning in 1979, Shell Oil expanded exploration in the Gulf, spurred on by the sense of missed opportunity in previous years and dimmed prospects in other offshore areas. Meanwhile, the Bureau of Land Management accelerated its lease sales. In 1981, there were a record seven offshore sales held in the United States. "We had a lot of lease sales. We went through a lot of lease sale reviews," remembered Charlie Blackburn. Competition for leases in the Gulf became fiercer than ever. The oil-price shock of 1979 (see chapter 7) and the perception that offshore prospects were declining created a feeding frenzy for what was left. Bonus bids skyrocketed, shattering all previous records. "The bidding just got ridiculous," said Blackburn. "The whole business got ridiculous!" The September 1980 sale in New Orleans brought in $2.8 billion; Shell Oil purchased sixteen tracts for a whopping $316 million, second highest in the sale. "I got a three-letter description: W-O-W!" said John Rankin, manager of the BLM's New Orleans OCS office, after the sale.[74]

Shell's exploration managers became increasingly dissatisfied with the direction of BLM's leasing program in the Gulf. First, there was the question of steeply increasing costs, as Blackburn indicated. Bonus bids, even those by Shell Oil, the most accurate and cost-efficient explorer in the industry, were too high for the volume potential available. During 1979–82, the company's bonus per BOE discovered soared to $3.94, from well under $1 for the previous eight years, while the ratio for the top companies in the industry increased by a factor of at least four or five. Shell tried to maintain its advantage by bidding on deeper, more subtle traps rather than compete only on the few bright spots nominated, yet most discoveries were on bright spot prospects such as Roberto, Hornet, Cougar, Boxer, Glenda, Wasp, Peccary, Hobbit, and Cheetah. And the company made only two geopressured discoveries at Onyx and Persian.[75] Although all nice discoveries, these were still predominantly gas deposits containing lower average volumes than those discovered by Shell Oil in the preceding years.[76]

In Shell's view, the second problem, which contributed to escalating costs, was the federal government's method of rationing leases through the nominating process. The relatively small amount of nominated acreage actually offered in sales was creating an artificial shortage of exploration opportunities. "Tract selection," as the BLM method was called, offered tracts or blocks in a piecemeal fashion, which hindered more efficient exploration strategies involving basin-wide assessments or the pursuit of structural trends that transcended tract boundaries. The Department of Interior's policy of stipulating a two-year time limit before the release of well logs compounded the problem. Often, when a company had a discovery on a given tract, it would fail to get a promising offset tract nominated before having to surrender its well logs on the discovery. This policy both increased the cost and inhibited the development of prospects that spanned across multiple blocks. Billy Flowers remembered Picaroon and Cougar as two important discoveries with open offset tracts that were not being followed up in 1980.[77]

Cougar was particularly important in that it held clues to finding petroleum in deepwater—depths beyond the record one thousand feet set by Cognac. At Cougar, Shell had found hydrocarbons in the "turbidite sands" associated with deepwater geology. The company had been focused on so-called "deepwater" since the late 1950s. But the definition of the concept had changed over time—first deeper than sixty feet, then deeper than two hundred feet, deeper than six hundred feet, deeper than one thousand feet. The only constant definition of deepwater over time has been "the depth of the water just past the deepest platform." The modern concept of deepwater, in use since about the early 1980s, refers to depths deeper than one thousand to fifteen hundred feet, the maximum depth for a conventional six-leg platform, although every company has had their own definition.

Deepwater was still largely uncharted territory in 1980. The soaring price of bonuses, the small amount of acreage offered in the sales, and the short time-horizon of leases stipulated by the BLM prevented pioneering moves into these depths. But Shell geologists believed such depths held interesting possibilities.

In the early 1960s, Shell seismic vessels had shot some reconnaissance probes off the edge of the continental shelf and down the slope. "These big structures popped up out there," remembered John Bookout. "Everybody was just astonished. 'Look at all the different kinds of possible traps and variety of configurations.'"[78] The structures were huge salt pillars but different from the conventional Gulf Coast salt dome. These pillars had squeezed up through the rocks, usually less than a couple miles across at the seafloor surface. During 1964–67, the Bellaire lab had conducted a core drilling program, using the self-positioning *Eureka* drillship, to find out if any oil had been generated in deepwater. The original *Eureka* party chief was Leighton Steward, and the two other key figures from the lab who worked on the project were Peter Lanier, a regional geologist, and Bob Chuoke, a brilliant theoretical physicist who would go on to help quantify bright spots in the early 1970s. The *Eureka* drilled some thirty, one-thousand-foot cores around the edge and on top of salt structures in six hundred to four thousand feet of water. "In fact, we got cores of salt with inclusions of oil in it," recalled Steward. "That was really a surprise to everybody because you could take the salt out of the core, you could put a little of it in a glass of water, and as it dissolved, you could hear it pop as the pressure was released. Then, oil came out with it. You could physically see the little black droplets of oil in the salt. That answered the question as to whether any oil had been generated in deep water."[79]

Over time, geologists speculated that ancient "turbidity currents," underwater rivers formed by suspended sediment, might have carried significant amounts of sand out into deeper water, forming reservoirs to trap oil. Whereas reservoirs on the shelf were highly faulted and required numerous wells to develop, deepwater reservoirs, if they were there beyond the edge of the shelf, might be large and continuous. Urged on by Pres. John Bookout to look for better exploration opportunities in the Gulf, Billy Flowers and Bob Nanz came up with the idea of shooting a set of long, regional seismic lines.[80] Shell, nor any other company for that matter, really understood the Gulf of Mexico as a basin. There had been little prior incentive to go out far enough to gain that kind of knowledge. With prospects in shallower water declining, however, Flowers and Nanz now found ample reason to do so. In the 1979 and 1980 Gulf lease sales, Flowers and Bill Broman, exploration manager for the Offshore Division, nominated some of these prospects, which ranged out to 1,300-feet-plus depths. But because the industry as a whole was not yet concerned about those depths, the Department of Interior would not put them up for sale. Large areas had no calls for nominations, and some were not even blocked out yet.

Shell exploration managers decided that they needed a wide-open lease sale, similar to that in 1962, to bring those areas into play. They also realized that deepwater development would require lease terms longer than five years. Flowers, Broman, Nanz, Jack Threet, and Lloyd Otteman, the Offshore Division's production manager, put together a traveling road show of talks and presentations to high-level Department of Interior, BLM, and U.S. Geological Survey officials to persuade them to open up the deep water for leasing. Instead of maximizing bonus bids in small sales, they argued, the government could take in more aggregate revenue in the form of royalties through larger, broad-area sales. But lease terms would have to be revised to provide incentive to the companies. They told the officials that the standard five-year leases and one-sixth royalty would not promote deepwater development. Something on the order of ten-year leases and one-eighth royalty might generate interest. They also pointed out the need for a safe supply for the country and the effect it would have on the U.S. balance-of-payments situation. "And we did something we had never done before," remembered Flowers. "We showed them prospects." Flowers and Broman were careful not to give away crucial information or overstate the potential, but they wanted to let government officials know that there *was* potential out there. They presented seismic data on some of these deepwater structures that made tracts in shallower water, which had been put up for sale, pale by comparison. One prospect in particular, code-named "Bullwinkle," showed three likely oil pays.[81]

Lobbying by Shell planted a seed with Interior officials that grew after the 1980 election of Ronald Reagan as president. Shell officials found a much more receptive audience in the new administration. Reagan's secretary of the interior, James Watt, believed fervently in letting the market determine energy outcomes and in releasing federal lands for exploration. Executives from other oil firms also lobbied for reforms to the leasing program, but according to J. Robin West, assistant secretary of the interior for policy, budget, and administration under Watt, none were as effective or forthright as representatives from Shell Oil. "Charlie Blackburn was the one who tried to really work with us and help us understand what were the pros and cons, what was reasonable, what was not reasonable," remembered West. "Some of the other guys . . . would come in like potentates with vast entourages and they would lecture us about what they wanted and leave." Lloyd Otteman remembered making a presentation with Billy Flowers to the undersecretary of the interior, Don Hodel, laying out their proposal for broad-area leasing with a slew of maps and view graphs. In the middle of the meeting, Hodel received a call from

Secretary Watt. "He said he needed to go see Jim and he said he needed what we got," said Otteman. "And he just gathered everything up and went off! Later, he came back, and it wasn't too long after that they came out with 'area-wide leasing.'"[82]

Area-wide leasing, which was part of a new five-year leasing program announced by Watt in May 1981, opened up the bidding on any unleased tracts in an entire planning area (e.g., the western, central, or eastern Gulf of Mexico). Millions of acres would be placed on the auction block at one time. For tracts in waters deeper than 900 meters (about 2,950 feet), the program also offered ten-year leases and one-eighth royalty. Watt's area-wide leasing plan aimed to allow oil companies to explore areas they believed to be most favorable rather than areas selected by the government through the nominating process. They could bid on any tract they wanted in a planning area, rather than have to choose from a limited number of carefully selected ones, and they would be more likely to acquire tracts in bunches, giving them greater control over large prospects. Area-wide leasing promised to reduce some of the competition and thus lower the costs of bidding; companies with independent data could submit smaller bids on deepwater tracts because the probability of another bid on a given tract was relatively low. It was the most effective way, on the other hand, of accelerating the pace of exploration in federal offshore waters. After years of vocally advocating such a leasing program, Shell Oil could take some credit for helping bring about this major policy change.

The new policy, along with the outspoken and confrontational style of Secretary Watt, were not universally popular, however. They drew protests from small oil firms and renewed political opposition at both the state and federal levels. Critics complained that the new system would give the majors, who had superior capital and technological capabilities for plying deepwater, a substantial edge over the independents. Environmentalists worried about a new wave of environmentally risky offshore development. The Pacific coast states, Florida, and several environmental groups went to court to block Watt's program. While the process was under litigation, Watt combined all offshore leasing, regulation, and royalty management functions in a new Minerals Management Service (MMS) within the Department of Interior, streamlining the leasing process and concentrating the growing pressure against the OCS leasing program in one agency. Legal and legislative challenges to the program failed in the U.S. Court of Appeals, and in May 1983, the MMS held the first big area-wide sale in the Gulf of Mexico, opening up over 37 million acres to bids, more than ten times what had normally been offered previously.[83]

To prepare the company for the new, big deepwater play, Shell had already embarked on a program to establish the viability and safety of deepwater drilling. Up to that point, nobody had drilled deeper than 1,500 feet in the Gulf of Mexico, and there were only a handful of wells in the world deeper than 3,000 feet, none of them in the United States. In 1981, after Watt's announcement of the new leasing program and as Shell exploration managers geared up for the play, Shell Oil president John Bookout had gathered top management together and told them: "I cannot in good conscience fund and launch this kind of program unless we can develop it. You've got to give me confidence you can get to 3,000 feet, and I want something on the drawing board saying you can get beyond 6,000 feet." Head office then assigned Carl Wickizer, manager of Production Operations Research, to conduct a feasibility study of "ultra deepwater" drilling and development in water depths beyond 6,000 feet in the Atlantic Ocean. After earlier exploration failures in the shallow waters of the Baltimore Canyon, Shell decided to see what the different geology of the deeper water in that area held. In the December 1981 lease sale, Shell obtained tracts in water extending to 7,500-foot depths in the Baltimore Canyon and Wilmington Canyon areas.[84]

Many critics of deepwater offshore leasing claimed that a technology barrier existed at 6,000 feet. Shell was determined to prove them wrong. In 1982, the company contracted with SONAT for the dynamically positioned drillship *Discoverer Seven Seas,* one of four vessels in the world rated for 6,000 feet of water. Shell then spent over $40 million extending its water-depth capability to over 7,500 feet, adding a new, large marine riser, a new long baseline dynamic positioning system with enhanced software and hardware, a new ROV designed for greater depths, and other modifications. Before the *Seven Seas* could begin drilling, however, Shell had to disprove the previous conclusion of the U.S. Geological Survey that the ocean floor in the area was too unstable for safe drilling.[85]

The company did this in 1981–82 by deploying its proprietary "deep-tow" technology, developed by an Offshore East Division engineering team led by staff civil engineer Earl Doyle. Deep-tow was a side-scan sonar that produced high-resolution images of the ocean floor and accurately revealed geological or man-made hazards. The deep-tow survey produced a new perspective on the seafloor geology of the area, showing a generally stable bottom topography and thus paving the way for deepwater drilling on Shell's leases. In late 1983, one hundred miles southeast of Atlantic City, New Jersey, the *Seven Seas* drilled an exploratory well in a world record water depth of 6,448 feet in the Wilmington Canyon. Although the drilling program in the Atlantic, which included two other deepwater wells, did not discover oil, the successful demonstration of drilling at such extreme depths established the industry's capability to drill in water depths beyond 6,000 feet. Just as importantly for Shell, it inspired confidence by the company's senior management about exploring in any deepwater frontier.[86]

Although the *Seven Seas* did not drill the first ultra-deepwater well until late 1983, Shell Oil was confident enough in the early feasibility study to bid aggressively in the first area-wide lease auction in May 1983. Gulf of Mexico Sale 72, as it was called, shattered all records. The industry leased 656 tracts for $3.47 billion. Under the leadership of Billy Flowers, offshore vice president, and Doug Beckmann, exploration general manager, and with the enthusiastic senior management support from John Bookout, Charlie Blackburn, and Jack Threet, Shell Oil put together an ambitious bidding strategy and spent $270 million for sixty blocks.[87] Several of the prospects it bought—Bullwinkle, Tahoe, Popeye— were in 1,300 to 3,000 feet. In October, Shell made a promising discovery on Bullwinkle, one of the prospects Nanz and Flowers had shared with government officials, in 1,350-foot waters of the Green Canyon area. Producing from this depth would be an extreme challenge, but Shell's civil engineering department believed that the fixed platform concept could be stretched to that limit. As engineers began to design such a structure, exploration managers were already thinking about venturing farther out. In 1984, the *Seven Seas* moved into the Gulf to drill on the deeper leases obtained in the 1983 sale.

As with each historic step into deeper water, production lagged behind exploration. Fixed platform technology could not be extended much beyond the depth of Bullwinkle. Either subsea wellheads or some kind of floating or compliant structure would be required. Concepts for all kinds of deepwater producing systems were being translated into practical designs. In 1981, Conoco installed in the Hutton field in the North Sea a tension-leg platform (TLP), which was an innovative concept using large steel tendons to tether a floating platform to the seafloor. But the costs of all these concepts presented serious questions. Subsea completions were still expensive and not yet perfected. And Conoco's Hutton TLP had experienced giant cost overruns. Shell Oil needed significant technological development and favorable economic scenarios to produce oil from 1,500 to 2,000 feet of water, let alone in anything much deeper.

One of the leading methods Shell's production department considered for those depths was subsea wellheads linked by pipeline back to a fixed platform in shallower water. But offshore pipelining faced distance limitations and economic constraints. Shell's production managers had a rule: the company could explore no farther than fifteen miles past 600 feet of water, the practical depth limit and distance for installing marine pipelines at the time. Due to this policy, Shell Offshore's exploration managers made only a few bids in the April 1984 sale, the second major area-wide sale in the Gulf. They were caught off guard, however, when other companies, notably Exxon and Placid Oil, acquired acreage in water deeper than Shell had been prepared to go.[88]

The results from this sale prompted a flurry of meetings and discussions in Shell about what its deepwater strategy should be. Exploration managers wanted to eliminate the fifteen-mile rule and probe the extreme depths. Upon transferring from the Alaska Exploration Division in 1984 to become general manager of exploration for Shell Offshore, Mike Forrest told the production managers in New Orleans: "We just spent millions of dollars on prospects in the Bering Sea where there is no infrastructure and there is no proven oil source rock. And yet, in the Gulf of Mexico, we are not willing to take risks going out into deeper water? This is a proven oil province!"[89] Shell's problem in the other U.S. offshore basins was unfavorable geology. The problem in the Gulf was water depth. The geology problem could not be solved. But the water-depth problem could, as Shell had proven again and again.

Shortly after the April lease sale, Billy Flowers met with top company officials Jack Threet, Charlie Blackburn, and John Bookout to make the case for pushing farther into the Gulf. All three appreciated the urgency, given the competition, and they resolved that Shell would drop the fifteen-mile rule and begin gathering seismic data from ultra-deepwater, using a $45 million, state-of-the-art seismic vessel, the *Shell America*, which had just been launched and outfitted. "We decided to drop everything we were doing on the shelf and put the *Shell America* to work in deepwater," recalled Flowers. As long as a football field and sixty feet

wide, the *Shell America* was one of the biggest, fastest, and most sophisticated seismic ships ever built. It housed a massive array of equipment and could deploy eight floats, all by a computerized launching mechanism. It not only could gather more specific data but did so with more speed and precision than ever. It also had the space to accommodate large processing systems, giving Shell the capability to do much faster processing of its offshore seismic data. The *Shell America* virtually revolutionized marine data acquisition.[90]

Time was short before the next Gulf of Mexico area-wide lease sale in July 1984. On the auction block, the government placed large tracts in the western Gulf. The *Shell America* immediately set out to gather as much data as possible. Because of the time constraint, however, the vessel had to focus on specific locations. Tom Velleca, general manager of geophysics in Houston, urged the Offshore Division to organize a team to search quickly for prospects in the Garden Banks area. Located in waters ranging from 1,000 feet to 4,000 feet, the prospects they worked on were considered very speculative. The geophysicists did not have as much seismic coverage as they would have liked. The *Shell America* only had time to shoot one seismic line across some of them. In fact, the areas they prepared bids for were more accurately classified as "leads" rather than "prospects." Previously, Shell had only bid on prospects, for which the company had good data. But the exploration managers thought that now was the time to take calculated risks. The introduction of area-wide sales had opened up huge swaths of virgin territory that could be purchased very cheaply. The lease sale team in New Orleans—Billy Flowers, Doug Beckmann, and Don Frederick—poured over the surveys in preparation for the July sale. Shell bid on ten leads in the sale and won seven of them, and for minuscule bonus prices compared to what the company had been paying prior to the introduction of area wide sales.[91]

Two blocks acquired in the sale covered a potential field called Auger, located in 2,900 feet of water. Further exploration identified it as the prospect with the most potential. The outline of the bright spots extended west into two open blocks. In those blocks, Shell geophysicist Mike Dunn mapped an amplitude anomaly at a subsurface depth of 19,000 feet, well beyond the horizon of conventional thinking about bright spots. Dunn and other geophysicists were convinced, however, that the amplitude effects were real. In the next area-wide sale, held in May 1985, Shell leased the two open blocks. "We were so afraid that other companies would go after the blocks," said Mike Forrest, "that we bid $5 million on one and $2 million on the other. It turned out we didn't have any competition at all."[92]

In the May 1985 sale, Shell expanded into water depths of 5,000 to 6,000 feet. Critical to economic operations at these depths was the need for thick, continuous oil sands that could yield large fields and large reserves per well. Because turbidity currents off the continental shelf dumped such immense quantities of sands in one place, geologists had reason to believe that reservoirs would tend to be far larger than shelf reservoirs. Some of Shell's earlier geological research predicted that, unlike in the deltaic setting, where oil was found near the crests of salt dome structures, the turbidite sands deposited beyond the shelf would have largely avoided such crests.[93] The seismic probes, therefore, were shot across the flanks of these structures, down dip from the crests. Shell's geoscience team mapped the salt ridges and the regional synclines where turbidite sands might have funneled into deep water. Meantime, Flowers and Forrest pressed the production managers on what size oil fields in water depths between 3,000 and 6,000 feet, using "to be designed" technology, would make deepwater producing operations economical. Gene Voiland and Carl Wickizer, production department managers, finally stated that if the exploration group discovered fields of at least one hundred million barrels, the engineers would find a way to make the discoveries pay.[94]

While these discussions were taking place, Shell drilled an exploration test well in 3,000 feet of water on Prospect Powell that had been leased in the April 1984 sale. Drillers located the well to penetrate a very strong, shallow bright spot anomaly, plus a deeper, poor-quality bright spot. Drilling indicated that the shallow anomaly was not associated with oil or gas. However, Don Frederick, division exploration manager, excitedly reported the discovery of a 40-foot-thick oil pay at the deep level. Further drilling and seismic surveys showed that the trap was entirely stratigraphic, likely to contain huge amounts of oil, certainly enough to meet the economic criteria set by the production department.[95]

Armed with this bit of intelligence, Shell Oil dominated the May 1985 sale. With partners or alone, the company was the high bidder on 86 of 108 blocks for which it submitted a bid, in a variety of areas. Its share in the high bids totaled more than $200 million. While most other deepwater lessees did not show interest in acquiring additional deepwater acreage, Shell took a giant plunge. It obtained tracts in the Green Canyon area ranging out to 7,500 feet of water.[96] It acquired prospects code-named Mensa and Ursa, among others. Combined with the tracts leased in the 1983 and 1984 area-wide sales, Shell now had huge areas of deepwater acreage in the Gulf of Mexico, acquired at very low bonus prices. Although nobody at the time knew the exact extent of what this acreage held, they would soon see Shell's deepwater play open the most spectacular new offshore frontier ever encountered.

Deepwater Treasures in a New Era of Oil

The late 1970s and early 1980s were golden years for Shell Oil. Under the leadership of John Bookout, Shell responded to the energy crisis by adopting a corporate strategy almost single-mindedly focused on developing new energy resources in the United States. Although profits on cheaper foreign oil buoyed other major oil companies, Shell outpaced all competitors in the more costly U.S. oil province. The company benefited from the huge run up in oil prices, but its performance was due in large part to its technological advantage and commitment to efficiency. Shell Oil had few equals in the business of offshore development and enhanced recovery. As the *Wall Street Journal* wrote, Shell "awed rivals with an offshore platform taller than the Sears Tower and rose to No. 1 among U.S. gasoline retailers with 9,400 stations sporting its yellow seashell signs."[1]

During the 1980s, however, the yellow luster wore off those signs in the wake of an unexpected and devastating turn of events. Numerous legal and operating problems in chemicals and refining dealt the company serious blows. To make matters worse, in 1985 Royal Dutch/Shell bought out the minority shareholding interest in Shell Oil, a move that caused considerable controversy within the company and angered some public shareholders who felt they had been forced to accept a lowball price. Most devastating was the effect of oil prices that began falling in 1981 and virtually collapsed in 1986. In retrospect, the post-1973 rise from three dollars to twelve dollars per barrel was an appropriate adjustment in the historical trend for the price of oil, which in real terms had declined in the decades after World War II. The second price spike after 1978, on the other hand, far overshot this trend. Many investments and strategic decisions made during 1978–85 proved disastrous when the intoxicating boom turned to sobering bust.

Given this context, Shell Oil's decision to press on with deepwater developments in the Gulf of Mexico was quite daring, even foolhardy in the eyes of some. But those who were running the company saw this decision as both a necessity and an opportunity.

It was a necessity in the sense that Shell had to continue finding large reserves to replace production and supply its downstream assets in order to remain an ongoing E&P concern. Deepwater Gulf of Mexico was one of the few frontier areas left in the United States where the discovery of relatively large reserves could be anticipated. If the reserves were there, Shell was better prepared than any company to exploit them. Few, it is safe to say, utilized exploration technology any better, or foresaw the possibilities of deepwater more clearly, than Shell Oil.

Deepwater Gulf of Mexico was the culminating achievement of Shell Oil's exploration organization in the post–World War II era. Carrying on the work of the pioneers who preceded them, Shell's geologists and geophysicists embraced advanced seismic technologies and decoded the complicated geology of deepwater Gulf of Mexico. With unwavering faith in their technical staff and in the oil and gas potential of this proven oil province, Pres. John Bookout, his second-in-command Charlie Blackburn, and their exploration managers, including Tom Hart, Billy Flowers, Bob Nanz, Jack Threet, and Bill Broman, persuaded the company to take a calculated but nevertheless substantial risk in leasing and drilling tracts in unprecedented water depths. They did this despite the distractions of serious problems downstream, a major battle for financial control of the company, and the collapse in oil prices.

Even after Shell Oil made a string of impressive discoveries in deepwater, however, the economic viability of this new frontier remained in question. The depths were so far beyond anyone's realm of experience. According to one estimate, Shell Oil spent $1.5 billion dollars on deepwater Gulf of Mexico before having the full confidence that the play would be profitable. How Shell Oil's production engineers designed systems to turn these discoveries into commercial reserves is the sequel to the story.

Corporate Crises

When John Bookout took over in 1976, Shell Oil ranked seventh among the U.S. oil companies in net profits. By 1985, it had moved into fourth place, with net profits of $1.65 billion, and first place in net profits per employee. This despite the fact that rivals like Mobil and Chevron had increased their profits through huge acquisitions—Mobil of Superior and Chevron of Gulf.[2] Shell reinvested most of its profits in the hunt for new reserves. During 1978–85, the company found enough new reserves to replace

94 percent of its production, second only to Amoco. If the purchase of Belridge Oil is included in this figure, however, Shell replaced 138 percent of its reserves. This finding rate looks even more impressive when one considers that Shell sold many smaller, older, and less productive fields during this period. During 1982–87, Shell sold oil fields containing about 165 million barrels of oil for $5.65 per barrel, while buying fields with 600 million barrels for $3.61 per barrel.[3] In addition to having one of the best reserve replacement rates, Shell's oil-finding costs were the lowest in the industry. From 1978 to 1985, Shell Oil discovered domestic oil and gas for an average $7.37 per crude-equivalent barrel, compared to an average of $11.51 per barrel among the other majors.[4]

No wonder analysts and journalists rushed to pronounce Shell as one of the best-managed companies in the industry, if not in all of corporate America. In the early 1980s, Shell Oil was the envy of the U.S. petroleum industry. Observers used adjectives such as "efficient," "innovative," "lean," "forward-looking," even "bold" and "arrogant," to describe the company. Shell had accomplished extraordinary things, finding new sources of oil and improving profit margins through a better allocation of each barrel of crude. The board of directors was so pleased with what John Bookout had achieved that in 1979 they extended his presidency two years beyond Shell's mandatory retirement age of 60 and continued to renew it until he eventually retired in 1988 at the age of 65.

Even the best and the brightest could not escape misfortune, however, as just about everyone in the oil industry found out in the 1980s. In the latter years of Bookout's presidency, slumping crude prices and other unfortunate developments tarnished the shining Shell. Troubles began with the 1981 recession, when higher feedstock costs undercut all commodity chemicals, forcing Shell Chemical to reduce capacity and shut down some operations. Higher environmental expenditures and refinery maintenance increased operating costs, as did a prolonged legal and political battle over environmental damage at the Rocky Mountain Arsenal, where Shell Chemical had manufactured pesticides for decades. Under the new low price regime, the Belridge acquisition and late-1970s refinery modernizations made to process heavy oil failed to live up to their earlier promise. The tremendous amount of money sunk into the unsuccessful hunt for oil offshore Alaska, and the successful but costly developments in deepwater Gulf of Mexico, also strained the bottom line.

On top of all this was intensified competition in the oil market. Trouble began with the Iranian Revolution and cutoff of oil from that nation in early 1979, which triggered crude shortages and price increases around the world. Nations suddenly without Iranian oil hustled to replace it with foreign production that U.S. oil companies had counted on importing. "Spot" market crude prices rose steeply, followed by general increases from oil-exporting nations. Oil traders entered the market in force, taking advantage of the enormous arbitrage between the lower prices in the term-contract market and the higher and more volatile prices in the spot market (a difference of about fifteen dollars per barrel in 1979), a situation that had been created by the loss of integrated control by the majors over international supplies. Following a wave of oil nationalizations, state oil companies now accounted for ever-increasing shares of total production. But they had no downstream outlets of their own, so they sold to a wide array of buyers. The Royal Dutch/Shell Group's international supply coordinator explained the difficult position of the majors: "Every negotiation with a producer government was a finger-biting exercise: there was one dominant thought in the mind of company presidents and negotiators alike—to hold on to term oil and limit the need for spot acquisition. The suppliers, of course, sensed this, and a cat-and-mouse dialectic began. . . . Both the terms of contracts and the prices that had to be paid became continuously worse."[5]

As an entity that produced less than half the crude oil it normally used, Shell Oil felt the pinch of tight world supplies. The company depended on growing imports of foreign oil, both through ongoing contracts and spot purchases. Even though Shell in the United States had imported very little oil from Iran, its refineries found themselves taking in 120,000 barrels per day below what they had expected to process. The nightmare of 1973 returned. Panic buying ensued, and long gas lines formed once again, extending for blocks around service stations. Again, the system of U.S. price controls and allocations exacerbated the shortage. This forced certain refiners to sell oil to other refiners short of supplies. Regulations froze refining and distribution patterns on a historical basis, preventing companies from moving supplies around to meet demand and discriminating against growing urban areas. Throughout 1979, Shell Oil had to sell increasing amounts of oil to other refiners, even when the company was unable to obtain enough for its own plants. The shortage not only disrupted motorists' daily lives but generated price increases that contributed to galloping inflation. It was a frustrating, confusing and infuriating period for gasoline consumers and producers alike.

Forced to take some kind of action, Pres. Jimmy Carter promoted an ambitious synthetic fuels plan and encouraged Americans to strive for energy conservation in all aspects of their lives. High prices and good times for oil producers continued for several years, but this obscured a tectonic shift in the underlying fundamentals of the oil market. Surprisingly effective and ongoing conservation measures (such as the production of more fuel-efficient vehicles), combined with the 1980–81 recession, softened fuel demand and thus real oil prices. "The massive twentieth-century march toward higher and higher dependence on oil within the total energy mix was reversed by higher prices, security considerations, and government policies," writes Daniel Yergin.[6] Coal and nuclear power stole market share from crude oil in electricity generation. Meanwhile, a surge of non-OPEC oil from the North Sea, Mexico, and Prudhoe Bay weakened the cartel's control over prices. When, in 1981, newly inaugurated Pres. Ronald Reagan lifted the federal government's seven-year price and allocation controls on both gasoline and crude oil, the seven-year trend of steeply rising oil prices reversed itself.

In the early 1980s, production from non-OPEC sources began to outpace that of OPEC. The official OPEC price (based on Saudi light crude) was still $34 per barrel, but in the buyer's market, discounts could be easily obtained. In February 1983, North Sea crude fell to $30 per barrel. The cartel tried to restrict output and hold individual producing nations to quotas, but cheating and discounting continued. Saudi Arabia, the world's largest producer and holder of one-third of the world's oil reserves, took on the role as swing producer, varying its own output to support the OPEC price. But the kingdom would only tolerate and underwrite quota violations for so long at the expense of declining market share for its own oil.

In the summer of 1985, Saudi Arabia decided it would no longer defend the OPEC price. Saudi oil production had dwindled to 2.2 million barrels a day, half its quota level, while quota-breaking among other members of the cartel was rampant. Finally, the kingdom moved to reclaim its market share, offering "netback" deals with its Aramco partners and other buyers in key markets. By late 1985, other OPEC exporters had followed suit, each striving for greater market share and forgetting the cartel's quotas. World production surged. "With the taps turned on and the cartel's artificial buttress removed," writes Shell Transport & Trading historian Stephen Howarth, "the only way for anyone to retain a market share, let along rebuild one, was to sell cheap."[7] In November 1985, the price for bellweather West Texas Intermediate crude was $31.75 a barrel. By spring 1986, it had dropped to $10 a barrel, with spot prices as low as $6. The great twelve-year boom in oil prices had finally turned to bust.

Against the backdrop of an emerging buyer's market around the world, the U.S. oil industry underwent a wave of corporate reorganizations. Deregulation of oil in 1981 lifted protection and increased competitive pressures, thus ripening the industry for consolidations, mergers, and breakups. At the same time, institutional investors, anxious for higher returns on their portfolios, began to exert pressure on the oil industry, whose performance was believed to be lagging in the aftermath of the boom. Many analysts also perceived a gap between the value of oil company shares and what oil and gas reserves would command in the open market. In the United States, adding new reserves through exploration and production was becoming expensive, as Shell Oil was finding out in Alaska and the Gulf of Mexico. But in the face of nationalistic barriers to oil investment elsewhere, oilmen still coveted reserves in the United States, where politics were stable and taxes comparatively low. The logical solution, it seemed, was to "explore for oil on the floor of the New York Stock Exchange"— to purchase undervalued companies, in other words.[8]

Many oil companies had plowed huge profits made from the two oil shocks into exploration in the United States, with disastrous results—most notably, the great Mukluk fiasco in Alaska's Beaufort Sea. Rather than risk millions of dollars exploring for oil in forbidding frontier locations, why not acquire or merge with another company and gain known reserves on the cheap? "The value gap, like a geological fault," writes Yergin, "facilitated a great upheaval throughout the oil industry. The result was a series of great corporate battles, pitting company against company, with a variety of Wall Street warriors mixed in and sometimes in command."[9]

At Shell, a corporate battle broke out over Royal Dutch/Shell's bid to buy out Shell Oil's minority shareholders. On the night of January 23, 1984, Sir Peter Baxendell, the chairman of Royal Dutch/Shell, telephoned John Bookout to announce that he and Royal Dutch Petroleum president Lodewijk van Wachem were boarding a plane and would be arriving in Houston in the morning. They did not disclose the purpose of their visit in the phone call but suggested that Bookout suspend trading of Shell Oil stock when the market opened the next day (Bookout found no basis for such action and thus did not suspend trading). When Baxendell and van Wachem arrived the next morning at his office, they announced that the Group was making a tender offer for the minority stake in the Shell Oil Company.

It was one of the longest-lived rumors in the U.S. oil industry: that the Group was going to buy up the stock in Shell Oil it did not already own. For years, the rumors had reappeared, but the purchase attempt never happened. The changing oil business and regulatory environment in the United States, however, provided a new opportunity. The merger wave in the U.S. oil industry was under way (Texaco-Getty, Chevron-Gulf, Mobil-Superior), and the Reagan administration's unwillingness to challenge these mergers and generally permissive antitrust stance signaled a green light for takeovers, even by foreign companies. "Royal Dutch would have done this years ago, except for its fear of antitrust problems," said a former chief of strategic analysis for the Group in 1984. "This was always a closed continent" for foreign ownership of energy assets. Now, he added, "you can buy without criticism." [10]

Full ownership of the highly profitable and technologically sophisticated Shell Oil Company was an investment too attractive for the Group to pass up.[11] The Group had an estimated $8 billion in cash on hand and was looking for some place to put it. The remaining interests were a natural corporate fit, matching Shell Oil's strengths in E&P with the Group's strengths downstream. The "merger" also promised to ease tensions between the Group and Shell Oil, which undeniably had been building for years. These tensions stemmed mainly from Shell Oil's independent international presence in the form of Pecten International and the restrictions on the flows of technology and capital between Royal Dutch and Shell Oil, which were necessary to protect the rights of minority shareholders and guard against the threat of antitrust action. With the public shareholders out of the way, as Baxendell noted at the time, integrated decision-making could "take place without inhibitions." [12]

The value of Shell Oil's outstanding stock quickly became a disputed issue. The Group initially offered $55 per share, significantly higher than the $44 per share at which Shell's stock sold before the offer but not high enough to match the $80–85 per share estimated value of Shell's crude reserves as determined by a Goldman

Sachs study commissioned by Shell Oil's outside directors. Other outside estimates ranged even higher.[13] The Group's share-value estimate, according to an appraisal by Morgan Stanley, was based on a $3.80-per-barrel price for Shell's reserves, which was roughly what Texaco had paid for Getty and Chevron for Gulf. But Getty's and Gulf's reserves had been shrinking for years, while during 1978–85 Shell Oil was a leader in U.S. reserve replacement and exploration efficiency.[14] The varying estimates also reflected disagreement over the net present value of Shell's "proven" reserves and the extent of its "probable" reserves. Shell Oil managers spent nearly all of 1984 trying to help put a value on the company for Goldman Sachs and the outside directors. But it was simply impossible because the value was what the market would pay, yet there was only one buyer. Exxon, for example, probably would have paid a lot more per share had the whole company been offered for sale. But it was not. So Shell eventually dropped work on valuing the company in the open market.

The tender offer raised a chorus of objections from Shell Oil's public shareholders. From their perspective, Royal Dutch was not only buying the company's reserves and physical assets but Shell Oil's talent for replacing them. At stake was the self-worth of Shell Oil executives and engineers and their fierce pride in the independent American operation that they had built up over decades. "Seventy-five dollars is to Shell what $128 was to Getty," said oil analyst Sanford L. Margoshes of Shearson/American Express Inc.[15] The Group sweetened its bid to $58 per share, but this seemed like an insult to some minority shareholders, many of whom were Shell Oil employees (who owned roughly nine percent of all Shell stock).[16] A "family feud," as the press dubbed it, soon erupted.

John Bookout found himself in an awkward position. He was being forced to choose between his corporate bosses and Shell Oil employees and retirees, whose strong loyalty he had nurtured for eight years. If he or the other members of the board publicly objected to the Group's offer without taking defensive measures—such as suing the parent company or attempting a self-tender at a higher price—they would be liable for failing their fiduciary duties to the minority shareholders. Faced with this predicament, Bookout considered resigning. Upon further reflection, however, he decided he owed it to the shareholders to remain in the job. The board counseled Bookout to keep quiet about the merger and focus on running the company's day-to-day operations. This was not easy for someone who had become a proud and outspoken leader of the industry, accustomed to meeting frequently with the press, securities analysts, government officials, community representatives, and the employees. Everyone closely scrutinized Bookout's every move for hints of what his position was. "We the employees felt the man carrying the banner for us was John Bookout," explained Tom Stewart, a Shell Oil public affairs manager. "He represented the employees. If John said it, we believed it."[17]

At the company's annual meeting in April 1984, rebellious shareholders looked to Bookout for guidance. "As he stood at the podium, the range of Bookout's diplomatic skills—and the depth of his dilemma—were on full display," wrote *Fortune*. He defended the Group's right to make a tender offer but refused to comment on the price. In the coming weeks, he chose a course of silent acquiescence. On May 2, a week before the Group's deadline for acceptance of the $58 offer, Bookout and four of his top executives sold their personal holdings in Shell Oil to the Group. Whatever they privately thought about the price, it was the best deal they were going to get. Controlling nearly 70 percent of Shell Oil's shares, the Group was not going to offer $80 a share for stock selling at $44. And it was prepared to enter into private deals for large blocks of stock to achieve the 90 percent needed to initiate a short-form merger. Said Baxendell: "A haggle, or auction over price, which seems par for the course in cases where people are contesting for control, is certainly not an option we have in mind. Indeed, we believe it would be inappropriate."[18]

Bookout declared at the time that his decision to tender his shares "best serves the operational needs of the company." Legally, he simply could not make a recommendation to others, saying, "all shareholders should make their own decisions." Most of them grudgingly sold their shares. Some held out, however, and brought class-action suits against Shell charging unfairness of price, unfair dealings, and inadequate proxy disclosure of relevant information. In early 1985, a settlement was reached that provided cash payments to class members of $190 million, reportedly the largest class-action settlement ever at the time. The settlement sum gave an additional $2 per share for all members of a subclass of Shell shareholders who had accepted the $58 tender offer and the same additional $2 per share for members of another subclass consisting of non-tendering stockholders if they waived their right to a court appraisal of their shares in the short-form merger at the $58 per share merger price.[19]

A few shareholders still pressed the issue in court. In June 1990, the Delaware Chancery Court ordered Shell Oil to pay $71.20 plus interest to holders of one million shares of the company who did not tender their shares in 1984, ruling that an "inadvertent error" by Shell Oil in calculating the discounted future net cash flow from estimated proven oil and gas reserves constituted a failure to disclose a material fact. Although this price was significantly lower than the $89 per share maintained by the plaintiffs as the fair value, Shell Oil (under full ownership of the Group) appealed, arguing that discounted cash flow analysis was "soft" data, based on numerous assumptions and estimates. However, in May 1992, Delaware's Supreme Court upheld the 1990 decision, resolving once and for all the minority shareholder issue at Shell Oil.[20]

All the wrangling and acrimony aside, the buyout turned out favorably for most everyone involved. Despite the money paid out in the settlement, the Group achieved good value for the rest of Shell Oil. "That was one of the best deals we ever made, if not the best," said the legendary John Loudon in 1989. He was long retired at the age of 84 but still sharp and attentive to the Group's business. "If we'd waited longer, we would have paid much more." While some Shell Oil stockholders called the deal a "stab in the back," Shell Oil employees actually fared pretty well.[21] Everyone in the company received huge bonuses that more than offset the tax effect of tendering Shell Oil stock. Executives exchanged Shell Oil stock options for more valuable Royal Dutch and Shell T&T stock options, and a special incentives program was implemented for senior staff.

Furthermore, John Bookout negotiated an important concession from Royal Dutch/Shell on the way to the "merger." He insisted upon continued operational autonomy for Shell Oil, and the Group complied, at least for a while. In September 1986, Lodewijk van Wachem formalized this agreement in a memorandum to all Group coordinators and division heads, which stated, "it is intended that Shell Oil shall maintain a board of directors on which non-executive directors of high quality serve, and that it shall continue to raise and manage its own finances, take its own decisions about the remuneration and deployment of staff, make its own decisions about investments and operations, and therefore manage its own affairs without the involvement of the Service Companies, thus perpetuating the position obtaining when there were minority shareholders."[22] This was an unusual arrangement to allow such a large operating company to function outside of the

Service Companies, which provided "advice and services" to all other operating companies around the world. Only someone with the clout and character of John Bookout could have preserved the organizational independence of Shell Oil. The board of directors and management did not immediately change. Bookout served as president until 1988. A feeling that things would never be quite the same still permeated the organization, but at least for awhile, the buyout did not fundamentally alter the company's identity.

That identity was increasingly defined by the "offshore imperative." In the midst of the buyout, as misfortunes in refining and chemicals, busted crude oil prices, and exploration setbacks in Alaska and elsewhere afflicted the company, offshore Gulf of Mexico remained the one area that showed some promise. It was still a risky bet, but it was the only one Shell Oil had left to make.

Bullwinkle: Redefining "Big"

On the morning of May 27, 1988, citizens of Port Aransas, Texas, were treated to an awesome spectacle. Escorted by a small fleet of tugs, the 1,365-foot-high steel jacket for Shell Oil's Bullwinkle platform began a slow barge cruise through the city's ship channel on its way out to sea. Hundreds of spectators lined the banks to gawk at what various reporters referred to as a "leviathan," a "behemoth," and a "Moose of a contraption." The most striking description came from the *Engineering News Record*. In view of the jacket's unimaginable human scale, the publication wrote that "the structure is so big, it evokes thoughts of the fallen Gulliver." [23] The spectators could have been easily mistaken for Lilliputians.

The search for multiple superlatives conveyed a single message: Bullwinkle was BIG. It was the tallest man-made offshore structure in the world. With deck and drilling rig in place, Bullwinkle became the tallest manned structure in the world period, nearly two-hundred feet taller than Chicago's Sears Tower. As a fixed platform, it set a new world water depth record of 1,353 feet, eclipsing the old record of 1,025 feet held by Shell's Cognac platform.

The story of Bullwinkle began with the landmark area-wide lease sale of May 1983, when Shell acquired leases for roughly thirty million dollars on Green Canyon Blocks 65 and 109. In October of that year, a discovery well was drilled on Block 65. It was a promising find of 82 million barrels of oil equivalent, but the water depth posed serious questions: Could a platform be installed in water that deep? [24] If so, what kind of structure would it be? A subsea system? Some kind of floating system, such as a new-

The Bullwinkle plat-
form before launch-
ing on May 31,
1988. *Photo courtesy
Peter Marshall.*

fangled tension-leg platform? Or a traditional fixed platform? If it
were a fixed platform, would it have to be built in multiple pieces,
like Cognac, or was it physically possible to build a single-piece
structure on land, put it on a barge, and launch it at sea?

Shell's civil engineering group, led by Pat Dunn, decided upon
a fixed platform. Some industry experts criticized the decision,
arguing that a floating platform or subsea system would be a better choice. "We have been involved in the subsea area a long time,
and we had already built Cognac so we knew what to expect,"
said Dunn. "We were working on preliminary designs for tension-
leg platforms; our analysis indicated that they were feasible but
too expensive for application at Bullwinkle." The task of producing a conceptual design fell to Shell's Development Group, headed
by Peter Marshall, who in 2006 received the OTC Distinguished
Achievement Award for his work on fixed platforms. Marshall's
team of engineers, draftsmen, and computer specialists spent a
year converting oceanographic and engineering mechanics into a
workable design. They produced two sets of bid drawings, one for
a monstrous one-piece platform and one for a two-piece structure
similar to Cognac. They then let competitive bidding help make
the final choice. The bidders unanimously selected the one-piece
option. "They all wanted to go with brute force and build a massive launch barge," said Marshall.[25]

The key consideration in this decision was installation. Cognac had been installed in three pieces because marine equipment big enough to install it in one piece had not existed. Also, engineers had feared that the long time required for installing and securing the foundation would leave the structure exposed to a potential calamity in the event of a hurricane. By the mid-1980s, however, marine equipment had increased in size, and Shell engineers were more confident in their ability to get a large jacket installed quickly. Underwater pile-driving hammers, first developed for Cognac, had come along way, as had launching techniques for large jackets. Shell's Eureka platform in California and Boxer in the Gulf of Mexico provided good experience for a large, one-piece installation. "We could also depend on fairly conventional technology with a one-piece jacket," said Marshall.[26]

Although Bullwinkle shaped up differently than Cognac, the experience of designing the earlier platform was invaluable. Many of the people who contributed to Cognac's revolutionary design, including Marshall, Lee Brasted, Gordon Sterling, George Uppencamp, Bobby Cox, and Charlie Stuart, were instrumental in the Bullwinkle project. Cognac also had been outfitted with many instruments that measured oceanographic conditions and provided ten years of fatigue data. "We monitored everything from stress on the platform to crack growth in 'early warning' specimens," noted Marshall.[27]

This data improved Shell Oil's capabilities for doing the detailed design. It produced a new tool—the random directional wave force theory—that introduced a new way of analyzing the energy that waves produce when striking a platform. This allowed Shell engineers to design Bullwinkle, which would have a much larger natural sway period than the Cognac structure, in such a way as to cancel out wave forces that might cause the platform to sway dangerously. Another important innovation pioneered on Cognac but implemented more completely on Bullwinkle was an advanced computer-assisted drawing and design (CADD) system, which dramatically expedited the design process. Hundreds of different load cases were analyzed to ensure strength in the steel members of the jacket. "We couldn't have gotten the job done without CADD," claimed Kris Digre, the engineering supervisor on the project. "The speed in getting the drawings out was just amazing. Take the joints, for example. We used to hand-draw each joint. It would take a draftsman one day to draw one joint, maybe even a *week* if it was a complex joint. The computer did it in a few *seconds*."[28]

Speed in design was crucial because the two contractors who were finalists in the bidding process submitted two completely different plans for fabrication. "It was the equivalent of preparing the drawings for a couple of high-rise buildings," explained Digre. "You have to take into account all the things that go into the structure, miles and miles of piping, tons and tons of steel—with a detailed view of everything."[29] Shell ultimately awarded the construction contract to Gulf Marine Fabricators, a subsidiary of Peter Kiewit & Sons, to build the jacket in one piece. In February 1986, construction began at Ingleside Point near Corpus Christi. Although Bullwinkle was "conventional" in the sense that it was a single-piece jacket, there was nothing conventional about the methods used to build it. Construction crews spent six months just preparing the foundation for the jacket, which would weigh fifty thousand tons. Gulf Marine incorporated several new innovations in fabrication, the most striking of which was an assembly technique that employed cranes and computer-controlled winches to "roll up" individual sections of the jacket. Meanwhile, Heerema Marine Contractors built the world's largest barge—the length of three football fields lined up end to end—to haul the steel giant on its side across the Gulf to Green Canyon.[30]

Managing the fabrication of Bullwinkle's jacket was a major industrial operation. The construction project employed six hundred people, making it one of the largest industrial companies in the area. Its scope, however, extended way beyond the Gulf Coast. Vendors came from thirty-three different states. Shell's team of construction and project engineers, led by Gordon Sterling, oversaw fabrication and installation. Capitalizing on their past experiences, Sterling's team managed the job with such precision as to make the construction seem almost uneventful compared to Cognac.

As fabrication proceeded over the next two years, Shell Oil's design engineers turned their attention to the most difficult aspect of the project: installation. They methodically worked out the details of the load-out, the transport, and the launch to make sure no part of the jacket was overstressed during those operations. "Getting Bullwinkle onto the barge was no trivial exercise," said Pat Dunn. The load-out itself took more than four days as cranes and jacks slowly pulled the jacket on greased skid timbers sliding over Teflon-coated steel on the concrete skidway. Three days after floating out past the awed spectators along the waterfront, Bullwinkle arrived at the installation site. Now came the critical moment. The installation crews only had one chance. "We had to know at what angle and at what speed it would hit the water, and we had to ballast the barge correctly or else the platform could have been severely damaged during launch," remembered Dunn. "It was mandatory that we be right."[31]

First-rate engineering and careful double-checking ensured that they were. At 6:30 P.M. on May 31, 1988, Heerema activated the hydraulic jacks that nudged Bullwinkle off the barge. Pete Arnold and a small group of co-workers were nearby in a work barge as the jacket began its slide. "It started out very slowly, very quietly," Arnold recalled. "I couldn't get over how quiet it was. It was almost *soundless*. By the time it was halfway off the barge it had picked up speed and you couldn't keep up with it running." The jacket hit the water with a giant, roaring splash but in almost perfect alignment. The force of the entry was best experienced on the launch barge, which was tipped up and pushed about one thousand feet from the point of entry. For a moment, the jacket disappeared under the water, which produced a "sinking" feeling among Shell executives witnessing the launch. A few seconds later, however, it popped back through the surface and everyone breathed easy again. Soon after the jacket settled in the water, crews climbed aboard and began to flood the compartments in the lower jacket, slowly bringing it first into an upright position and then gently down to the sea floor. Giant mud mats, specially designed for Bullwinkle, stabilized the legs on the bottom. "We set it down within an eyelash of being level, and 2.9 feet from the location where we wanted to be," said Gordon Sterling.[32]

After the successful launch, however, Shell encountered problems completing the installation. First, the main crane failed just as crews had begun to drive the piling to secure the jacket to the sea floor. Then in late June, labor problems on the Heerema barge caused further delays. Because of time lost over the summer, the project was exposed to two dangerous hurricanes, the project manager's greatest fear. In September, Hurricane Florence blew across the Gulf of Mexico toward Bullwinkle, bringing with it ten-foot seas and forcing the evacuation of the construction barge. When the barge returned, Gilbert invaded the Gulf. It was a so-called "hundred year storm" that generated 200-mile-per-hour winds. Fortunately, Gilbert turned due west and hit a remote area of Mexico rather than colliding with more populated areas along the Gulf Coast. Although the Bullwinkle platform was never in danger, the bad weather added frustrating delays. In the end, though, the installation was finished superbly. The jacket was topped off by a platform deck manufactured by McDermott, and the first production from the record-setting platform reached the surface on July 23, 1989.[33]

With a capacity of forty wells and a final price-tag of roughly $250 million (the total development cost was about $500 million), Bullwinkle established a landmark in the history of offshore oil. It symbolized both the end of one era in the Gulf of Mexico and the beginning of a new one. Shell and other companies no longer looked at one thousand to fifteen hundred feet as the practical limit of deep water. Henceforth, as new production technologies became commercially viable and unique geologic conditions were discovered, offshore development in the Gulf would take another quantum leap into two-to-three-thousand-feet-plus water depths. Bullwinkle was the last of the record-setting fixed platforms. Other giant fixed platforms would be installed in the Gulf but none equal the size of Bullwinkle. At greater depths, the technology was still feasible, but costs became prohibitive with the increased requirements of steel and larger launching barges. "Bullwinkle doesn't really represent a new technology," Gordon Sterling commented at the time. "It's *extending* our understanding of metallurgy, *extending* our understanding of oceanography, and so on . . . simply because of its size."[34]

By approaching the limits of fixed-platform technology, Bullwinkle took a major stride into deep water. Unlike Cognac, it was built and installed in a single piece, which allowed for greater cost savings than if it had been pieced together. Structural engineering at Shell Oil took a massive leap forward with Bullwinkle, moving into computer-assisted design and operations. For many Shell people, it was the project of a lifetime. In 1989, the American Society of Civil Engineers gave the Bullwinkle project its Outstanding Civil Engineering Achievement Award, making Shell Oil, which had won the coveted award for Cognac in 1980, the only organization twice honored in this category by the ASCE. Most significantly, however, Bullwinkle took the company into a new frontier, providing hints of what the water just beyond had in store and a base from which to reach it.

By the time Bullwinkle started producing in 1989, Shell Oil had a string of discoveries in this ultra-deepwater. During 1983–86, Shell Oil won 252 of the 327 tracts bid (77 percent) in the

Gulf of Mexico, leading the industry in finding 388 million BOE at very attractive bonus prices. Shell's bonus to BOE discovered ratio dropped to $1.84, again topping all competitors.[35]

The company made strikes on prospects Tahoe, Popeye, Mensa, Ram-Powell, Mars, and Auger, names that have since become famous within the industry. At the time, however, the leases on these prospects were still considered highly speculative. These discoveries were in two to five thousand feet of water. Buoyed by the success of Bullwinkle, Shell E&P pressed on to find out if the deeper water could be made profitable.

Auger: A Quantum Leap

Deepwater was the most daunting development challenge Shell E&P had ever faced. Having achieved such abundant exploration success, Shell did not have adequate resources (manpower, money, risk capacity) to undertake development of all its discoveries at once. Instead, managers devised a strategy to space out major platforms by about two years, allowing time to apply lessons and make improvements from one project to the next, spread out manpower and cash requirements, and mitigate against investment risk. Simultaneously, they laid out a similar strategy for subsea developments, in anticipation that subsea satellite wells would be widely used to develop smaller discoveries.

Auger rose to the top of the list of Shell's deepwater prospects. In 1984, Shell had obtained the Auger leases in the Garden Banks area 136 miles off the Louisiana coast but delayed exploratory drilling because the company's engineering analysis showed that no vessel, including the *Discoverer Seven Seas,* was capable of drilling to the formation depth and pressure required at Auger. So Shell entered into an agreement with Global Marine to build and operate a new, giant semi-submersible, the *Zane Barnes,* to meet its needs. Drilling at Auger began at an inauspicious time—in 1987, right after the price of oil had collapsed. This placed considerable pressure on the engineers and geologists who would have to find a huge field to cover the enormous cost of a tension-leg platform (TLP), which had emerged as the favored technology for production operations at the Auger prospect. Operations project leader Bill Sullivan and his staff prepared a drilling recommendation that would test all bright spot intervals, including a speculative one mapped by Mike Dunn at a subsurface depth of 19,000 feet, which was well beyond the horizon of conventional thinking about bright spots.

Early drilling using the *Zane Barnes* proved disappointing. Results indicated that two bright spot targets at a subsurface depth of about 15,000 feet were associated with poor quality sands. These initial tests cost over $20 million, and a pivotal debate ensued over whether to drill deeper. Some Shell geoscientists thought the deep bright spot amplitude at 19,000 feet was merely a salt reflector. Others, however, were convinced that the amplitude effects were real. In late July 1987, upon the strong recommendation of Jim Funk, division exploration manager, drilling continued.[36]

Shell had invested too much in deepwater technology and leases, and it had staked too much of its E&P identity on its offshore capabilities, to turn back now. The Gulf of Mexico had been kind to Shell Oil for decades, and once again it yielded buried treasure. In late September, drilling in 2,860 feet of water and at a subsurface depth of about 16,550 feet, the *Zane Barnes* struck oil. Drilling even deeper, Shell found the "field-maker" oil-bearing sand at 19,000 feet. It was a handsome find for the Gulf of Mexico. The Auger field contained an estimated 220 million BOE, the third largest discovery offshore by Shell Oil up to that point, behind South Pass 24 and South Pass 27. It gave new promise for deepwater development in the Gulf. Auger might herald bigger and better things. And if it did, the future would be bright indeed for Shell Oil, which held the majority of deepwater leases in the industry. In December 1987, the company announced it would develop the field and solicited bids for a tension-leg platform.[37]

Despite the size of the Auger field, this decision was not arrived at easily. The cost of developing the field was estimated at over $1 billion. Oil prices were down to $15 per barrel from highs of nearly $40 per barrel earlier in the decade and staying down. Nobody knew when, or even if, oil prices would recover. Shell Oil owned 100 percent of the leases at Auger and so would shoulder 100 percent of the costs of development. The company was really "out there" on its own. No other oil company was even considering such water depths. At Auger, the depth was twice the record-setting depth of the Bullwinkle platform, which had not even been installed yet. TLP technology was proven but still quite new. Getting consent from other parts of the company and approval from the board of directors to go ahead with Auger, therefore, was a hard sell to say the least.

Shell Oil had gone through a similar exercise in soul-searching over "deepwater" development back in the late 1950s. E&P managers faced the same question their predecessors faced thirty years earlier: Even with viable technology, would deepwater ever pay? Once again, the final determination boiled down to an abiding faith in Shell's technical staff by top management and the supreme confidence that Shell Oil could do something that no other company had ever done. Selling this idea within the company, however, took concerted effort from managers in E&P, from deepwater division production manager John Krebs, to general manager Wade Dover, to exploration vice president Bill Broman, to production vice president Larry Smith, to president of Shell Offshore, Inc., Bob Howard. "You talk about people sticking their necks out," recalled Rich Pattarozzi, later president of Shell Deepwater. "That was an unbelievable undertaking by any company, I don't care how big you are, to go out there and basically put $1.2 billion on the line."[38] In the end, the board was persuaded to go forward with Auger. John Bookout, who was just finishing his tenure as president of Shell Oil, credited Shell's outside directors for supporting this strategic thrust. It took courage on their part, but had the proposal been made in another company without Shell's peerless reputation offshore, it is hard to imagine the directors going along.

Auger was yet another unprecedented undertaking for Shell. It bridged the gap between the company's ability to explore in deepwater and its ability to produce there. The gap had grown quite large as seismic teams moved out into thousands of feet of water. But the tension-leg platform concept promised to close it rather quickly. Unlike conventional fixed platforms, resting pyramid-like on the sea floor, the TLP consisted of a production facility situated on a floating hull and held in place by long tendons made of tubular steel, which connected the hull to templates on the ocean floor. The tendons were placed under tension to keep the hull from bobbing like a cork in choppy seas, but allowed side-to-side motion caused by wind or wave forces. In Shell's unique design, a lateral mooring system would provide the means to position and hold the structure directly over the wells, which were spaced widely and elliptically on the sea floor.

As with Bullwinkle, Shell Oil did not invent radically new technologies with Auger but took existing technologies to new limits. Auger was not the first platform to employ a floating, tension-leg design. In 1984, Conoco installed the first full-scale TLP in the North Sea's Hutton field in 485 feet of water. In mid-1988, Placid Oil installed a floating system in 1,540 feet of water in the Gulf of Mexico. And in 1989, Conoco placed its Jolliet tension-leg well-platform (not a full-blown TLP) in 1,760 feet of water in the Gulf. While the TLP was considered leading-edge technology, Shell's engineers had seriously studied the concept at least since 1977, when the head office formed a special Marine Systems Engineering Group led by Carl Wickizer. This group had initiated ongoing development work on floating production systems and subsea systems, all with a design depth of at least 3,000 feet. With

help from the civil engineering group and Shell Development, they had looked at various alternatives, including the TLP, and actually did the first conceptual design of a TLP at the time. These preliminary designs could be applied directly to Auger. But making these designs come to life was a serious undertaking. At 2,800 feet, Auger was extending TLP technology over a thousand feet deeper than Conoco's Jolliet. Its design and construction were by far the most sophisticated yet attempted. With the largest design-to-capacity ratio of any of the company's production facilities, the Auger TLP would be an equipment-heavy platform supporting full drilling and production with thirty-two wells.[39]

A key design consideration at Auger was cost. While this was true for any major project, it was mandatory in the context of persistently low oil prices in the 1980s and early 1990s. Newly sophisticated three-dimensional computer modeling techniques aided this effort. They lowered the cost of shooting data and sped up its processing. Commonly employed by the early 1980s to define proven oilfields and by the late 1980s to explore for new prospects, "3-D seismic" allowed seismic readings to be rendered into a detailed, three-dimensional picture of the subsurface geology. At Bullwinkle, Shell geophysicists used the technology very effectively for the development drilling phase of the project. At Auger, 3-D seismic data increased Shell's confidence about the extent of the field and provided enough definition to limit the number of development wells, each of which cost millions of dollars.

The other key design considerations were safety and system integration. Again, such considerations were important to all major projects, but they took on new significance at Auger. Design engineers faced so many new aspects of safety that risk assessment and mitigation became a major part of all design decisions. Shell could not afford a failure! Because Auger was a completely self-contained floating drilling and production platform, weight distribution was a critical factor that had to be managed at all stages of the TLP's life—at float out, installation, drilling, and production.

In addition, Shell engineers incorporated innovative well systems, subsea components, and specialized tensioning mechanisms into the TLP design. It was thus essential that all structural, topside, and well systems be integrated in an unprecedented manner.

Consummating this design required a brand-new organization that combined specialists of all disciplines from Shell Oil's various engineering groups, contractors, support groups, operators, technicians, and consultants into a unified design and construction team. This organization emphasized teamwork, with each team multiskilled in various tasks. For example, a mechanic who in the past only worked on compressors now found himself helping ready oil wells. "The new organization has changed things," said one worker. "We're expected to speak up if we see something we think we should be doing differently. Let me tell you, that's a big, big change for the oil patch."[40] Shell engineers designed the entire project and worked closely on every detail of fabrication, installation, and drilling, but contractors came from far and wide. Using a complicated contracting strategy that relied on multiphased bidding, Shell Oil contracted with nine hundred companies in the United States and thirty-three foreign companies on the project. McDermott, Inc. fabricated the 23,000-ton deck in Morgan City, Louisiana, and the Italian firm Belleli S.p.A. built the 21,000-ton hull. Shell personnel also worked with various subcontractors to develop advanced, deeper-water versions of underwater remote operating vehicles (ROVs), which assisted with the installation and maintenance work on the sea floor.[41]

As construction of the Auger TLP was just getting underway, problems arose. These included cost overruns, delays, and strained relationships with contractors. Fabrication of the topsides, in particular, fell behind schedule. Auger was the first of its kind. Design and fabrication involved a great deal of trial and error. Most distressing, however, was the continuing slump in oil prices, which threatened the viability of projects like Auger. In 1991, E&P learned Shell Oil was the highest-cost oil producer in the United States, and the company registered its first quarterly loss since the Great Depression (see epilogue). Shell E&P's 1990 "long-range plan" forecast continuing low oil prices, making Shell Oil's grand strategy for deepwater uneconomical. Furthermore, deepwater oil tended to be sour—containing sulfur, which had to separated out at the refinery—and thus its price would have to be discounted. Shell Offshore would have to make major changes in its strategy for deepwater or give up on the entire play. It seemed quite possible that Auger would be Shell's first and last deepwater project.

This was the situation facing Rich Pattarozzi in 1991 when he was promoted to general manager of E&P for deepwater. Searching for ways to make the project profitable, he and his managers realized they needed to improve well productivity. Auger wells were forecast to produce 3,000 to 4,000 barrels per day, similar to what Bullwinkle wells were producing. "This isn't going to work," Pattarozzi told his people. "The economics aren't there. We've got to do better than that." Shell geologists and reservoir engineers had a possible answer. They had done extensive outcrop work around the world, looking at rocks they felt were comparable to those found in the turbidite reservoirs of deepwater Gulf of Mexico. They discovered that turbidite sands might behave differently from the deltaic sands found in shelf reservoirs. Turbidity currents had deposited massive amounts of sand, making the reservoirs potentially much larger. These sands also tended to be less faulted and unusually porous, thanks to the sifting of the sands as turbidity currents traveled long distances. Finally, deepwater sands were more tightly sealed below layers of dense mud and therefore highly pressured. "It's like putting a brick on a balloon full of water," explained Shell exploration geologist Alan Kornacki. "The water wants to burst upward. If you put a straw into that balloon, the weight of the brick will push the water out at a tremendous rate."[42]

This was the theory anyway. But where was the proof? Shell's deepwater engineers thought the wells at Bullwinkle might offer some clues. Bullwinkle was on the outer-continental slope, just off the shelf, and had turbidite reservoirs similar to those at Auger. In the spring of 1992, Pattarozzi's engineers approached the production superintendent at Bullwinkle to see if he would test producing a well at higher rates. It took some cajoling. Bullwinkle's wells were producing at about 3,500 barrels per day, which was considered highly productive. On the shelf, a good well produced 1,000 b/d and an excellent well flowed at 2,000 b/d. Understandably, the superintendent was not thrilled about increasing the choke on one of his wells and possibly damaging it. But Pattarozzi's engineers convinced him that the draw-down pressure in the well would not increase enough to do any damage. They were right. The first well they tested increased its production from 3,500 barrels per day to 7,000 barrels per day with hardly any change in bottom-hole, draw-down pressure.

It was a divine breakthrough. If this was any indication of the general productivity of the wells at Auger and other prospects Shell had under lease, it not only radically changed the cost structure of Auger but gave the company a whole new economic model with which to approach deepwater projects. The results were so encouraging that Shell Offshore mangers were able to obtain relatively swift approval from Shell Oil's board of directors to move ahead with developing another TLP on the company's Mars prospect (see below), even though Auger had not yet been installed. After that, said Bob Howard, president of Shell Offshore, Mars was a "pretty easy sell."[43]

In the fall of 1993, the various components of the Auger TLP were finally brought together. But it was a trying ordeal for the managers of the project. McDermott finished fabricating the deck just in time for it to be trapped in Amelia, Louisiana, when the great Mississippi River flood silted up the Atchafalaya River. After six weeks of dredging by the Army Corps of Engineers, the deck was finally freed to be transported through the channel to the Gulf. The delay, however, forced Shell to gamble on the weather and schedule the mating of the hull and deck during hurricane season. In October, McDermott performed the tricky mating operation, which involved submerging the hull, bringing the barge carrying the giant deck in between the four legs, and then gently deballasting the hull up to the deck. "When the process was explained to me," remembered Pattarozzi, "I was so scared I could hardly stand it. The things that could go wrong in that whole mating process were just phenomenal." Pattarozzi was so edgy as he awaited the outcome of the mating operation, his wife described him as a "cat on a hot tin roof." His sense of relief was just as palpable when he received the phone call informing him that it had been successfully accomplished.[44]

In December, the unit was towed to site. Under tow, Auger rose twenty-six stories above the water line, a monstrous structure that looked like nothing before seen on earth. "I was reminded of the spaceship in the movie *Close Encounters of the Third Kind*," said Senior Construction Superintendent Pete Arnold.[45] Over the winter, work crews connected the mooring system and tendons, fortunately with very few weather delays. Many observers thought that such a precise operation could not be done during that time of year, but the Auger crew pulled it off with careful patience and planning. On February 12, McDermott officially turned the platform over to Shell.

In April, Auger's first well came in. Shell managers had expected 8,000 b/d or more, but the well flowed at only 2,200 b/d. Something was wrong. Production engineers figured it was probably a downhole or a mechanical problem. While they tried to sort it out, the second well came in at only 1,500 b/d. "Now, things were getting fairly tense," remembered Pattarozzi. "Here we were already building Mars. We've got two wells which should be producing 15,000 b/d, and they're producing 4,000 b/d. Did we just blow a billion dollars?" It was not yet time to panic, however. No inquisition came down from head office. But as engineers from New Orleans and the lab people at Bellaire joined forces and started looking at the problem, tension mounted. "I couldn't even find people because they kept hiding," said Pattarozzi. "They didn't want to talk to me."[46]

Finally, in June, the engineers thought they identified the problem. The fluid used to complete the wells, they surmised, had combined with the gas emitted during the completion to form calcium carbonate, a precipitate which blocked the formation. If this were the case, the simple solution would be to dissolve the precipitate with acid. Crews pumped hydrophloric acid down one of the wells. It hit the formation and broke through, opening up the well. Within a couple weeks, the two wells were each producing over 10,000 b/d. At that rate, deepwater projects would be very profitable, even with oil under $20 a barrel. Now, Rich Pattarozzi was a very popular man. His platform at Auger had opened up a whole new deepwater frontier, and Shell had a string of deepwater discoveries waiting to be developed.

News soon trickled out about the productivity of the Auger wells, setting the industry abuzz about deepwater. Derided by oilmen as the "dead sea" after a twenty-year decline in overall production, the Gulf of Mexico suddenly became one of the hottest areas in the world. After the news from Auger, companies scrambled for deepwater leases. The number of lease sale bidders in the Gulf of Mexico went from less than ten companies prior to 1994 to anywhere from thirty to forty different companies afterward. Of course, they were ten years behind Shell Oil, which had leased many of the most promising prospects in the early area-wide sales of the mid-1980s. After its string of discoveries in the late 1980s, Shell Offshore quietly accumulated more deepwater leases while bemused skeptics insisted that deepwater would never pay off. By 1995, Shell controlled one-third of all Gulf of Mexico leases in depths greater than fifteen hundred feet, making it the envy of the entire U.S. oil industry.[47]

The development of Auger gave the industry more reason to be envious. Although Shell Oil's management had bet right on Auger, they nevertheless greatly *under*estimated the rate at which the field would flow. Shell designed the Auger TLP to handle 42,000 barrels of oil (and 100 million cf/d of gas) a day from twenty-four wells, but by July the first three wells were already producing nearly 30,000 b/d. By the end of 1994, a short-term debottlenecking operation had increased total capacity to 60,000 barrels a day, yet the wells were still capable of generating more. Several more stages of debottlenecking in the mid- to late 1990s eventually raised the TLP's capacity to 105,000 b/d of oil and 420 million cf/d of gas.[48]

In more ways than one, Auger laid the groundwork for what followed in deepwater Gulf of Mexico. An early recognition of this was the American Society for Civil Engineer's decision to give Auger the 1995 award of Outstanding Civil Engineering Achievement, honoring Shell Oil for an unprecedented third time. "What Shell has done out there is truly extraordinary," said John Kingston, editor in chief of *Platt's Oilgram News*. "They basically opened up a new vista."[49]

Subsea Comes of Age

Not only did Auger pioneer a new frontier for deepwater platforms in the Gulf of Mexico, it heralded subsea technology's coming of age. In a subsea completion, the wellhead is located on the ocean floor rather than on a production platform at the surface, and production is pipelined to a nearby platform. First developed by Shell in the early 1960s (see chapter 3), subsea wells could never quite stay commercially competitive with fixed platforms in the Gulf of Mexico, though they were used with increasing frequency to develop high-volume production in the North Sea. With the discovery of highly productive fields in the deepwater, however, subsea strategies began to make economic sense for the Gulf as well, especially for gas fields and smaller fields that did not warrant a platform. "Auger triggered our thrust into tension leg platforms, which were really a hybrid between subsea and structural technology," explained Shell Facilities Engineering Advisor Bill Petersen. "Most of our TLP studies had included a subsea component, either as part of an early production system or as a remote subsea development producing to the floating system. Auger was the first time that we actually brought these two disciplines together."[50]

Shell's first deepwater subsea project actually took place at prospect Tahoe, a gas field in 1,500 feet of water in the Viosca Knoll area. First drilled in 1989, the Tahoe project officially went forward in the fall of 1992. In January 1994, using a refurbished subsea tree and a control system that had been proven in the North Sea, Shell crews successfully brought Tahoe on-stream producing roughly 30 million cf/d of gas. The importance of Tahoe was not in the gas recovered but in getting Shell established in deepwater subsea operations. Said Project Engineer Ken Orr, "we looked at Tahoe as a stepping stone to bigger and better projects." Indeed, it was an important first step toward installing a larger and deeper system in the Popeye gas field in 2,000 feet of water in the Green Canyon area. In early 1996, two subsea wells at Popeye began flowing, ultimately reaching a production rate of 120 million cf/d of gas. With Popeye, Shell achieved a number of technological firsts, including the Gulf's first diverless, cluster production system, the first guideline-less, 10,000 psi subsea trees, and the Gulf's first concentric completion/workover riser, a design that extended riser water-depth capabilities beyond 3,500 feet. Using a sleeve that pivoted pipe into place without divers or ROVs, Shell eventually connected the two wells to a central manifold, from which production was pipelined 24 miles to the Cougar platform.[51]

This was the longest tieback from a subsea well in the Gulf of Mexico. By the mid-1990s, a host of newly developed technologies—new riser technology, horizontal trees, new methods for preventing oil from cooling and clogging in deepwater pipelines, and more capable umbilicals for hydraulic and electronic control systems—offered improved abilities for installing production equipment from the surface and for making long tiebacks between subsea wells and production platforms. In 1996, Shell pushed the boundaries of subsea technology by installing a cluster system modeled on Popeye at the Mensa gas field in a record-setting depth of 5,400 feet of water. The sixty-eight-mile tieback from the manifold to the West Delta Block 143 platform also set a world record. In 1998, the Mensa system, consisting of three wells linked to the manifold five miles away, reached a production rate of 280 million cf/d of gas, boosting by almost 25 percent the company's overall natural gas production in the Gulf.[52] Mensa served notice that subsea technology would be integral to future deepwater Gulf projects.

Indeed, after Mensa, the ratio of subsea to surface projects in the Gulf of Mexico would grow, bringing in oil as well as gas. In 1997, Shell Oil, in equal partnership with Marathon Oil and British Petroleum (operator), started production of oil and gas from subsea facilities in the Troika field, located in 2,700 feet of water in Green Canyon, and began pipelining both back to the Bullwinkle platform. In preparation for Troika production, Shell increased the processing capacity of Bullwinkle from 55,000 b/d to 200,000 b/d.[53] The development of satellite subsea wells around platform "hubs," such as Bullwinkle, extended the life of existing infrastructure and increased the cost-effectiveness and flexibility of deepwater production.

The hub concept came to play a central role in Shell's development strategy for deepwater TLPs. Discoveries at Tahoe, Popeye, Mensa, and Troika led Shell to marry subsea technology to the TLP concept at Auger. In the late 1990s, the company developed two fields adjacent to Auger—Macaroni and Cardamom—with subsea wells and tied them back to the TLP. Much of the new capacity at Auger was added to take production from these fields, which would not have been commercial on their own. The hub strategy further emboldened Shell's approach to its deepwater play. "It's meant that we continued our forward-looking exploratory tactics, explained Dave Lawrence, vice president of exploration and development for Shell Exploration & Production Co. (SEPCo), the Royal Dutch/Shell Group's worldwide E&P organization. "It's meant that we built our acreage position . . . and have continued to look for satellite opportunities around the Augers of the deepwater Gulf of Mexico."[54] In the early 2000s, Shell tied other satellite subsea fields, such as Serrano and Oregano, into Auger and made the platform available to produce additional fields as they are discovered and as production from the Auger field itself declines.

Mars: A New Paradigm

One of the most important achievements of Auger was in providing lessons for improving future TLP designs and operations, Mars in particular. Mars started out as a risky venture for Shell. In the spring 1985 lease sales, the company acquired acreage in the Mississippi Canyon area covering the Mars lead—it was too speculative even to be considered a prospect. In fact, managers

added Mars on at the last minute in the meeting to decide the bidding strategy. Shell had access to a limited amount of data—some proprietary seismic and some well information from two older wells drilled in the basin by Getty and ARCO.[55] The first round of new two-dimensional seismic logs, shot in 1986 after the acreage was acquired, did not impress Shell geologists. They estimated the chance of success at only 10 percent, because the prospect was way down the flank of a shallow salt dome, where there seemed to be less probability of discovering hydrocarbons, and because the pay sands were thought to be thin. "We didn't see enough potential there based on early two-dimensional seismic logs," remembered Jim Funk, division exploration manager at the time. "We saw some hydrocarbons there, but we didn't have a good enough picture yet."[56]

A second round of 3-D seismic, shot in 1988 after Shell leased further acreage in the area, gave Funk's team a brighter outlook. It provided a framework with which to understand the depositional pattern of the turbidite sands in the basin. Shell geologists came up with a model for analyzing the reservoir that indicated the possibility of many deeper pay sands with strong stratigraphic trapping components. The few wells that had been drilled out in deepwater, such as those by Getty and Arco, had been drilled on the tops of structural crests, where the sands were poor. But Shell's model showed that the big-pay sands were down off the crests in the basin. "It was initially hard to grasp that this was as big as the seismic suggested," said Funk. "It looked like it could be a one-of-a-kind field."[57]

Yet the economics remained dubious in the years before Auger changed the nature of the game. Shell's development teams initially assumed the wells at Mars would produce 3,000 b/d. At that rate, the project would need many wells, almost too many to be profitable, especially during a period when collapsed oil prices severely strained Shell Oil's bottom line. In response to orders from head office in 1988 to slash his budget, and with two drilling vessels already under contract, deepwater exploration manager Jim McClimans sought a partner to help spread the risks and costs if the project went forward. He led a team of Shell negotiators that brought in British Petroleum with a 28.5 percent interest in Mars, a tactical decision that would later come back to haunt Shell. In 1989, Sonat's *Discoverer Seven Seas* drilled the first exploratory well and struck oil bearing zones at 14,500–18,000 feet. Appraisal drilling over the next two years identified more than twenty-four individual pay sands within fourteen intervals, confirming the validity of the depositional model developed by Shell geo-scientists. "The seismic was just working perfectly," marveled Bill Broman, general manager of exploration at the time.[58]

The implications of this discovery were enormous, in two respects. First, Mars turned out to be a world-class field. Although some of the zones contained heavier oil and higher sulfur than Auger, the estimated reserves of Mars were 500 million barrels, the largest discovered in the Gulf of Mexico in twenty-five years. Second, Shell Oil's new model for analyzing deepwater depositional patterns and classifying turbidite sands proved to be a significant geologic breakthrough, not only for the deepwater Gulf of Mexico but for other areas of the world with similar geology. "It doesn't take a mental giant to go from that point to thinking through several spots in the world where other situations ought to be there," said Broman. The Mars discovery led Broman to expect to find three to five billion barrels of oil equivalent on Shell Oil's acreage in the deepwater Gulf.[59]

Despite Mars' handsome potential, the economics were still questionable. Recounted Dan Godfrey, Mars project manager: "The general manager told me, 'Dan, you're going to have to find a way to improve the earning power by about 3 percent if you want us to carry this forward to the board. We will have to find ways of doing it differently (from Auger)." This was in 1991, when cost overruns and delays at Auger had dimmed the outlook for deepwater development. Godfrey and his Mars team realized they would have to take a new approach to the design and construction of a TLP for Mars. They considered a floating production system with subsea wells as an alternative to a TLP. The overall economics of the two alternatives were not all that different. But in the final analysis, as Pattarozzi put it, "all else being equal, if I can have the trees at the surface versus on the ocean floor, I am going to go with the trees at the surface."[60] Still, things would have to be done differently than at Auger. The Mars team searched for a new method for managing the project that would bring about more collaboration with contractors and more shared responsibilities between the two partners, Shell and BP, and between the partners and the contractors.

They found a solution in forming "alliances." In 1992, Shell Oil and BP formed a joint project team for managing the project. In the early stages, both companies found they were doing a lot of the same work independently. Eliminating duplication would reduce costs. The alliance also broke new ground in the industry by establishing an arrangement for sharing technology and patents created on the project, although Shell Oil ended up giving away a lot more than BP, which had no experience in deepwater. Indeed, this was an unprecedented move for Shell Oil, which for the most part had always gone at it alone offshore. But the costs of deepwater were too staggering to continue this way. Although wary of each other at first, Shell Oil and BP personnel bought into the team concept. Said Shell Construction Superintendent John Haney: "In most cases we've taken the best of their practices and the best of our practices and molded them into a winning combination."[61] Team members from the two companies were even encouraged to socialize, spending Christmas together one year.[62]

Shell and BP carried the alliance concept over to their relationship with contractors. Rather than drawing up all the specifications for various parts of the TLP and then asking for lump sum bids, the Shell-BP project team proposed forming alliances with contractors who would share the risks and benefits and as well as participate in critical decisions. This was a radical idea. It replaced the traditional adversarial relationship of client and contractor with a new spirit of collaboration that Shell was already incorporating internally through its Continuous Improvement Initiative. "When we had our first workshop with contractors, the facilitator kind of paused and asked for participation—you could hear a pin drop in the room," remembered Godfrey.[63] Nobody could believe Shell was serious about it.

The key to such a relationship was instilling flexibility in the contract. If, for example, construction drawings differed from preliminary designs or if contractors encountered unexpected cost increases on their end, then the contract could be adjusted. With this mechanism in place and man-hour targets established, Shell and the contractors would agree to share cost overruns or savings. For every dollar a contractor went over budget, in other words, he would split the cost with Shell. For every dollar saved, the contractor received half the benefit. But the alliance concept was not only about costs and savings; it was about giving contractors a greater voice in the project. One lesson learned on Auger was that Shell engineers may be very good at designing something, but not necessarily expert at building it. The alliance concept would bring the contractors in on the process as early as possible, allowing them to help make design and fabrication more complementary.[64]

Another advantage of this approach was that it promised to reduce considerably the so-called "cycle time"—the period covering design, bidding, and contracting. Instead of finishing the design before moving on to fabrication, these two stages would overlap in order to slash time off the schedule. In a deepwater development, where the time from discovery to first production averaged about ten years, reducing any amount of time was invaluable. The alliance method of contracting took an estimated six to nine months out of the cycle.[65] On a platform producing 50,000 to 100,000 b/d, the time-value of the money made at the beginning rather than at the end of the platform's life was quite significant. This was the kind of improvement to earning power that Shell's top management wanted to see. In the summer of 1993, Shell Oil's board gave final approval to the Mars project.

In October 1993, Shell-BP awarded all its contracts based on the alliance concept. Initially, contractors hesitated to get involved, but they soon warmed to the new and innovative method of contracting, which proved to be advantageous to all parties. As on the Auger project, Belleli won the right to build the hull and McDermott became general contractor for the topside modules. Aker Gulf Marine fabricated the piles and tendons and integrated the hull and modules. In late 1993, construction began on the hull in Taranto, Italy, and on the topside modules in Morgan City, Louisiana.

Having some of the same major contractors from Auger on the Mars project allowed all parties to draw on past experience, good and bad, to make improvements on the design and construction of the TLP. The deck modules, for instance, were lifted individually by crane onto the hull rather than mated as one piece to the hull in the water. "We never did that mating again," said Pattarozzi. Although smaller and lighter than Auger, Mars carried twice the production capacity, originally designed for 100,000 b/d of oil and 110 million cf/d of gas. Saving weight was important to the builders—each extra pound added another five dollars to the cost of the structure.[66] With assistance from their contractors, Shell engineers designed Mars to be "fit-for-purpose." They factored little extra capacity into the design and maximized every square inch of space. The deck provided room for twenty-four well slots, separation and treatment facilities, and accommodations for 106 people. As fabrication was proceeding in 1995, however, pre-installation drilling by the Sonat semi-submersible *George Richardson* discovered that Mars wells were much more prolific than the expected 11,000 b/d, thanks to better well completion technology. Pleased with these results, the Mars team nevertheless scrambled to do a debottlenecking study while the TLP was already under construction and introduced several innovative technologies to increase design production rates to 140,000 b/d of oil and 170 million cf/d of gas (further tweaking after start-up boosted those numbers even higher). The flexibility of the alliance system helped pull off this crucial modification with little delay.[67]

The biggest difference between Mars and Auger was beneath the surface of the water. Shell spaced the twenty-four wells on Mars in a close, rectangular pattern, as opposed to the widely spaced oval pattern on Auger. On the first project, the engineers were not sure how close wells could be spaced working in 2,850 feet of water. So they opted for a conservative design, providing ample room to work between wells. Drilling experience at Auger, however, proved that spacing could be done more conventionally. Closer well spacing at Mars allowed Shell to use conventional guidelines instead of an expensive lateral mooring system like the one used at Auger, which was needed to move the TLP over the wells. With closer wells, moreover, engineers were able to design the structure without Auger's huge, expensive moon pool.

"Mars knocked down so many myths about deep water," said Steve Peacock, exploration manager for British Petroleum. "It's amazing how the industry gets caught, not so much by what it doesn't know, but by what it thinks it knows that just ain't so."[68]

On April 10, 1996, Aker Marine towed out the completed Mars TLP from its Ingleside, Texas, facilities across the Texas and Louisiana Gulf to its home in the Mississippi Canyon. Everyone credited the alliance approach for bringing the Mars TLP in under budget and ahead of schedule. With an initial production capacity more than double that of Auger, the Mars project cost a total of $1 billion, about $200 million less; and the time from discovery to first production was an impressive seven years. Installed at a water depth of 2,940 feet, Mars surpassed its sibling Auger in setting yet another world depth record for Shell Oil. In early July, Mars's first well came in producing as much as 15,000 b/d, the greatest sustained daily flow rate yet reported in the Gulf. McDermott laid a new, dedicated, $135 million deepwater pipeline to transport Mars oil forty miles to a hub platform at West Delta Block 143, the same platform handling Mensa natural gas. In September, Shell put into service the $62 million Mississippi Canyon Gathering System to move gas from Mars and two other deepwater projects, Mensa and Ursa (see below). Like Auger, Mars in the late 1990s became a hub for subsea tiebacks from adjacent fields. In fact, Mars was the first TLP to host a subsea tieback, installed in 1996. In 2000, oil and gas from another subsea field, Europa, was linked up to Mars as the platform's capacity was extended to 200,000 b/d of oil and 185 million cf/d of gas.[69]

While Auger opened a whole new vista for deepwater Gulf of Mexico, Mars established a new paradigm for working in those depths and demonstrated Shell's signature facility for technological and organizational innovation. It represented a culmination of nearly a half-century of effort by Shell Oil's offshore pioneers to extend oil exploration and production into a frontier that was often viewed as unconquerable. The discovery and analysis of the field resulted from cutting-edge geophysics and geology. The TLP was an engineering marvel. And the organization of the project changed the industry's thinking about how to manage deepwater projects.

This great advance, however, came at a price for Shell Oil. To cut costs and risk, the company had sacrificed some of its hard-earned competitive advantage in deepwater exploration and production technology by forming a partnership with British Petro-

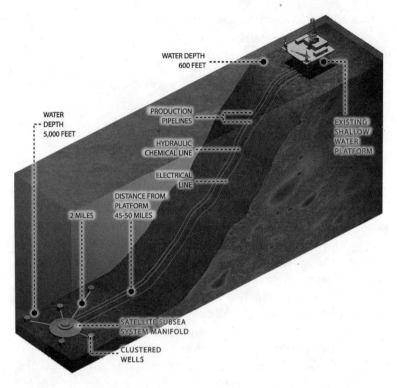

31
Schematic diagram
of a subsea well
system and plat-
form tieback.

WATER DEPTH
600 FEET

WATER
DEPTH
5,000 FEET

PRODUCTION
PIPELINES

EXISTING
SHALLOW
WATER
PLATFORM

HYDRAULIC
CHEMICAL LINE

ELECTRICAL
LINE

DISTANCE FROM
PLATFORM
2 MILES   45-50 MILES

SATELLITE SUBSEA
SYSTEM MANIFOLD

CLUSTERED
WELLS

leum, a longtime and bitter rival of Royal Dutch/Shell around
the world. The deal let BP in on the deepwater Gulf of Mexico
business, giving its managers and engineers a close-up view of all
aspects of Shell Oil's operations, from its exploration and reservoir evaluation models to its drilling and production techniques.
With nothing in deepwater, BP "went to school" and subsequently
staked out a big position in the deepwater Gulf. By 2004, the British oil giant was the largest leaseholder and, after the Shell E&P
Company, the second-largest producer in deepwater. The partnership with BP was understandable given the financial constraints
on management in 1988, and Jim McClimans cut a good financial
deal for Shell. But in crystal-clear hindsight, it was a major strategic blunder.

For many years, however, Shell Oil remained ahead of the
game. Journalists who wrote awestruck feature stories on Auger and Mars had a hard time keeping up with Shell Offshore
afterwards, as the company continued to perform encores, forg-
ing ahead in deepwater with one major project after another. Although there were some deviations from the initial grand plan
calling for a major deepwater project every two years, the over-

all timing and capital expenditures followed the original strategy quite closely. Fortunately for Shell, the only major deviation was in well production rates, which far outstripped the original forecast.

In 1997, the TLP Ram/Powell went on-stream in a record-setting depth of 3,200 feet in the Viosca Knoll area about eighty miles south of Mobile, Alabama. Shell acquired its lease on the Powell prospect in the April 1984 area-wide sale, while Amoco and Exxon held adjoining leases on what they code-named the Ram prospect. When Shell drilled the discovery well on Powell in March 1985, the logical decision was to form a joint venture in which all three companies agreed to pool their resources rather than develop the field on their own. The Ram/Powell joint venture, in which Shell Oil had a 38 percent operating interest, was one of the company's first deepwater projects on the drawing board, but it was also in significantly deeper water than Auger and with less reserve potential than Mars, so the other projects received the inside track.[70]

Ram/Powell greatly benefited from Shell's experience and achievements with Mars. Managers often referred to Ram/Powell as a "clone" of Mars. "This is the first project I've ever been involved with in which we've tried to duplicate the work we've done before as much as possible," said Structural Design Team Leader Bill Luyties.[71] Knowledge and experience were transferred directly from one project to the other, and Ram/Powell adopted Mars' alliance contracting arrangement and overlapping design-build approach and employed the same major contractors and many of the same crews. But Ram/Powell was unique as well. Shell Oil worked more closely with its partners and its contractors than ever before, especially in the design phase. This was partly born out of necessity. Working on Mars and a host of other deepwater projects, Shell simply did not have the manpower to tackle Ram/Powell on its own. Ram/Powell also incorporated new innovations, such as large-bore wells and new horizontal drilling techniques, that reduced the number of wells needed on the platform. Costing some $50 million less than Mars, Shell's third TLP in the Gulf shaped up as another stunning achievement.

Three other TLPs rounded out Shell Oil's deepwater hub system as the century came to an end. In March 1999, production came on-stream from a $1.5 billion TLP in the Mississippi Canyon called Ursa, a joint venture between Shell (operator and holder of a 45 percent interest), BP (23 percent), Conoco, and Exxon (each with 16 percent). At 63,300 tons, the Ursa platform was the largest structure in the Gulf, nearly double the weight of Mars. Its giant size was dictated by the need to accommodate astounding initial well-production rates of up to 30,000 b/d and by the fact that it was installed in nearly 4,000 feet of water, shattering Ram/Powell's record of the deepest drilling and production platform in the Gulf. On September 8, 1999, the A-7 well at Ursa broke previous Gulf of Mexico records with a daily production rate of 39,317 barrels of oil and 60.67 million cubic feet of gas, or a total of nearly 50,000 barrels of crude oil equivalent per day. Ultimate recovery for the field was estimated at about 400 million barrels of oil and gas equivalent, the second largest deepwater field in the Gulf after Mars. In late 2003, Shell began production in the nearby Princess field (estimated ultimate recovery of 150 million BOE) with a three-well, expandable subsea system tied back to the Ursa platform. In the process, it set a drilling depth record of 33,200 feet, including the water column of 4,695 feet. Shell Offshore also has a 25 percent interest in Marlin, a TLP operated by BP-Amoco in 3,200 feet of water, and a 100 percent interest in Brutus, the latest TLP, whose first production came in August 2001 from a 200-million-barrel field at 3,000-foot water depths in Green Canyon.[72]

Shell's deepwater activities in the 1990s included one other important and innovative dimension: the marketing and transportation of natural gas, an increasingly important resource in the United States. In 1995, to make the most of all the gas coming in from deepwater offshore, plus the gas being produced on the shelf, Shell formed Coral Energy, L.P., a joint venture with Tejas Gas Corporation, the largest intrastate gas pipeline company in the country, to market and trade natural gas along the Gulf Coast and in other markets. A year later, Shell expanded its presence in natural gas by forming Shell Midstream Enterprises to provide third-party production handling and gas processing among other services. Through this unit, Shell built its infrastructure of natural gas pipelines in the Gulf of Mexico, such as the Mississippi Canyon Gathering system installed to serve Mars and Ursa, and two other major gas pipelines, centered around Bullwinkle and Auger. Shell expanded further into downstream gas in 1997 by acquiring Tejas Gas. Moving down the "value chain" in natural gas was an entirely new endeavor for Shell E&P and one that was not as immediately rewarding as originally planned, due to rising natural gas prices. But it demonstrated the new opportunities opened up by deepwater production.[73]

Shell's pioneering deepwater activities continued into the new century with the discovery and development of six widely spaced oil and gas fields in ultra-deepwater. In September 2000, the company announced plans to develop its Na Kika project. This was a subsea development of five independent fields (Kepler, Ariel, Fourier, Herschel, and East Anstey around Mississippi Canyon 474a), four predominantly oil and one natural gas, tied back to a centrally located semi-submersible floating production facility, an industry first for deepwater Gulf of Mexico. A 50–50 joint venture between Shell and BP, with Shell operating pre-production and BP post-production, Na Kika will produce an estimated 300 million barrels of oil equivalent (100,000 b/d of oil and 325 million cf/d of gas) in 5,800 to 7,600 feet of water, the deepest for any installation of its type in the world. A sixth field, 100 percent owned by Shell Exploration & Production Company, was tied back in 2005. With total capital expenditures of $1.26 billion, the unit producing cost of Na Kika will be 50 percent of Auger, and Shell estimates that future floating production systems will have a unit cost of one-half that of Na Kika.[74] This project established many drilling and pipelining firsts, in addition to being the deepest permanently moored floating production system, and won the prestigious Distinguished Achievement Award from the Offshore Technology Conference in 2004. "Developing the Na Kika project in these record-setting water depths is one of the most substantial steps forward in the application of deepwater development technology since SEPCo's [Shell Exploration and Production Company] Auger tension leg platform was installed in 1993," said Walter van de Vijver, SEPCo's president at the beginning of the project.[75]

Shell Oil's deepwater adventure that began in the 1980s was the crowning achievement of a half-century of work and innovation in the Gulf of Mexico. Beginning in the 1950s, visionary Shell executives such as Ned Clark and Mac McAdams recognized that the only way for Shell Oil to survive in the United States as an E&P company was to push aggressively offshore in the Gulf of Mexico. Early successes were not forgotten as each generation of leaders expanded on Shell's technological ability to make offshore a paying proposition. Shell was always one or two steps ahead of the competition, in production as well as exploration technology, and more than once almost single-handedly revived the fortunes of offshore Gulf of Mexico when most people had written it off as a dead sea. In the process, Shell Oil E&P helped postpone the day of reckoning for U.S. oil production on its inevitable decline. As of 2004, the 300,000 barrels a day of oil and 1.2 billion cubic feet of natural gas per day produced by Shell in the Gulf accounted for 80 percent of the company's hydrocarbon production in the United States. With an interest in more than seven hundred primary leases, most in deepwater, and other projects coming on-stream, that percentage is rising.

Looking back in time from the perspective of Auger and Mars to Shell's first bold steps into thirty-foot waters of the Gulf with the submersible barge *Mr. Charlie*, or even to the pathbreaking voyage of the *Bluewater 1* semi-submersible out to three hundred feet, the technological achievements made by Shell's engineers and geoscientists were collectively unparalleled. It is especially obvious now, in view of developments such as Na Kika, that Shell's pioneering innovations with the *Bluewater 1* and the RUDAC and MOBOT subsea systems back in the late 1950s laid the foundation for deepwater production today. Shell Oil names dominate oil company inductees into the Offshore Energy Center's Offshore Hall of Fame. The surviving pioneers looked upon the succession of deepwater milestones in the 1990s and 2000s with well-deserved pride in knowing that their contributions helped open offshore oil and gas frontiers, in the United States and elsewhere, that are becoming increasingly vital to world energy needs.

As visionary as they were, however, they could not have foreseen how far Shell and the industry would travel. Said Rich Pattarozzi: "If we had been talking thirty years ago, when we were in 200 feet of water, and you had said we would be drilling in 10,000 feet of water and producing in 5,000 feet, I would have replied, 'there is no way in hell.'"[76]

Epilogue: The Globalization of Shell Oil

Nineteen ninety-four was a pivotal year for Shell Oil. The company's deepwater play in the Gulf of Mexico blossomed with the installation of Auger and the spudding of its prolific wells. It appeared that the Gulf, once again, had made the company. The next year, cost-cutting and corporate reforms helped earnings and profits rebound. "Today, no one worries about Shell fading away," wrote *Business Week* in May 1995. "Thanks to a string of discoveries of huge reserves in the gulf's deep waters and technological advances that have made it easier and cheaper to get that oil out, Shell's once bleak fortunes have been revived."[1]

Even with this revival, however, Shell Oil would never return to the pre-1990s status quo. Ironically, just when its deference to the offshore imperative was beginning to pay off richly, the company was being restructured and reincorporated into the larger worldwide Royal Dutch/Shell. Shell Oil Company's loss of financial independence through the 1985 buyout of minority shareholders had signaled the beginning of the end for the old "American" Shell. By the end of the 1990s, Royal Dutch/Shell had asserted complete organizational control over American operations, in violation of the agreement made with John Bookout at the time of the buyout.

Two developments set this in motion. First, a harsh, competitive business environment imposed difficulties, despite the bright future for deepwater production. Volatile trends in the supply and demand for oil and gas, increasing societal and environmental expectations, mounting frustrations with persistently soft oil prices, and the continuing underperformance of Shell Oil's heavy oil properties and downstream assets forced Shell leaders to look for new ways to reshape the company. Shell Oil management tried both traditional and creative ways of adjusting, slashing jobs and adopting a new management structure, but returns on invested capital, though improved, still did not reach acceptable levels. And Shell's big finds in deepwater Gulf of Mexico still did not add enough reserves to replace declines in production, as the United States continued its long trend toward depletion.

The structure of the world oil industry, and the position of the United States within it, had changed. Political liberalization and deregulation on a global scale, a merger movement in the world oil industry, and the emergence of increasingly integrated global markets led to a new "globalization" strategy in which Royal Dutch/Shell reorganized businesses along functional rather than geographical lines. Beginning in 1997, the Group began weaving parts of its U.S. operations more deeply into its other international businesses. For Shell Oil, this meant another round of job cuts and the virtual elimination of its long-cherished managerial autonomy. Step by step, first in chemicals, then in services, then in exploration and production, the Group reduced Shell Oil's authority to act on its own. Managers in Houston increasingly found that their immediate superiors resided in Europe, bringing the history of Shell in the United States full circle. At the beginning of the new century, almost one hundred years from when Shell first entered the country, operations once again consisted of a collection of companies or businesses directly managed and coordinated from abroad.

Big-Time Trouble

The slump of the 1980s hurt Shell Oil worse than many of its rivals. While big companies concentrated resources on cheaper oil developments overseas, Shell was committed to expensive projects in the costly oil-producing environment of the United States. It had risked a lot of money in offshore Alaska and lost. Heavy spending on deepwater Gulf of Mexico would eventually pay off but not for years down the road. The Belridge acquisition, which had won wide praise for Shell, came back to haunt the company. The fact that enhanced recovery methods in the field were a technical success provided little consolation when the heavy oil produced there only fetched twelve dollars a barrel. Shell's great strength, its technical proficiency in exploration and production, became a handicap under the new price regime. The engineers and geologists who ran the company in the late 1980s clung too long to the belief that the world had not radically changed. Shell did not follow the rest of the industry with layoffs during the 1986 collapse. The price projections coming out of E&P Economics, as late as 1988, were still overly optimistic. Made not disingenuously but perhaps out of intellectual and bureaucratic inertia, such projections allowed E&P to continue doing what it did best.[2]

Shell's costs had gone up with the soaring crude prices, but they did not follow prices down as easily. In 1990, E&P managers were shocked to learn that Shell had the highest cost per barrel of oil and gas produced of any of its competitors. Return on investment that year was 5.5 percent, while the other major companies averaged 12 percent. "We put a lot of money into research," admitted Jack Little, executive vice president for E&P during the mid-1990s. "We had a large staff. We threw technology at everything we did. And we were not really focusing on the bottom line like we should have been."[3] The major underlying problem, as it had been since the 1960s, was that the company could not find enough crude to offset declining production and reserves. Between 1986 and 1991, its daily oil output declined by 160,000 b/d. The day of reckoning finally came in the summer of 1991, when Shell Oil reported its first quarterly loss since the Great Depression—$68 million. For the year, the company squeaked by with a paltry $20 million in net income, down from over $1 billion the year before. "This was big-time trouble," said Phil Carroll, then executive vice president for administration, describing the thinking of senior management at the time. "This cannot go on."[4]

After enjoying distinction as the U.S. oil industry's most admired and successful company, Shell Oil had taken a hard fall. "Now people are saying, 'Hey the emperor isn't wearing any clothes,' and the poor guy who's sitting on the throne is Frank Richardson," remarked one oil analyst in 1991, referring to the man who had replaced John Bookout as Shell Oil president three years earlier. "My heart goes out to him."[5]

Indeed, Richardson's timing could not have been worse. Things started to go badly, it seemed, the moment he took over, starting in 1988 with an oil spill at Shell's Martinez refinery near San Francisco, a pipeline rupture in Missouri, and the destruction of a catalytic cracker in a tragic explosion at the Norco refinery outside New Orleans. The following year, six workers were killed on a small platform in the East Bay, one of the worst accidents in the history of offshore operations at Shell. Events in the industry compounded matters. In 1988, 167 workers died in a fire on Occidental Petroleum's Piper Alpha platform in the North Sea, and the following year the *Exxon Valdez* tanker struck Bligh Reef in Alaska's Prince William Sound, spilling eleven million gallons of crude oil. Both events focused negative publicity on the entire U.S. oil industry, not just Occidental and Exxon alone, and added to the state of siege felt by Shell managers on environment and safety matters. Moreover, a boycott of Shell over the Group's involvement in apartheid South Africa was in full swing. Fortunately for the company, a proposed amendment to economic sanctions legislation prohibiting any company operating in South Africa from bidding on federal oil and gas leases in the United States never made it out of Senate Foreign Relations Committee.[6]

As another in a long line of technical men to lead the company, Frank Richardson had to fight many fires at once. He had started with Shell as an engineer in Midland, Texas, rising up through various E&P assignments to become general manager of production for the Western E&P Region in 1975. Two years later he was assigned to the strategically important position of Shell Oil Liaison to the Group, returning in 1978 to become vice president for Corporate Planning, a department that had accumulated tremendous power in the company under John Bookout. More reserved and measured than his assertive predecessor, Richardson became known for his "Mr. Fix-It" style of management.

Once environment and safety questions had been addressed through clean-ups, remediation, refinery rebuilding and upgrading, and a renewed commitment to safe operating practices, Richardson took aggressive steps to cut costs. The first and most painful step was reducing personnel. In 1991, Shell laid off 15 percent of its 31,500 workers. The cutbacks hit E&P in a big way. With the potential for layoffs every day and the general contraction of investment and budgets, company morale plunged.

Richardson also pared down the company's assets. In 1991, Shell sold off nonstrategic oil-producing properties and various downstream investments.[7] Phil Carroll, who replaced Richardson as Shell Oil president in July 1993, carried on the process of revitalizing the company. Operating improvements, asset "restructuring," and selective investment under Richardson had resulted in a leaner, more cost-effective business. But despite these gains, Shell's earnings were still less than a third of what they should have been, given the company's asset base. So restructuring continued in 1993. Carroll slashed employment by another 12 percent, down to 22,000 workers. And he sold off other nonstrategic assets, including most of Shell's producing interests in Syria and its working interest in the East Bay, Louisiana, oil and gas fields (South Pass 24 and 27), an historic offshore legacy of the company. But by the end of the year, Carroll realized that operating efficiency could not be increased simply by further cost-cutting and prudent investments. To achieve an acceptable rate of return while growing at the same time would require some kind of fundamental change. The oil price collapse of 1986 had dramatically altered the world of oil. It called into question the fundamental assumptions about the future that had underlain the strategy, structure, and culture of Shell Oil. Although E&P management had long encouraged open communication and teamwork, the company was still largely geared toward command-and-control management of vertically integrated assets. Entrepreneurial energy and more flexible management, it seemed, were needed to meet the challenges of a new era, which included increased global competition and the dawning of a new information age. "We can't win in tomorrow's world with yesterday's approaches," Carroll concluded.[8]

Carroll was an unlikely candidate to effect transformative change within Shell. "He's a dyed-in-the-wool Shell man," wrote *Business Week*, "steeped in the language and perks of corporate life. And the Shell Oil CEO's love of process and structure betrays the heart of a petroleum engineer."[9] The son of a New Orleans banker, Carroll received his master's in physics from Tulane before joining Shell in 1961. He took various E&P assignments in New Orleans, Midland, New York, Los Angeles, and eventually Houston, where he served as division production manager for California and Alaska from 1975 to 1979. During the 1980s, he held increasingly responsible positions, from vice president of public affairs to managing director of Shell International Gas for the Group in London (after the buyout), before moving into administration. A cerebral and intellectual man, with three decades of experience in the oil business, Carroll reasoned that Shell, while successful and respected, nevertheless needed to rewrite many of its traditional rules, not only to revive growth and profitability, but to restore the motivation and morale of the workforce.

In early 1994, Carroll launched what he called Shell Oil's "transformation journey." Taking advice from such management consulting gurus as Noel Tichy, Peter Senge, and Charles Handy, who counseled numerous companies to become more "entrepreneurial," "flexible," and "revolutionary" during the years when companies like Enron were the darlings of this set, Carroll attempted to discard the top-down model of corporate governance and move toward a less paternalistic structure in which employees theoretically had more responsibility and accountability for results.[10] He developed a new system of governance, which sought to disperse decision-making widely throughout the company. This involved dismantling the single national operating company, Shell Oil, and creating four separate operating companies—Exploration & Production, Oil Products, Chemicals, and Services—each with its own board of directors and financial structure. The results of the four companies were consolidated into a single overall balance sheet. This idea was to "free" the principal businesses to become more competitive in their respective markets.[11]

Rather than creating exhilaration and optimism for employees, however, "transformation" appeared to breed confusion, apprehension, and fear. They might have felt like they had greater shared input into what happened in their companies, but many employees also had much less job security since the primary objective was to reduce costs. By 1995, Shell and the business press hailed the improved financial performance that transformation had effected. Shell's earnings rebounded to $1.5 billion in 1995 from the low of $20 million in 1991, while return on capital jumped from 4.4 percent of revenues in 1993 to 10.2 percent in 1995. In 1994 and again in 1995, *Fortune* magazine, which annually ranked America's best and worst companies on eight key attributes of reputation, named Shell Oil as the most admired major oil company in America.[12] But rising oil, natural gas, and refined product prices between 1992 and 1996—crude oil rose from around thirteen dollars per barrel to nearly twenty-four dollars per barrel during this period—were no doubt responsible for most of the improved returns.

Shell Oil under Carroll continued to expand the application of its new business model, looking for other places to squeeze out costs. One striking feature of this business model was the new approach to portfolio management. The basic strategy here called for Shell to look beyond internal solutions and enter into joint ventures with competitors, both downstream and upstream, combining assets to achieve better operating efficiency. The first of these historic, ground-breaking joint ventures in the upstream business was a limited partnership launched by Shell Oil and Amoco in 1997 to combine and operate all the oil- and gas-producing assets of the two companies in the greater Permian Basin of West Texas and New Mexico. The joint venture, called Altura Energy Ltd. was the first time in U.S. history that two major oil companies created a freestanding organization to merge their operations in a major basin. Based on the relative value of assets brought into the new partnership, Amoco claimed 64 percent ownership of the new company, and Shell 36 percent. Once antitrust officials approved the joint venture, Altura took over a portfolio of 6,300 wells with a combined daily production of 170,000 barrels of oil and 220 million cubic feet of natural gas. This made Altura the largest oil producer in Texas and the third-largest independent producer in the United States.[13]

As the Altura deal approached the latter stage of negotiation, Shell proposed another joint venture with Mobil. If such an arrangement could work with Amoco in West Texas then it could work in California, where Mobil sat across the lease line from Shell on the giant Midway-Sunset and Belridge fields in Kern County. The mechanics of the deal were easier than with Altura, which was complicated by carbon dioxide transportation and injection considerations. In June 1997, Shell and Mobil agreed to merge their producing operations into a new company called Aera Energy LLC, in which Shell held a 59 percent interest and Mobil 41 percent. With proven reserves of more than 1 billion barrels of crude oil equivalent and production of 250,000 b/d, Aera became the largest oil producer in California. At the time of the merger, the two parties expected pre-tax benefits of $1 per barrel in lower production costs. Unfortunately but inevitably, part of the cost savings came from the elimination of 275 positions overall between Shell and Mobil in California.[14]

Some Shell people who had spent the better part of their lives working for the company had a difficult time coming to terms with these joint ventures. Alliances with longtime competitors seemed to compromise Shell's familiar corporate culture and identity. More shocking than either Altura or Aera was Shell Oil's 1997 announcement that it was merging all its downstream operations with Texaco. It was not a formal corporate merger but rather two joint refining-marketing arrangements: Equilon Enterprises, which covered the Midwest and West regions of the country, and Motiva Enterprises, which included Saudi Aramco in the East and Gulf Coast regions. These arrangements were shocking in two respects: First, although joint ventures in oil exploration and production were fairly common, a refining and marketing alliance was highly unusual in the United States; and second, it was shocking because it was Texaco. According to one financial journalist, Texaco had long been "the company that the rest of the industry loved to hate."[15] It was perceived as unusually closed, autocratic, and imperious. Texaco had built its reputation on its ability to sell gasoline in every state in the union, but in the postwar period it had been one of the highest cost and least successful exploration companies of all the majors. Shell Oil, by contrast, was a very accomplished and technologically oriented E&P company whose efficient marketing organization also had fiercely battled Texaco for years to claim the position of number one gasoline retailer in the nation. Many longtime Shell employees were surprised, to say the least, to learn that Shell and Texaco had decided on a multibillion-dollar merger of most of their nationwide refining-marketing assets as well as all of their domestic transportation, trading, and lubricants businesses.

The Shell-Texaco alliance was a troubled marriage from the beginning. It was created just as another sharp downturn hit the world oil industry in 1998. Amid worldwide economic turmoil, marked by the collapse of several Asian economies and a Russian debt crisis, crude oil prices plummeted from $18 per barrel to $12 per barrel by year's end. Shell Oil reported a net loss of $1.7 billion on the year, the worst since 1992. Shell E&P was forced to write down oil producing assets, primarily Altura and Aera, and sell others, such as South Louisiana onshore and remaining producing properties in Alaska. Despite lower crude prices, the Shell Oil products organization and the Texaco alliances barely broke even, handicapped by high product inventories and the lowest gasoline prices since the 1970s (adjusted for inflation). At the same time, Shell Chemical struggled through another trough in industry returns. "The oil industry is 'on the ropes' again," wrote Jack Little, who in 1998 succeeded Phil Carroll as president of Shell Oil.[16]

Globalization

On June 1, 1998, Shell Oil Company converted to using the international Shell pecten emblem as its trademark and dropped the slightly different yet distinctive pecten it had used for years.[17] A trivial event on the face of it, the adoption of the international logo nevertheless symbolized profound changes in the works for the company. Royal Dutch/Shell was taking Shell Oil global.

This was the most significant departure in the relationship between Shell Oil and the Group since World War II. Organizational and administrative integration had not immediately followed the Group's buyout of Shell Oil minority shareholders in 1985. Shell Oil's strong-willed president John Bookout conducted Shell Oil's affairs post-buyout in very much the same manner as before, according to the agreement reached with the Group. After Bookout retired as president and moved on to Royal Dutch Petroleum's supervisory board from 1988 to 1993, the American company continued to retain much of its autonomy. Houston still largely ran the capital budgets and determined staffing and development strategies for Shell in the United States. And Shell Oil continued to report its corporate earnings separately. In the late 1990s, this arrangement came to an end. Beginning in 1996, Royal Dutch/Shell underwent its own transformation, reestablishing all its businesses on a more global basis, reducing the operating autonomy of Shell Oil as well as that of all the Group's subsidiary companies. The agreement Bookout had negotiated with Royal Dutch/Shell, which amounted to nothing more than a handshake, apparently did not apply in the changed world of oil.

Royal Dutch/Shell's transformation took place amid radical restructuring in the international oil industry prompted by worldwide financial deregulation accompanied by a growing pool of equity capital flowing around the world. Shareholders placed increasing pressure on managers of oil companies to improve short-term financial results and demonstrate prospects for long-term profits. To achieve the first imperative, during the early 1990s, the world's leading oil and gas companies, like Shell Oil, refocused on core businesses of oil, gas, and chemicals, restructured their asset bases, and "downsized" to reduce fixed costs. This proved to be inadequate as oil prices swooned in the late 1990s. To pull themselves off the "ropes," oil companies then embarked on one of the greatest merger movements in the history of the industry. Mergers allowed them to bolster short-term profits by paring away overlapping functions and laying off personnel. Mergers also equipped the new firm with enough capital to undertake new growth and riskier ventures, especially in the hunt for large reserves and "elephant" fields, although most companies continued to slash exploration budgets and staff to bolster short-term earnings. First, in 1998, British Petroleum acquired Amoco. Then Exxon combined with Mobil, followed by BP-Amoco and ARCO, Total with Fina and Elf, Chevron and Texaco, and finally, Conoco and Phillips. In the new competitive environment created by these mega-mergers, the greatest rewards would go to those companies that organized their functions on a global basis and were able to take advantage of new economies of speed and scale.

Well before the downturn in 1998, Royal Dutch/Shell leaders recognized the need to reposition the Group globally, even though the organization had weathered the crash of the 1980s better than the other majors. In 1990, the Group passed Exxon as the world's largest oil company. Its $107 billion in revenues placed it second only to General Motors on the Fortune 500 list of largest industrial corporations, and its $6.4 billion in earnings ranked it as the world's most profitable corporation. With a century of worldwide experience, Royal Dutch/Shell had learned to deal with instability by using long-term "scenario planning." Having developed a scenario and made tentative plans for $15/barrel oil during the great oil boom, Shell performed better than its competitors when prices actually dropped to that level.[18] The weakening dollar and high margins in refining and marketing, where the Group was the strongest, generated huge cash flows.

Flush with billions in cash reserves, Royal Dutch/Shell could afford to think long-term. "Shell doesn't react, it commits," wrote *Financial World.* "One of the interesting consequences is that management must be forever contrarian in its thinking, for it frequently makes decisions that fly in the face of current conditions, knowing that a crisis today likely means a windfall tomorrow."[19] The Group retrenched in the early 1990s, shedding things like coal, metals, and agricultural chemicals that did not relate directly to oil, but it also made larger investments than many of its rivals. It expanded worldwide exploration, buying acreage that would not be developed for years. It spent heavily on advanced technology such as 3-D seismic and made big bets on the production and shipping of liquified natural gas. During 1987–92, the Group vastly outpaced the other major oil companies in two key measurements of success in the industry: reserve replacement (185 percent) and finding cost per barrel ($3.38).[20]

Still, Royal Dutch/Shell had its soft spots. Financial analysts never failed to point out that despite the Group's perennial leadership of the Global 500 in total profits, its return on invested capital trailed its big competitors. Struggling Shell Oil, which accounted for about one-quarter of the parent's assets but delivered only about 13 percent of its earnings, contributed to the problem.[21] Also troubling was the mounting pile of over $10 billion in cash that the oil giant seemed unsure of how to invest. Profitability questions soon moved to the center of discussion. In May 1994, Cornelius Herkströter, who had just succeeded Sir Peter Holmes as chairman of the Committee of Managing Directors, addressed the issue openly by convening an unprecedented meeting of the Group's top fifty men at a seventeenth-century English manor called Hartwell House in Buckinghamshire. The upshot of that meeting was that Shell's immediate returns were not good enough and that Shell's corporate culture was part of the under-lying problem. The company then initiated the most far-reaching internal review of the organization in thirty years, scrutinizing its structure, portfolio, leadership, and relations, both internal and external. Among the conclusions drawn from this review was that the Group was "bureaucratic, inward looking, complacent, self-satisfied, arrogant," said Shell director Sir John Jennings. "We tolerated our own underperformance. We were technocentric and insufficiently entrepreneurial."[22]

Creating a more fast-moving, profit-centered organization out of one of the world's largest, tradition-bound, and decentralized industrial organizations was no easy task. Two public events in 1995, however, helped pushed Royal Dutch/Shell in this direction. Widespread environmental opposition in Europe to Shell's plans to sink Brent Spar, a once-innovative but aging oil-storage instal-lation in the North Sea, and the Nigerian government's execution of Ken Saro-Wiwa, a prominent Nigerian author who protested Shell's environmental record, forced Group leaders to pause for self-reflection. "We had to look deeply at ourselves and say 'Have we got everything right?'" said Mark Moody-Stuart, the manag-ing director who later succeeded Herkströter.

In the wake of these developments, Royal Dutch/Shell tried to reinvent itself. In a radical new departure, the company began publicly reporting its environmental record and defending human rights wherever it operated.[23] Most significantly, the Group began to rethink its decentralized and internationalized management structure, which had long been based on strong, national-level CEOs who reported to regional coordinators and managing direc-tors. It hired American consulting firms such as McKinsey and Coopers & Lybrand, as well as Noel Tichy, who had advised Shell Oil during its transformation journey. In January 1996, based on a McKinsey plan, Royal Dutch/Shell reduced its central bureau-cracy by 30 percent and established five committees to run Shell's major businesses (exploration and production; oil products; gas and coal; chemicals; and central staff functions).[24]

Despite the reduction in Shell's central bureaucracy, the reor-ganization actually centralized decision-making. Falling barri-ers to trade and global movements of goods and services placed national-level organizations at a disadvantage. More than ever before, Shell's national-level CEOs had to deal with new forces spilling over national political boundaries. Technological and managerial expertise, service and supply arrangements, and procurement and sales could be run more efficiently and profitably by giving Shell Centre more power and authority to organize its affairs on a global basis. In effect, the Group had begun to reshape a loose international organization of national operations into a single international operation.[25]

Over the next several years, Royal Dutch/Shell vigorously pursued globalization and cost-cutting. After the Asian crisis blindsided both oil and chemicals, Mark Moody-Stuart, chairman of the Committee of Managing Directors, announced that the Group was selling 40 percent of its chemicals businesses, with a book value of $7.7 billion, and sought to prune $300 million a year in operating costs. The divestment was in fact part of a larger five-year restructuring program. Responding to criticism that the Group had been too slow to react to the mega-mergers of Exxon and Mobil and BP and Amoco, Moody-Stuart confessed that "our Group's reputation with investors is on the line" and pledged to attack costs at a rate of $2.5 billion a year by 2001, a sum later raised to $4 billion. To achieve these cuts, Shell took special charges and write downs on high-cost production and refining assets, which contributed to the largest annual loss in the Group's history—$2.47 billion.[26]

This "clearing out of the cupboard," as Moody-Stuart described it, involved large job cuts and fundamental management changes. During 1998–2001, Shell shed nearly 25,000 jobs from a global workforce of 105,000. CEO executive structures replaced the business committees set up in 1996, thus further centralizing authority and enabling the Group to make faster business decisions. The Group appointed CEOs, which already ran businesses such as Chemicals and Services, to take over Oil Products and Exploration and Production. By early 2000, restructuring and cost-cutting, combined with sharply rising oil prices, had turned the Group around. It was reporting strong profits again, consistently beating analysts' expectations.[27]

The restructuring of the Group, in effect, globalized oil products and exploration and production, further tightening the Group's control over Shell Oil and leading to sweeping management changes and cutbacks in the United States. In June 1999, as profit margins in refining stubbornly remained low, the troubled Shell-Texaco joint venture Equilon sold two of its Midwest refineries, El Dorado, Kansas, and Wood River, Illinois, another historic legacy of the old Shell Oil. Globalization of the Group's E&P strategy refocused on deepwater plays around the world and away from non-core projects such as tertiary oil recovery in the United States. In March 2000, Shell and BP-Amoco sold Altura Energy to Occidental Petroleum for $3.6 billion, ending a transitional chapter in the withdrawal of Shell Oil from onshore production in the United States. The sale of Shell's interests in the Wassan Field of West Texas marked the severing of yet other historic links to Shell Oil's past (Shell remained an active participant in thermal recovery through its joint venture with Exxon-Mobil in Aera Energy). American managers were forced to compete with other Group entities for a finite global budget, and returns to U.S. operations did not necessarily stay in the United States. In 1999, Shell Oil stopped publishing its own annual report, symbolizing the passing of the old American Shell.

Even Shell Oil's offshore operations could not escape cost-cutting and consolidation in a climate of depressed oil prices. In 1999, Shell Oil Exploration and Production eliminated 740 jobs, or 20 percent of its staff. It sold 40 percent of its mature producing fields and exploration prospects on the Gulf of Mexico Shelf to Apache Corporation, although it continued to invest selectively in shallower Gulf waters, in projects such as the company's first subsalt well in the Enchilada field. It also sold working interests in a few deepwater fields (Angus and Macaroni) to Santa Fe Snyder in order to "maximize the value of its deepwater portfolio and enhance capital efficiency." Finally, Royal Dutch/Shell moved its global exploration and production technology organization and its deepwater services group from New Orleans, which once housed the most powerful E&P area office in the old Shell Oil Company, to Houston.[28]

Shell continued to possess one of the largest lease positions in the deepwater Gulf, and the reorganization promised to maintain the organization's leadership in frontier developments such as the Na Kika project. But offshore Gulf of Mexico, including its deepwater frontier, became just one in the Group's global portfolio of frontier developments that included offshore Brazil, West Africa, and the Far East. "In the Gulf, we're looking for more robust, larger scale projects with a higher gas mix," explained Dave Lawrence, SEPCo's vice president, E&D, in 2000. "We will continue to drill and develop conventional prospects in the deepwater, especially where we can leverage off of our strong infrastructure position. Our technical knowledge and development experience positions us well for pursuit of subsalt and other less conventional plays. What we're not doing is focusing on small scope opportunities."[29] Large scope opportunities meant large investments, which were now prioritized on a global basis. Shell continued to be a major player in the Gulf of Mexico. But, in the flurry of deepwater leasing during the early 2000s, it surrendered the commanding lead it had maintained for decades, a process that started with the BP partnership at Mars. In 2002, Royal Dutch/Shell was only the fourth-largest leaseholder in the ultra-deepwater (3,500 feet or more) of the Gulf, behind BP-Amoco-Vastar, ExxonMobil, and Chevron (combined with Texaco's leases).[30]

In early 2004, a front-page news scandal raised troubling questions about Royal Dutch/Shell's E&P organization. In a series of announcements, the Group lowered its estimate of proven world petroleum reserves by an astounding 4.37 billion barrels, more than 20 percent of its total. Revelations that top officials had tried to conceal shortfalls in proven oil and gas reserves led to the resignations of chairman of the CMD, Sir Philip Watts, Vice Chairman Walter van de Vijver, and Chief Financial Officer Judy Boynton.[31]

Estimating reserves is a subjective and inexact science, and the reserves taken off the books may well become proven some day. Nevertheless, Shell's restatement was too large to be downplayed, and it was not followed by major revisions of reserve estimates by other major oil companies. A tougher accounting interpretation by the U.S. Securities and Exchange Commission prompted the restatement, as did, some analysts speculated, an SEC investigation into the methods of accounting for deepwater reserves by offshore Gulf of Mexico operators. A 1977 rule required companies to conduct a flow test to confirm discoveries before booking reserves. But many operators contended that this was too expensive and too dangerous in deepwater and that new technologies such as 3-D seismic surveying provided sufficient confirmation. In April 2004, the SEC agreed and exempted deepwater operators in the Gulf from the old rule. Shell's overstated reserves, however, appeared to be mainly offshore Australia, Nigeria, and Oman, and not the Gulf of Mexico.[32]

The scandal pointed to two problems in the Group's E&P organization. First, it revealed that Royal Dutch/Shell had not been as successful in discovering new reserves as the public had thought, or at least as successful as its major competitors had been. Second, the attempted cover-up was uncharacteristic of an organization, at least in the United States, that long prided itself on conservative financial accounting and above-the-board business practices. Exploration prowess and straight shooting management were characteristics of the old Shell Oil. Even though the reserves problem had nothing to do with the United States, the fact that the Group's E&P organization was not meeting traditional standards in these two areas provided another example of how the old Shell Oil had become a relic in the wider world of Shell.

The Group's emphasis on the downstream businesses in the United States was further proof that the E&P-oriented Shell Oil was history. Vertical integration by the major oil companies on a global scale made the United States, where oil production and reserves continued to decline, more important as a market for oil products than as a place to look for oil and gas. After Jack Little, who had been Shell Oil's fifth consecutive president promoted from E&P, retired in 2000, American Steve Miller, a Group managing director with worldwide responsibility for supply and marketing, became the first CEO of Shell Oil from the products side since Dick McCurdy in 1965. Miller's first action was to draft a new "Blueprint for Shell in the U.S." It was a plan to redefine the mission of Shell in America, to "ensure Shell continues to play an active, productive leadership role in its largest market." It laid out several aspirations that would guide the company to success: take passionate care of customers; be a model of diversity for corporate America; help people reach their full potential; achieve leading-edge financial performance; be a responsible corporate citizen; be among the leaders in environmental performance; and build a strong national profile and identity. What the blueprint did not say, strikingly, was anything about finding or developing energy in the United States.[33]

Personnel changes and restructuring continued at a bewildering rate for Shell companies in the United States. In September 2002, Dutchman Rob Routs succeeded the retiring Miller. Routs had been president and CEO of the Shell-Texaco alliance, Equilon. When, as a result of the Chevron-Texaco merger in 2000, the U.S. Federal Trade Commission forced Texaco to sell its Equilon and Motiva interests, Routs became president of Shell Oil Products, U.S.A. He stayed long enough to oversee the final disintegration of the national operating company. For good measure, Routs laid the history of Shell in the United States to rest, killing a corporate history of Shell Oil started by Phil Carroll in 1998 and dismantling the Shell Museum, a marvelous display of Shell Oil historical artifacts and exhibits located in the lobby of One Shell Plaza in Houston. After he moved on in July 2003 to become Group managing director in the Netherlands, another oil products executive,

47-year-old Lynn Elsenhans, became Shell Oil Products CEO and assumed the new title of "country chair" for the various subsidiaries operating in the United States. The first female chief executive for Shell in the United States, she oversaw the $650 million campaign to rebrand Texaco stations into Shell stations. Also on her plate was the absorption of Pennzoil–Quaker State Company, acquired in March 2002 for $1.8 billion and the assumption of $1.1 billion in debt. Indeed, one 2003 public relations piece described Elsenhans' main mission as "refining Shell" and "shaping the brand."[34]

Shell Oil's offshore imperative in the United States was replaced by the imperative to achieve greater market share in oil products, for which Americans, with their ever-larger sport utility vehicles and expanding network of freeways, had a seemingly unquenchable thirst. Given that fossil fuels are a finite resource, it was inevitable that the race against depletion in America would ultimately be a losing one, as M. King Hubbert had foreseen in the 1950s. To be sure, there is still a lot of oil and gas to be produced in the United States thanks to advanced technologies in drilling, seismic surveying, and enhanced recovery, all of which Shell Oil helped pioneer. But aside from deepwater Gulf of Mexico and perhaps parts of Alaska and other offshore areas closed by leasing moratoria, there are few substantial fields left to be discovered, at least not enough to offset declines in production and reserves. The United States as a whole has lost *strategic* importance in the search for petroleum as the industry has become increasingly globalized.

The dissolution of Shell Oil Company reflected another late-twentieth-century trend in the petroleum industry: the passage of technological leadership from the major oil companies to service companies, niche companies, and independents. In deepwater Gulf of Mexico, independents have become as technologically aggressive, if not more so, than majors. In the 1990s, the majors reduced spending on technology and slashed exploration budgets. Under constant financial pressure from investors, they have stripped away and outsourced research functions that they used to perform in house. The heyday of Shell Oil's Bellaire Research Center has long past. But this trend is not purely a financial one. It also reveals new geological and geopolitical realities in the world of petroleum. Increasingly, as a greater share of known world petroleum reserves belongs to state-owned companies, and as new reserves become harder to find, the ownership of oil production is becoming a smaller part of major oil companies' income. According to former Shell Oil geologist Kenneth Deffeyes, "The major oil companies are coming to resemble large service companies with attached merchant banks."[35] The old Shell Oil was essentially an E&P technology company. It became an anachronism in this new era.

Still, the old Shell Oil opened the way for the best E&P opportunities that now exist for private oil firms. In this globalized world, just about everyone is driven by the offshore imperative. With state oil companies controlling most onshore reserves and with the future of U.S.-occupied Iraq highly chaotic and uncertain, companies large and small have no choice but to look in the oceans for future oil and gas supply. Furthermore, Hubbert's methodology for estimating peak production in the United States is now being applied to world oil production. A great debate now rages over exactly when the world peak will be reached. But there seems to be consensus that it will happen in at least the next ten to thirty years, if not the next five or even last year. On the downside of the peak, whenever it comes, an increasing percentage of oil and gas will come from offshore.[36]

Although Shell Oil's offshore pioneers may no longer recognize the Shell they used to work for in the United States, they surely recognize the lasting impact of their work as the offshore imperative spreads to other companies and countries around the world. From the 30-foot water depths plied by *Mr. Charlie* in the East Bay in the 1950s to the 7,600-foot depths of Na Kika in the 2000s, it is a large legacy, and one that will live on until the last drop of oil is produced from the earth.

Notes

Introduction

1 James E. Akins, "The Oil Crisis: This Time the Wolf Is Here," *Foreign Affairs* 51 (Apr. 1973): 462–90.

2 Michael B. Stoff, *Oil, War, and American Security;* David S. Painter, *Oil and the American Century;* and Daniel Yergin, *The Prize.*

3 See, for example, Colin J. Campbell, *The Coming Oil Crisis;* Kenneth S. Deffeyes, *Hubbert's Peak* and *Beyond Oil;* Bob Williams, "Debate over Peak-Oil Issue Boiling Over, with Major Implications for Industry, Society," *Oil & Gas Journal* (July 14, 2003): 18–29; and Matthew R. Simmons, *Twilight in the Desert.*

4 Joseph A. Pratt's *Prelude to Merger* is the only one so far to address this theme directly.

5 Offshore Stats and Facts, Minerals Management Service, U.S. Department of the Interior, http://www.mms.gov/stats/OCSproduction.htm.

6 Robert Gramling, *Oil on the Edge;* Joseph A. Pratt, Tyler Priest, and Christopher Castaneda, *Offshore Pioneers;* Clyde W. Burleson, *Deep Challenge!;* and Tai Deckner Kreidler, "The Offshore Petroleum Industry: The Formative Years, 1945–1962" (Ph. D. dissertation, Texas Tech University, 1997).

7 For histories of Shell Oil and Royal Dutch Shell, see Kendall Beaton, *Enterprise in Oil;* Frederik Carel Gerretson, *History of the Royal Dutch;* R. J. Forbes and D. R. O'Beirne, *The Technical Development of the Royal Dutch/Shell, 1890–1940;* Stephen Howarth, *A Century in Oil;* Jack Doyle, *Riding the Dragon;* Ide Okonta, *Where Vultures Feast;* Steve Lerner, *Diamond.*
For a good review essay of oil company histories, see Paul Horsnell, "Oil Company Histories," *The Journal of Energy Literature* 5, no. 2 (Dec. 1999): 3–31.

8 Jonathan Rauch, "The New Old Economy: Oil, Computers, and the Reinvention of the Earth," *The Atlantic Monthly,* Jan. 2001, 35–49; and Helen Thorpe, "Oil & Water," *Texas Monthly,* Feb. 1996, 90–93, 140–45.

9 See, most recently, James W. Cortada, *The Digital Hand.* Cortada does look at the application of digital technology in the petroleum industry but mainly in the downstream or "process manufacturing" business of refining and petrochemicals.

10 See, for example, Manuel Castells, *The Rise of the Network Society.*

11 See, for example, John J. McCloy, Nathan W. Pearson, and Beverly Matthews, *The Great Oil Spill.*

12 R. H. Nanz, vice president, exploration, Shell Oil Company, "The Offshore Imperative—The Need for and Potential of Offshore Exploration," paper presented at Colloquium on Conventional Energy Sources and the Environment, University of Delaware, Newark, Apr. 30, 1975.

13 John Reilly interview with author, Oct. 30, 2002, Houston. Tapes and transcripts of most interviews cited in this book, except those conducted by James Cox in 1973, will eventually be archived at the University of Houston.

14 Chuck Edwards interview with author, Apr. 30, 2003, Houston.

15 See, for example, Robert Bryce, *Cronies.*

16 Thomas P. Hughes, *Human-Built World,* 5.

17 Peter Marshall interview by Joseph Pratt, Sept. 21, 2002, Houston.

18 See Philip Scranton, ed., *The Second Wave;* and Geoffrey Jones and Lina Gálvez-Muñoz, eds., *Foreign Multinationals in the United States,* which includes Tyler Priest, "The 'Americanization' of Shell Oil," 188–205. Also see Mira Wilkins, *The History of Foreign Investment in the United States, 1914–1945.*

Chapter 1
The Americanization of Shell Oil

1 Royal Dutch Petroleum Company, Summary Annual Report and Accounts 2004.

2 Yergin, *The Prize,* 56–77; Howarth, *A Century in Oil,* 33–47; and Beaton, *Enterprise in Oil,* 38–45. See Beaton for a detailed history of Shell Oil's first thirty-three years. For the standard

biography of Marcus Samuel, see Samuel Henriques, *Marcus Samuel.*

3 Sir Henri Deterding, as told to Stanley Naylor, *An International Oilman,* 54.

4 Yergin, *The Prize,* 114–27; Howarth, *A Century in Oil,* 49–71; Beaton, *Enterprise in Oil,* 46–55.

5 On the straight-line policy, see Deterding, *An International Oilman,* 50–51.

6 Quoted in Beaton, *Enterprise in Oil,* 58.

7 "Oil—Shell's Game," *Forbes,* Apr. 1, 1952, 21.

8 Beaton, *Enterprise in Oil,* 206–34.

9 Ibid., 361–73.

10 Ibid., 631–36.

11 Monroe Spaght interview by James Cox, 1973, Houston (copies of Cox interviews in author's possession); Shell Oil Company, *Annual Reports,* various years.

12 "Oil—Shell's Game," 25.

13 "Portraits of Two Presidents," *Shell News* 28, no. 12 (Dec. 1960): 3.

14 Bill Kenney interview by James Cox, 1973, Houston.

15 "Oil—Shell's Game," 18, 25.

16 Shell Oil Company, *Annual Report, 1951,* 1, and *Annual Report, 1960,* 26.

17 "Full Circle," *Forbes,* June 1, 1956, 25. In 1957, Shell T&T also launched stock on the NYSE, although most of the ownership remained in British hands.

18 John Loudon interview by James Cox, 1973, Houston.

19 L. L. L. Golden, *Saturday Review,* Dec. 9, 1967, 3.

20 Quoted in Howarth, *A Century in Oil,* 262.

21 "The Diplomats of Oil," *Time,* May 9, 1960, 98.

22 Quoted in Gilbert Burck, "The Bountiful World of Royal Dutch/Shell," *Fortune,* Sept. 1957, 140. In his old age, Deterding's disposition toward autocratic rule transferred to his political views, as he expressed admiration for Italian dictator Benito Mussolini and the German Nazi regime. After retiring, he married his former secretary, a German who strongly supported the Nazis, and bought an estate in Germany where he lived until his death. See Howarth, *A Century in Oil,* 187–88.

23 Burck, "The Bountiful World," 140.

24 According to William N. Greene, a major impetus behind the McKinsey study was to address problems with coordination, exemplified by the dumping of oil products in Rotterdam by the Group's Curaçao refinery while its other European affiliates were trying to maintain prices. Greene, *Strategies of the Major Oil Companies,* 227.

25 Burck, "The Bountiful World," 136.

26 Ibid., 140.

27 Ibid., 136.

28 Loudon's grandfather was a governor general of the Dutch East Indies, his uncle was Holland's foreign minister, and his father, Hugo Loudon, was one of the founders of Royal Dutch Petroleum and later a managing director and chairman. "The Diplomats of Oil," 94–98.

29 Kenney interview.

30 Loudon interview.

31 "Shell Oil Turns to Caribbean," *Business Week,* Oct. 30, 1948.

32 "More Dollars for More Oil," *Fortune,* Dec. 1948, 15. The Group was also greatly aided economically by a British Treasury agreement in 1946 that freed Shell from exchange controls and permitted oil payments in sterling instead of scarce dollars.

33 On the Group's postwar recovery program, see Howarth, *A Century in Oil,* 211–31; "International Training Program," *Shell News* 16, no. 10 (Oct. 1948): 4; "Royal Dutch/Shell and Its New Competition," *Fortune,* Oct. 1957, 139–42.

34 "The Diplomats of Oil," 98.

35 Burck, "The Bountiful World," 135.

36 "Ideas for Rent," *Shell News* 21, no. 1 (Jan. 1953): 21.

37 "Shell Research Laboratory Is Dedicated in Houston," *The Pecten* (Dec. 1947): 5.

38 Shell Development Company, *Bellaire Research Center,* 29–30.

39 "Aliens Welcome," *Business Week,* Feb. 21, 1953, 170–71; and "Refinery Site Chosen," *Oil and Gas Journal* 52, no. 10 (July 13, 1953): 76.

40 See Painter, *Oil and the American Century,* 75–127, for a well-researched discussion of the Middle East oil deals.

41 See Burton I. Kaufman, *The Oil Cartel Case.*

42 For more on the Timken decision, see Wyatt Wells, *Anti-trust and the Formation of the Postwar World,* 134–35.

43 See also the discussion on antitrust in James Bamberg, "OLI and OIL: BP in the U.S. in Theory and Practice," in Jones and Gálvez-Muñoz, eds., *Foreign Multinationals in the United States,* 172–73.

44 "Agreement for Research Services" between Shell Internationale Research Maatschappij N.V. and Shell Oil Company, Jan. 1, 1960. Shell Oil Company SEC Form 10-K, 1980.

45 Bob Nanz interview by author, Sept. 15, 1998, Houston.

Chapter 2
Testing the Waters

1 Raymond L. Lankford, "Marine Drilling," in *History of Oil Well Drilling*, ed. J. E. Brantly.

2 Beaton, *Enterprise in Oil*, 203; Edgar Wesley Owen, *Trek of the Oil Finders*, 504–505. For the history of the origins of LL&E and Texaco in South Louisiana, see Owen, *Trek of the Oil Finders*, 760–61; and George Elliott Sweet, *The History of Geophysical Prospecting*, 135–38.

3 "The Other Fellow's Job: The Seismograph Shooter," *Shell News* (Mar. 1939): 15.

4 Ibid.

5 Randall A. Detro, "Transportation in a Difficult Terrain: The Development of the Marsh Buggy," *Geoscience and Man* 19 (June 30, 1978): 93–99.

6 F. C. Embshoff, "Floating Derricks: Modern Drilling in the Swamplands," *Shell News* (July 1938): 4.

7 Lankford, "Marine Drilling."

8 Other significant south Louisiana salt dome fields developed farther inland by Shell Oil in the 1930s included Black Bayou (Cameron Parish), Iowa (Calcasieu Parish), and White Castle (Iberville Parish).

9 Shell paid $6.27 per acre in bonus payments for this lease. Louisiana State Mineral Board, *Biennial Report* (May 1942): 15.

10 C. H. "Steve" Siebenhausen Jr., "Notes on Shell's Early Offshore Engineering and Operations Experience," 1999, in author's possession.

11 "Salt, Sugar, and Oil," *Shell News* 18, no. 6 (June 1950): 6–9; Beaton, *Enterprise in Oil*, 645; A. J. Galloway interview by James Cox, 1973, Houston.

12 Beaton, *Enterprise in Oil*, 783; "Salt, Sugar, and Oil," 9.

13 James Hebert interview by Andrew Gardner, Apr. 17, 2001, New Iberia, La.

14 For the classic critique of U.S. domestic oil policies, see John M. Blair, *The Control of Oil*, especially chapter 7 on the "domestic control mechanism," 152–86.

15 "New Home for Shell," *The Pecten* (May 1947): 3–9.

16 "Midland, Houston, and New Orleans Are Named Area Headquarters," *The Pecten* (Nov. 1946): 2; "Houston's Regional Staff Coordinates Shell's Exploration and Production Activities," *The Pecten* (Jan. 1950): 4–5.

17 Bob Ferris interview by author, Dec. 17, 1998, Austin, Tex.; Bouwe Dykstra interview by James Cox, 1973, Houston.

18 Shell Oil Company, *History of Shell Offshore Seismic Operations in the United States*, 1.

19 "Disputed Louisiana Lease Looks Hot," *The Oil and Gas Journal* (May 5, 1958): 72–73; John Bookout interview #2 by Tom Stewart, Sept. 24, 1998, Houston.

20 James M. Parks, "Recollections of Shell's Bellaire E&P R&D Lab," *Oil Industry History* 4, no. 1 (2003): 118.

21 Quoted in Kenneth S. Deffeyes, *Hubbert's Peak*, 2–3.

22 Shell Development Company, *Bellaire Research Center*, 33–35; "'Doc' Adkins Retires From E&P Research to Devote Full Time to Many Research and Hobby Activities," *The Pecten* (Apr. 1951): 5; Nanz interview.

23 Parks, "Recollections of Shell's Bellaire E&P R&D Lab," 18–22; "Achievements Belie His Modesty," *AAPG Explorer* (Mar. 2002), www.aapg.org/explorer/.

24 Ed Picou Jr., "A History of Shell Paleontology in the Gulf Coast: A Personal Perspective," 2003, draft in author's possession.

25 John Abbott, "Back to Back," *Shell News* (Jan. 1989): 6; "Oil—Shell's Game," *Forbes*, Apr. 1, 1952, 25; Abbott, "Back to Back," 7.

26 Miner Long interview by author, July 17, 2003, Houston.

27 For more on the early history of geophysical prospecting, see Sweet, *The History of Geophysical Prospecting*; Owen, *Trek of the Oil Finders*; and L. C. Lawyer, Charles C. Bates, and Robert B. Rice, *Geophysics in the Affairs of Mankind*.

28 For a good survey of early Texas oil, see Diana Davids Olien and Roger M. Olien, *Oil in Texas*.

29 Shell Development Company, *Bellaire Research Center*, 39–45; "With Shell 25 Years, Dr. Merten is Seismic Pioneer," *The Pecten* (Dec. 1951): 8; Sam Bowlby interview by James Cox, 1973, Houston.

30 CDP is sometimes referred to as common mid-point or common reflection point.

31 W. Harry Mayne, *50 Years of Geophysical Ideas*.

32 John Redmond's notes on chapter 2, Feb. 2001, in author's possession.

33 Shell Development Company, *Bellaire Research Center*, 46–56.

34 A log is a detailed record of rock properties obtained by lowering a tool with sensors into the wellbore on electrical conductive wireline.

35 E. C. Thomas, Shell Oil Company, "50th Anniversary of the Archie Equation: Archie Left More than Just an

Equation," *The Log Analyst* (May–June 1992): 199–205.

36 Galloway interview.

37 Shell Oil Company, *History of Shell Offshore Seismic Operations*, 2–6; "Shrimp Boats Is A-Comin,' Did They Find Any Oil?" *Shell News* 3 (1984): 3.

38 "Shrimp Boats Is A-Comin,'" 3.

39 Aubrey Bassett interview by author, Jan. 16, 2003, New Orleans; Shell Oil Company, *History of Shell Offshore Seismic Operations*, 6.

40 Shell Oil Company, *History of Shell Offshore Seismic Operations*, 19.

41 Ibid., 10–19.

42 Dr. Milton B. Dobrin, "Introduction to Geophysical Prospecting," *The Oil and Gas Journal* (Mar. 24, 1952): 124–32; Curtis A. Johnson and John W. Wilson, "Marine Exploration Comes of Age," *World Petroleum* (Mar. 1954): 76–77.

43 William H. Newton, "Doodlebuggers Afloat," *World Petroleum* (Feb. 1954): 35.

44 Louisiana State Mineral Board, *Biennial Report* (July 1954); "Mobile Drilling Platforms Seen as the Next Possible Step in Offshore Exploration," *The Oil and Gas Journal* (Nov. 22, 1947): 57; Galloway interview.

45 Boone Williams, "Shell Moves in at New Orleans," *The Pecten* (Aug. 1947): 8.

46 "Gulf of Mexico Oil Play," *Shell News* (Oct. 1949): 7.

47 Ibid., 8.

48 Howard Shatto interview by author, Oct. 2, 1999, Houston; "Gulf of Mexico Oil Play," 8–9.

49 Frank Poorman interview by author and Sam Morton, Dec. 8, 1998, Houston.

50 Director, U.S. Geological Survey, to Department of Interior Solicitor, Jan. 28, 1952, Box 3283, Central Classified Files, 1937–1953, Records of the U.S. Secretary of Interior, RG 48, National Archives and Records Administration (NARA), College Park, Md.

51 "Island on Stilts," *Shell News* (Feb. 1952):11–13; Keith Doig interview by James Cox, 1973, Houston.

52 Cliff Hernandez interview by Andrew Gardner, May 1, 2000, New Iberia, La.; Galloway interview.

53 "Bouwe Dykstra Looks Ahead," *The Oil and Gas Journal* 57, no. 41 (Oct. 5, 1959): 245.

54 Dykstra interview.

55 "Offshore Activity Is Booming Again," *Shell News* (Nov. 1953): 3–4; Ferris interview; James W. Calvert, "Louisiana Offshore Operations," *World Petroleum* (Nov. 1954): 68.

56 Doig interview.

57 Tyler Priest, "Claiming the Coastal Sea: From the Tideland's Controversy to the Landmark 1962 Sale," in *History of the Offshore Oil and Gas Industry in Southern Louisiana: Interim Report*, vol. 1, *Papers on the Evolving Offshore Industry,* http://www.gomr.mms.gov/homepg/regulate/environ/studies/2004/2004-049 .pdf.

58 Ibid.

59 Martin G. White, Solicitor, Department of Interior, to Shell Oil Company, Jan. 22, 1952, and Secretary of Interior to Shell Oil Company, Feb. 4, 1952, Box 3283, CCF, 1937–1948, RG 48, NARA; Galloway interview.

60. The SLA and the OCSLA essentially divided the continental shelf into two areas, one belonging to the coastal states and the remaining area set aside for the United States. But for whatever reason, political, legal, diplomatic or otherwise, these pieces of legislation left two major questions unanswered with respect to state ownership: 1) which states bordering the Gulf of Mexico were entitled to a historical boundary of three leagues; and 2) how would the coastlines of the states be determined.

61 Gramling, *Oil on the Edge*, 50.

62 Ben C. Belt, "Louisiana and Texas Offshore Prospects," *Drilling* (Mar. 1956): 119; "Offshore Payout Still Years Away," *The Oil and Gas Journal* 58, no. 18 (May 2, 1960): 78; Dean A. McGee, "Economics of Offshore Drilling in the Gulf of Mexico," *Offshore Drilling* (Feb. 1955): 16.

63 Kreidler, "The Offshore Petroleum Industry," 157–62, 171–73.

64 "A Pioneer Keeps Pioneering," *The Oil and Gas Journal* (Jan. 31, 1966): 259–60.

65 The Provident Fund was one of the hallmark creations of Henri Deterding. It was a pension plan that allowed salaried employees to contribute up to 10 percent of their annual salary to the fund, which the company matched and invested conservatively. In 1944, Shell extended eligibility to hourly employees.

66 L. A. "Pete" Rogers interview by Andrew Gardner, Jan. 16, 2002, Patterson, La.; Keith Viator interview by Andrew Gardner, Apr. 30, 2001, Erath, La.; and Hernandez interview. Also, see "Oil Sentinels at Sea," *Shell News* 28, no. 7 (July 1960): 11–15.

67 "Strange Armada in the Gulf," *Shell News* (May 1955): 10–13.

68 Dykstra interview.

69 Galloway interview; Alden J. Laborde, *My Life and Times,* 166–67; Bruce Collip interview by author, Aug. 21, 1999, Houston.

70 Laborde, *My Life and Times,* 174.

71 Derrell A. Smith, "Geology of South Pass Block 27 Field, Offshore, Plaquemines Parish, Louisiana," *Bulletin of the American Association of Petroleum Geologists* 45, no. 1 (Jan. 1961): 51–71.

72 Oil and gas companies discovered thirty-four new fields in 1954, fifty-seven fields in 1955, and seventy-two fields in 1956. The success rate for wildcat exploratory wells was exceptionally high (34 percent in 1956), much higher than onshore. During 1949–56, the increase in U.S. domestic reserves from offshore development was nine times the average for onshore wells. U.S. Department of Interior, "Petroleum and Sulfur on the U.S. Continental Shelf," internal study, Aug. 1969, box 134, CCF, 1969–1972, RG 48, NARA.

73 See Garry Boulard, *Huey Long Invades New Orleans,* 18–25.

74 "The Roosevelt Story," *Roosevelt Review* 29, no. 1 (Dec. 1965): 2–3; Ed Picou Jr. interview by author, July 8, 2003, New Orleans.

75 B. B. Hughson interview by author, June 26, 2002, New Orleans.

76 Galloway interview; Kenney interview; "Majors Speed Offshore Efforts," *World Oil* (June 1954): 97.

77 "First Offering of Continental Shelf Leases Brings High Bonuses," *World Petroleum* (Nov. 1954): 86; Joe Reilly, "Millions Gambled on Gulf Tidelands," *The Oil and Gas Journal* (July 18, 1955): 62–65.

78 Galloway interview.

79 Robert O. Frederick, "Marine Drilling: The Future Remains Bright," *Drilling* (Dec. 1959): 55–56; Joe Zeppa, "What is the Outlook for Drilling in the Gulf of Mexico?" *Drilling* (Dec. 1959): 59.

80. Charles A. Stuart, *Geopressures,* 4; James A. Hartman, Edward B. Picou Jr., "A Memorial: Charles Allison (Chuck) Stuart, 1918–2001," *Nogs Log* 42, no. 9 (June 2002): 11.

81 Hartman and Picou, "A Memorial: Charles Allison (Chuck) Stuart."

82 Stuart, *Geopressures,* 43–44, 107; C. H. "Steve" Siebenhausen Jr., "Management's Significant Contributions to Production Department's Rise to Technical and Operating Excellence, 1952–1981," unpublished manuscript provided to author.

83 Siebenhausen, "Management's Significant Contributions."

84 Stuart, *Geopressures,* 4.

85 "Transportation and Supplies," *Shell News* 16, no. 4 (Apr. 1948): 12; Shell Oil Company, *Annual Report, 1951,* 8; "Transportation and Supplies," *Shell News* 18, no. 4 (Apr. 1950): 10.

86 "Crude Oil 'Round the Bend," *Shell News* 20, no. 6 (June 1952): 9; "Big Ditch for Barges," *Shell News* 22, no. 11 (Nov. 1954): 1–3; "The Norco Story," *Shell News* 15, no. 3 (Mar. 1947) 12–15.

87 Kenney interview.

88 "Underwater Gathering System," *Shell News* 24, no. 12 (Dec. 1956): 22–24.

89 "Pipeline on the Delta," *Shell News* 27, no. 1 (Jan. 1959): 25.

90 "Outer Continental Shelf Lands Act, the Act of August 7, 1953, Chapter 345, as Amended," Minerals Management Service, U.S. Department of the Interior, http://www.mms.gov/aboutmms/pdffiles/ocsla.pdf; Ann Hollick, *U.S. Foreign Policy and the Law of the Sea,* 34–37.

91 American Embassy, The Hague, to Department of State, Embassy Despatch 90, July 28, 1954, FCN Treaty Negotiations (23–25 meetings), State Department Decimal File 611.564/8–354, RG 59, NARA.

92 American Embassy, The Hague, to Department of State, Embdesp. 601, Jan. 20, 1956, Treaty of Friendship, Commerce, and Navigation, State Department Decimal File 611.564/2–1056, RG 59, NARA.

93 "Early Structures Marked Several 'Firsts,'" *Offshore* (Sept. 1979): 113; "Largest Mobile Drill Barge Joins ODECO Offshore Fleet," *Drilling* (Apr. 1957): 115.

94 "Strange Armada in the Gulf," 11; "Dry Hole in Deep Water," *Shell News* (Nov. 1956): 19.

95 "Steel Island in the Mississippi Mud," *Shell News* (Oct. 1955): 6–9; Dr. Richard J. Howe, "Offshore Mobile Drilling Units," *Ocean Industry* (July 1968): 61.

96 "Operation Hurricane," *Shell News* (Sept. 1957): 12–15.

97 "Insurance Problems Mount with Offshore Operations," *Drilling* (Aug. 1956): 74–75, 124; Kreidler, "The Offshore Petroleum Industry," 219–25.

98 Kenney interview.

99 Belt, "Louisiana and Texas Offshore Prospects," 119.

Chapter 3.
Betting on Technology

1 Yergin, *The Prize,* 535.

2 D. B. Kemball-Cook, "'The Challenge

Today of Finding and Producing,'"
Shell News (Apr. 1959): 6.

3 Bouwe Dykstra, "Costs, Allowable Rate
Hinders Offshore Work," *Drilling* (Aug.
1959): 114.

4 Don Russell interview by author,
June 18, 1999, Tyler, Tex.; Jerry
O'Brien interview by author, June 5,
2002, Conroe, Tex.

5 Marlan Downey interview by author,
Aug. 24, 1999, Dallas; Jack Threet
interview by Bruce Beauboeuf, Dec. 5,
1997, Houston.

6 "Clark Has Varied Background," *The
Oil and Gas Journal* (Sept. 16, 1963):
174.

7 Bruce Collipp interview by author,
Aug. 21, 1999, Houston.

8 Bookout interview #2; John Redmond
to author, Dec. 12, 1999.

9 John Redmond interview by author,
Nov. 12, 1999, Houston.

10 Redmond to author, Dec. 12, 1999.

11 "Kemball-Cook Faces Tough Job," *The
Oil and Gas Journal* (Oct. 20, 1958): 167.

12 Redmond to author, Dec. 12, 1999.

13 Ibid.

14 John Rankin, "History of Federal Offshore Leasing, 1986," 48, unpublished
paper provided to author by Rankin;
"Offshore Sale Again Breaks Records,"
The Oil and Gas Journal 57, no. 34 (Aug.
17, 1959): 91–92.

15 Ferris interview.

16 John Rankin interview by author,
Sept. 30, 2000, Houston; "Federal
Lease Sale Nets Record $285 Million,"
Offshore 12, no. 3 (Mar. 1960): 15–16;
"Rivalry for Underwater Land," *Shell
News* 28, no. 5 (May 1960): 18–19.

17 R. F. Bauer, A. J. Field, and Hal Strat-
ton, "A Method of Drilling from a
Floating Vessel and the 'Cuss I,'" paper
presented at the spring meeting of the
South District, Division of Production,
American Petroleum Institute (API),
Houston, Feb. 1958, in API, *Drilling
and Production Practice* (1959): 124–29.

18 Bruce Collipp interview by Joseph
Pratt, Oct. 17, 1998, Houston; Collipp
interview by author.

19 Collipp interview by Pratt.

20 "They Call Him the Father of the Semi-
submersible," *Shell News* 1 (1980): 6;
Collipp interview by author; Bruce G.
Collipp, "Offshore Industry 1950–
1965: Invention," in Harry Benford
and William A. Fox, eds., *A Half Century of Maritime Technology 1943–1993*,
590.

21 Collipp interview by author.

22 Russell interview.

23 Collipp interview by author.

24 "A Giant Step to the Deeps," *Shell News*
(Oct.–Nov. 1962): 10; "They Call Him
the Father of the Semi-submersible," 6.

25 "They Call Him the Father of the Semi-
Submersible," 6.

26 Douwe (Dee) DeVries interview by
author, Sept. 30, 2000, Houston; Collipp interview by author; "Sand in His
Boots," *Shell News* 5 (1984): 10.

27 Ron Geer interview by Joseph Pratt
and Bruce Beauboeuf, Dec. 12, 1997,
Houston; "Sand in His Boots," 10.

28 DeVries interview.

29 "First Underwater Completion in the
Gulf," *Offshore* (Feb. 1961): 11; Collipp
interview by author.

30 Geer interview; Collipp interview by
author; "They Call Him the Father of
the Semi-Submersible," 7; "Spies Pry,
but Shell Keeps Secret of New Deep-
Water Oil Drilling Rig," *Wall Street
Journal*, Aug. 13, 1962.

31 Collipp interview by author; "Offshore
Drilling Rig Gets Better Sea Legs,"
Business Week, Aug. 18, 1962, 101.

32 "Federal Government Puts 3.67 Million Offshore Acres on the Auction
Block," *Offshore* (Mar. 1962): 13–14.

33 O'Brien interview.

34 Ibid.

35 Jim Lampton interview by author,
July 2, 2003, Cold Spring, Tex.; Picou
interview; Picou, "A History of Shell
Paleontology"; O'Brien interview.

36 O'Brien interview.

37 Hughson interview; Harry Hasenpflug
interview by author, Aug. 21, 2002,
Sugarland, Tex.

38 Lawyer, Bates, and Rice, *Geophysics in
the Affairs of Mankind*, 82–87; and W.
Harry Mayne, *50 Years of Geophysical
Ideas*.

39 Gene Bankston interview by author,
Dec. 3, 1999, Houston.

40 Picou interview; O'Brien interview.

41 The big spenders on cash bonuses in
the sale were Humble ($63.1 million)
and Gulf ($46.6 million). "What the
Biggest Offshore Sale Means," *The Oil
and Gas Journal* (Mar. 26, 1962): 79.

42 "Sand in His Boots," 10.

43 "A Robot Goes to Sea," *Shell News*
(Jan.–Feb. 1963): 10.

44 Shatto interview; Shatto, comments on
chapter draft, in author's possession.

45 Bill Petersen interview by Joseph A.
Pratt, Oct. 2, 1999, Houston; "Robot
'Diver' Developed," *Offshore* (Dec.
1962): 11–12; "A Robot Goes to Sea,"
8–12.

46 C. W. Swanlund Jr., "Economics of Deep Water Hydropressured Exploration and Development Drilling—Louisiana Offshore," Shell Oil New Orleans Area Production Department, Jan. 1964, copy provided to author by Swanlund.

47 "Five Giants for the Gulf," Shell News 33, no. 4 (July–Aug. 1965): 24–26; Pratt, Priest, and Castaneda, Offshore Pioneers, 62.

48 Shell Oil, Annual Report, 1966, 6; "The Gulf: Offshore Oil's Major Leagues," Shell News (Sept.–Oct. 1967): 2.

49 F. P. Dunn, "Deepwater Production: 1950–2000," OTC 7627, paper presented at the 26th Annual Offshore Technology Conference (OTC), Houston, May 2–5, 1994, in Proceedings (1994): 921–28.

50 Gene Bankston interview by James Cox, 1973, Houston.

51 Dunn, "Deepwater Production," 923; "Hurricanes—A Hazard for the Oil Industry," Shell News (Nov.–Dec. 1964): 21–24; "A Bad One Named Betsy," Shell News (Nov.–Dec. 1965): 1–4. On the impact of the 1960s hurricanes on the offshore oil industry, see Joseph A. Pratt, "The Brave and the Foolhardy: Hurricanes in the Early Offshore Industry," in History of the Offshore Oil and Gas Industry in Southern Louisiana: Interim Report, Vol. 1, http://www.gomr.mms.gov/homepg/regulate/environ/studies/2004/2004–049.pdf.

52 Collipp interview by author; "A Bad One Named Betsy," 1–2.

53 Billy Flowers interview by author, June 18, 1999, Tyler, Tex.; "Shell Erects World's Largest Fixed Platform," Offshore (Aug. 1967): 28.

Chapter 4
The Trials and Triumphs of Exploration

1 "Changes in Shell Leadership," Shell News 33, no. 3 (May–June, 1965): 2; "Monty Spaght's Royal Dutch Treat," Fortune (June 1966): 156;"A Visit with Monty Spaght," Shell News no. 3 (1980): 22.

2 Blair, The Control of Oil, 207.

3 Shell Oil, Annual Reports, 1961–1963; "Royal Dutch Treat," 191. On the strike, see "Why Strike Shell?" Union News 19, no. 2 (Apr. 1963): 9; "Shell Strikers 100% Solid in 8th Month," Union News 19, no. 2 (Apr. 1963): 8; "Oil Strikers Get Global Support," Business Week, May 25, 1963, 62; "Honorable Settlement at Shell," Union News (Sept. 1963): 3; and Milden J. Fox Jr., "The Impact of Work Assignments on Collective Bargaining in the Petroleum Refining Industry on the Texas Gulf Coast" (Ph.D. dissertation, Texas A&M University, 1969).

4 Galloway interview; Spaght interview.

5 Shell Oil, Annual Report, 1968, 4; "Putting in More Chips," Forbes, Apr. 15, 1968, 43.

6 Shell Oil, Annual Reports, 1959–1969.

7 "Shell to Develop Oil Leases in Deep Louisiana Waters," Offshore (July 1965): 28–30; "Blue Dolphin Line to Go on Stream," Offshore (Oct. 1965): 34, "From the Bayous to the Heartland," Shell News 35, no. 4 (July–Aug. 1967): 20–23.

8 "Approval Asked for Red Snapper Line," Offshore (Jan. 1966): 40–42.

9 Bowlby interview.

10 "Accent on Alaska," Shell News 26, no. 2 (Feb. 1958): 7; Donald Worster, Under Western Skies, 184. On the role of Anchorage businessmen in the leasing, see David Postman's eight-part series, "Inside Deal: The Untold Story of Oil in Alaska," Anchorage Daily News, Feb. 4–11, 1990; and Jack Roderick, Crude Dreams, 59–71.

11 "Oil Hunt Warms Up in Alaska," Shell News 32, no. 1 (Jan.–Feb. 1964): 20–25.

12 "Conquering Cook Inlet," Shell News (Nov.–Dec. 1966): 16. For more on the Cook Inlet, see R. C. Visser, "Platform Design and Construction in Cook Inlet, Alaska," Journal of Petroleum Technology (Apr. 1969): 411–20, and Pratt, Priest, and Castaneda, Offshore Pioneers.

13 On the history of oil exploration on the North Slope, see H. C. Jamison, L. D. Brockett, and R. A. McIntosh, "Prudhoe Bay—A 10-Year Perspective," in Giant Oil and Gas Fields of the Decade 1968–1978, ed. Michel T. Halbouty, 289–313; and R. N. Specht, A. E. Brown, and C. H. Selman, "Geophysical Case History, Prudhoe Bay Field," Geophysics 51 (1986): 1039–49.

14 Herrera quoted in Roderick, Crude Dreams, 133; Bert Bally to author, Feb. 3, 2005.

15 Bowlby interview.

16 Bob Sneider interview by author, Nov. 10, 2004, Houston; Bert Bally interview by Sam Morton, Nov. 29, 1999, Houston; "Oil Hunt Warms Up in Alaska,"22; Charlie Blackburn interview by author, Aug. 23, 1999, Dallas; Threet interview by Beauboeuf; Ron Geer phone conversation with author, Feb. 27, 2004; Bowlby interview.

17 Roderick, Crude Dreams, 187–89; Specht, Brown, and Selman, "Geo-

physical Case History, Prudhoe Bay Field"; Sneider interview.

18 "Western Waters: The New Oil Frontier," *Shell News* 32, no. 6 (Nov.–Dec., 1964): 19–20; Collipp interview by author.

19 "Shell Will Test Federal Acreage off Oregon Coast," *Offshore* (Mar. 1965): 44.

20 "Western Waters," 20.

21 For the full story of the Prudhoe Bay discovery, see Charles S. Jones, *From the Rio Grande to the Arctic;* Kenneth Harris, *The Wildcatter;* and Roderick, *Crude Dreams.*

22 Blackburn interview; Bally interview.

23 "Report from the North Slope," *Shell News* 1 (1970): 16–23; Ron Geer, phone conversation with author, Feb. 27, 2004.

24 Bert Bally letter to author, May 6, 2001.

25 "For Bally, Geology Is an Adventure," *AAPG Explorer* (May 1998): 30–31; Downey interview.

26 Bill Broman interview by author, Dec. 15, 1999, The Woodlands, Tex.

27 Downey interview.

28 Nanz interview.

29 Marlan Downey, "Reefs Hid 'Between the Lines,'" *AAPG Explorer* (Mar. 2000), www.aapg.org/explorer/.

30 Ibid.; "The Sound of Michigan," *Shell News* 4 (1969): 9–11; "Shell Is Leader in Michigan," *The Oil and Gas Journal* (May 19, 1975): 147.

31 Shell Oil, *Annual Report, 1973,* 10.

32 Flowers interview.

33 "Wilson Is Youngest Shell V.P.," *The Oil and Gas Journal* 57, no. 53 (Dec. 28, 1959): 257. After retiring from Shell in 1973, Wilson became president of the AAPG and began a second career as an international consultant. In 1987 he won the AAPG's Sidney Powers Award for distinguished service to the organization, the year before Rufus LeBlanc won it, which made Shell Oil the only company to have winners in back-to-back years. In 1999, he published a celebrated book, *Terroir,* on the geology of French vineyards.

34 "Niobe: A Search Offshore," *Shell News* 5 (1972): 18.

35 Milton B. Dobrin, "Geophysics Offshore Is a Must," *The Oil and Gas Journal* (Oct. 30, 1967): 77–81.

36 Flowers interview; "Lease Sale Nets $60 Million," *Offshore* (June 1964): 13–14.

37 Long interview; Flowers interview; Russell interview.

38 "Offshore Sale Smashes Records," *The*

Oil and Gas Journal (June 19, 1967): 71; "Shell is Big Spender in OCS Auction Off Magical Louisiana," *Ocean Industry* (July 1967): 53.

39 Flowers interview; "'67 Lease Sale No Bargain to Producers in Gulf of Mexico," *Offshore* (Aug. 1972): 29–31.

40 Long interview; Lampton interview.

41 Long interview.

42 Ibid.

43 "Oil Goes Big for Offshore Texas," *Offshore* (June 5, 1968): 17–20.

44 Mike Forrest interview by author, June 29, 1999, Houston; Mike Forrest, "Bright Idea Still Needed Persistence," *AAPG Explorer* (May 2000), www.aapg.org/explorer/.

45 Flowers interview; Mike Forrest, "Evolution of 'Bright Spot' Technology," June 12, 1974, in author's possession; "Bright Spots—A Sure Way to Find Oil and Gas?" *Shell News* 41, no. 3 (1976): 26.

46 Forrest interview.

47 David DeMartini interview by author, July 9, 2004, Houston; DeMartini personal communication to author, July 28, 2004; Forrest, "Bright Idea Still Needed Persistence;" Bookout interview #2; Flowers interview.

48 Marlan Downey, "Carrots, Sticks Are Good Tools," *AAPG Explorer Online* (Oct. 2000), *www.aapg.org/explorer/;* Mike Forrest, personal communication to author, Apr. 30, 2004; Forrest interview.

49 D. A. Holmes, interoffice memorandum, "1970–1986 Lookback of Offshore Lease Sales Gulf of Mexico Cenozoic," Aug. 24, 1987, Shell Offshore Inc., copy provided to author by Mr. Holmes.

50 Forrest interview; E. Baskir, D. C. DeMartini, Shell Development Company, "Do Competitors Utilize 'Bright Spots'?" memorandum to J. H. Robinson, Jan. 13, 1971, copy provided to author by Mr. DeMartini.

51 Flowers interview.

52 "Stampede for Gulf Blocks Smashes All Bonus Records," *The Oil and Gas Journal* (Dec. 25, 1972): 37–41; "The Super Sale of Offshore Leases," *Business Week,* Dec. 19, 1972, 22.

53 Mike Forrest, "'Toast' Was on Breakfast Menu," *AAPG Explorer* (June 2000), www.aapg.org/explorer/; "Win Some, Lose Some," *Shell News* 6 (1972): 13; Blackburn interview.

54 Forrest interview; *Wall Street Journal* quote in "Bright Spots—A Sure Way," 26.

55 Elmer L. Dougherty, Lawrence A. Bruckner, and John Lohrenz, "Cumu-

lative Bonus and Production Profiles with Time for Different Competitive Bidders: Federal Offshore Oil and Gas Leases," paper presented at the 48th Annual California Regional Meeting of the Society of Petroleum Engineers of AIME, San Francisco, Apr. 12–14, 1978, SPE Preprint 7134.

56 Holmes, "1970–1986 Lookback."

57 Ibid.

Chapter 5
The End of Business as Usual

1 John Redmond, notes to author, Mar. 2001.

2 Ibid.

3 Ibid.

4 Alexander Stuart, "Texan Gerald Hines Is Tall in the Skyline," *Fortune,* Jan. 28, 1980, 105.

5 "Tall Texas Tower," *Shell News* 34, no. 3 (May–June, 1966): 20–21.

6 "Shell's $25 Million Trip to Houston," *Business Week,* Sept. 19, 1970, 68.

7 "Let Me Tell You about My New Home Town," *Shell News* 41, no. 6 (1973): 2.

8 Harry Bridges interview by Sam Morton, July 15, 1998, Geneva, Switzerland.

9 Russell interview.

10 M. D. Reifel, "Offshore Blowouts and Fires," in ETA Offshore Seminars, Inc., *The Technology of Offshore Drilling,* 239–57.

11 Doyle, *Riding the Dragon,* 151; "The Wicked Witch Is Dead," *Shell News* 39, no. 2 (1971): 6–7.

12 "The Wicked Witch Is Dead," 3.

13 Ibid; Doyle, *Riding the Dragon,* 151.

14 "The Wicked Witch Is Dead," 8–9; O. J. Shirley interview by Tom Stewart, June 8, 1999, Houston.

15 Harold H. Davenport, et al., "How Shell Controlled Its Gulf of Mexico Blowouts," *World Oil* (Nov. 1971): 71–74.

16 Lucius Trosclair interview by Jamie Christy, July 2, 2004, Bayou Vista, La.; Alden Vining Jr. interview by Jamie Christy, Mar. 16, 2004, Morgan City, La.

17 Hallowell, *Holding Back the Sea;* Mike Tidwell, *Bayou Farewell;* and Don Davis, "From the Marshes to Deepwater, Louisiana's Hydrocarbon Infrastructure Is at Risk," http://www.epa.gov/oilspill/pdfs/d_davis_04.pdf.

18 Tidwell, *Bayou Farewell,* 117; Joel K. Bourne Jr., "Gone with the Water," *National Geographic,* Oct. 2004, 88–105.

19 The casualty rates of the jack-up and semi-submersible were not that far apart when compared to the number of units built. During 1955–74, jack-ups suffered forty-seven casualties out of 143 vessels built, with an estimated total value of damages at $122 million. By comparison, semi-submersibles suffered twelve casualties out of 72 vessels built, with $50 million in total damages. One of out every three jack-up units experienced a casualty, compared to one out of every five semi-submersibles. Ralph G. McTaggart, "Offshore Mobile Drilling Units," in ETA Offshore Seminars, Inc., *The Technology of Offshore Drilling, Completion and Production,* 24–29.

20 Ken Arnold interview by author, May 10, 2004, Houston.

21 Viator interview; Hernandez interview.

22 For a good summary of the problems of enforcement offshore, see Neil R. Etson to President Nixon, Mar. 18, 1970, Box 71, Central Classified Files, 1968–1974, RG 57, NARA.

23 Trosclair interview; for information on the legal history of the Jones Act and the LHWCA, see the Steinberg Law Firm, Offshore Injury Litigation, *http://www.offshoreinjury.net/.*

24 E. W. Standley, interior liaison representative, to Jack W. Boller, assistant executive secretary, Marine Board, National Academy of Engineering, Feb. 15, 1972, Part 13, Box 136, Central Classified Files, 1969–1972, RG 48, NARA; Richard B. Krahl and David W. Moody, "Gulf Coast Lease Management Inspection Program," OTC 1714, paper presented at the 4th Annual OTC, Houston, May 1–3, 1972; Donald Solanas, "Update—OCS Lease Management Program," OTC 1754, paper presented at the 5th Annual OTC, Houston, Apr. 29–May2, 1973.

25 K. E. Arnold, P. S. Koszela and J. C. Viles, "Improving Safety of Production Operations in the U.S. OCS," OTC 6079, paper presented at the 21st Annual OTC, Houston, May 1–4, 1989; Arnold interview.

26 E. L. Pace and C. R. Turner, "Producing Operations Personnel Training—A Program to Help Meet Modern Needs," OTC 1992, paper presented to the 6th Annual OTC, Houston, May 6–8, 1974. The offshore oil industry's safety record in the Gulf improved significantly after the introduction of new regulations and practices. Both the reported incidence and rate of fatalities and injuries in the OCS declined. Although the total number of drilling man-hours reported for the OCS increased from 26 million to 105 million between 1962 and 1977, the

reported accident frequency for the same period declined from 14.9 to 9.3 accidents per 100 man-years. In other words, there was a fourfold increase in exposure hours but a 38 percent decrease in accident frequency. The rate of fatalities in the Gulf of Mexico was also much lower than in the North Sea. During the 1970s, there were 187 fatalities in OCS activities. This averaged about 0.05 fatalities per 100 man-years, whereas the North Sea average was closer to 0.2. According to a 1981 National Research Council study, the frequency of injuries in oil and gas operations in the OCS was comparable to other industries. The injury and illness rates per 100 full-time workers in all oil and gas extraction activities totaled 13.9 in 1978, only slightly higher than general manufacturing. Committee on Assessment of Safety of OCS Activities, *Safety and Offshore Oil*, 134–36.

27 "Shell's Game," *Forbes*, Sept. 15, 1972, 68.

28 Siebenhausen Jr., "Management's Significant Contributions."

29 Shell Oil, *Annual Report, 1975*, 23.

30 Logan Fromenthal interview by Jamie Christy, July 6, 2004, Morgan City, La.; Shell Oil, *Annual Reports*, various years.

31 "Women in Shell," *Shell News* 40, no. 1 (1972): 20–23; "The Career Woman: Anatomy No Longer Determines Destiny," *Shell News* 44, no. 5 (1976): 14–19. For more on women in the offshore industry, see Diane E. Austin, "Women's Work and Lives in Offshore Oil," in Dannhaeuser and Weiner, eds., *Markets and Market Liberalization*, 163–204.

32 Quoted in Yergin, *The Prize*, 588.

33 "Time to Break Out the Umbrella," *Shell News* 41, no. 3 (1973): 2.

34 "He Predicted the Oil Shortage 19 Years Ago," *The National Observer*, date unknown, 1975, 1E.

35 Ibid., 20E.

36 Richard K. Vietor, *Energy Policy in America since 1945*, 195; Shell Oil, *Annual Report, 1974*, 19–21; Barran quoted in Howarth, *A Century in Oil*, 302.

37 Yergin, *The Prize*, 590.

38 "Oil Decontrol: Who Would be Helped, Who Hurt," *Forbes*, Sept. 1, 1975, 15.

39 "How to Think about Oil-Company Profits," *Fortune*, Apr. 1974, 98–103; Yergin, *The Prize*, 659; Anthony Sampson, *The Seven Sisters*, 322.

40 Quoted in Yergin, *The Prize*, 657.

41 Robert Sherrill, *The Oil Follies of 1970–1980*, 213, 217.

42 Sampson, *The Seven Sisters*, 324.

43 Shell Oil, *Annual Report, 1974*; S. G. Stiles, Presentation—Public Relations Meeting, Mar. 21, 1973; Stan Stiles interview by James Cox, 1973, Houston.

44 Stiles interview by Cox.

45 Bridges interview.

46 "The Bridges of Shell," *Shell News* 39, no. 6 (1971): 4, 5.

47 Bridges interview.

48 John Bookout interview #1 by Tom Stewart, July 16, 1998, Houston.

49 Spaght interview.

50 Bookout interview #1.

51 *Robert Halpern et al., v. D. H. Barran, et al.*, Court of Chancery of Delaware, New Castle, 272 A.2d 118; 1970 Del. Ch. Lexis 93, Dec. 15, 1970; Shell Oil, *Annual Report Form 10-K*, 1977, 7.

52 On the growing disparity between "posted" and "arms-length" prices, see Blair, *The Control of Oil*, 261–75.

53 *Robert Halpern et al., v. D. H. Barran, et al.*

54 Bridges interview.

55 Bridges interview; John Redmond interview by author.

56 "Pecten Companies," *Shell News* 4 (1980): 1.

57 Michael Tanzer and Stephen Zorn, *Energy Update*, 113–17.

58 Bridges interview.

59 "Once over Lightly," *Shell News* 44, no. 1 (1976): 16.

60 "Shell and the World: An Update," *Shell News* 2 (1982): 3.

61 Ibid., 2–3; "That's Some Move Pecten International," *Shell News* 1 (1984): 8–10.

62 Downey interview.

63 "A Visit with Marlan Downey," *Venture* 1, no. 3 (1984): 14.

64 Ibid., 12; "Shell and the World," 1.

65 Forrest interview.

66 Russell interview.

67 Yergin, *The Prize*, 665.

68 "'It's Time to Take Risks,'" *Forbes*, Oct. 6, 1986, 127.

69 Bookout interview #1.

70 Bridges interview.

71 Nanz interview; "Shell's New President Must Find More Oil," *Business Week*, Dec. 22, 1975, 19.

72 Blackburn interview.

73 "Born with a Little Somethin,'" *Newsweek*, June 15, 1981, 77; Jack Little

interview by author and Sam Morton, Feb. 14, 2000, Houston.

74 "Shell Oil's Man in the Middle," *Fortune*, May 28, 1984, 182; "Born with a Little Somethin,'" 77.

75 "For Big Units, Gulf Coast Is Still No. 1," *Chemical Week* (Sept. 14, 1977): 16–17.

76 Bookout interview #2.

77 In August 1975, Shell Oil, through its Pecten Arabian Limited subsidiary, and Petromin, the Saudi Arabian national oil company in charge of refining, began negotiations for installing a jointly owned complex that would produce ethylene and other petrochemicals in return for granting Shell a long-term contract for crude. After several years of negotiations, Pecten Arabian Limited and the Saudi Basic Industries Corporation together formed the Saudi Petrochemical Company—nicknamed "Sadaf," which is Arabic for "sea shell"—to build and operate the complex. The $2.7 billion petrochemical complex, in which Shell provided $400 million in equity capital, was the first and largest project in Saudi Arabia's industrial program and one of the most ambitious petrochemical construction projects ever attempted. In 1981, Shell Oil began lifting crude based on a contract that gave it the right to buy up to one billion barrels of Saudi oil over a fifteen-year period. But the company imported only a few cargoes, as new and cheaper production from around the world started pouring onto the market. "Coming into the Kingdom," *Shell News* 6 (1980): 17; "Putting a Petrochemical Plant Together Piece by Piece," *Shell News* 3 (1982): 26–29; "The Saudi Experience," *Shell News* 6 (1985): 1–2.

78 Ibid; "The Oil Companies Are Sloshing in Ethylene," *Business Week*, July 10, 1978, 84.

79 John Bookout, Overview of Presidency notebook, provided to author by Bookout.

80 John Bookout, Presentation to the Los Angeles Society of Financial Analysts, Oct. 12, 1978, 3, provided to author by Bookout.

81 "Aggressive Shell on Hunt for Oil as Rivals Retreat," *Los Angeles Times*, Mar. 29, 1987, 4.

Chapter 6
The Offshore Imperative

1 "Rising above the Crowd," *Shell News* 6 (1989): 28.

2 "A Visit with Charlie Blackburn," *Venture* 1 (1984): 3.

3 "The Offshore: Heading for Deep Water," *Shell News* 4 (1981): 3.

4 Blackburn interview.

5 "The Significance of One Shell Square," special issue of *The Times-Picayune Dixie Roto Sunday Magazine*, Sept. 3, 1972.

6 B. G. Collipp and C. A. Sellars, Shell Development Co., "The Role of One Atmosphere Chambers in Sub-sea Operations," OTC 1529, paper presented at the 4th Annual OTC, Houston, May 1–3, 1972; "Offshore Operators Looking Deeper," *The Oil and Gas Journal* (Apr. 30, 1973): 118–20; "Breaking the Deepwater Barrier," *Shell News* 1 (1977): 23–25.

7 Carl Wickizer interview by Bruce Beauboeuf, Nov. 27, 1997, Houston.

8 Pat Dunn interview by Joseph Pratt and Bruce Beauboeuf, July 1, 1996, Columbus, Tex.

9 Pat Dunn, "Deepwater Production: 1950–2000," OTC 7627, paper presented at the 26th Annual OTC, Houston, May 2–5, 1994.

10 "Offshore Technology Meeting Spotlights Progress, Problems," *Offshore* (June 5, 1969): 23–24.

11 Dunn, "Deepwater Production."

12 Ibid.

13 "Camille Knocks out 300,000 b/d and Costs Industry $100 Million," *Offshore* (Sept. 1969): 33–35; "How Sea-Floor Slides Affect Offshore Structures," *The Oil and Gas Journal* (Nov. 29, 1971): 88–92.

14 E. G. "Skip" Ward interview by Joseph Pratt, Oct. 17, 1998, Houston; Dunn interview.

15 Dunn, "Deepwater Production."

16 Lloyd Otteman interview by author, May 17, 2001, Houston; Ward interview.

17 Mike Forrest, "'Toast' Was on the Breakfast Menu," *AAPG Explorer* (June 2000), www.aapg.org/explorer/.

18 John Redmond notes on the Cognac Project, in author's possession.

19 See explanation in "On the Block: One Million Acres," *Shell News* 1 (1982): 6.

20 Redmond notes on the Cognac Project.

21 The Shell group consisted of Shell (41.67 percent), Conoco (33.33 percent); Sonat Exploration (10.42 percent); Drillamex (4.17 percent); Barber Oil (4.17 percent); Florida Gas Exploration (4.16 percent); and Offshore Co. (2.08 percent). "Bidders Snub Most Deepwater Tracts," *The Oil and Gas Journal* (Apr. 8, 1974): 36–40.

22 Redmond interview.

23 "Deepwater Tracts Will Be Offered on

Continental Slope for the First Time," *Offshore* (Oct. 1973): 52–53; "Bidders Snub Most Deepwater Tracts," 36–40.

24 Otteman interview; Forrest, "'Toast' Was on the Breakfast Menu."

25 Sam Paine interview by author, June 8, 1999, Houston.

26 Redmond notes on Cognac Project; Paine interview.

27 "Cognac Goes Down Smoothly," *Shell News* 6 (1977): 2–3.

28 Dunn, "Deepwater Production"; "Cognac Rises into a Class of Its Own off Louisiana Coast," *Offshore Engineer* (July 1978): 32.

29 Paine interview.

30 "Tallest Offshore Oil Platform Sets Water Installation Records," *Engineering-News Record* (Aug. 24, 1978): 20.

31 "Cognac Goes Down Smoothly," 5.

32 "Shell's Cognac Yields New Technology," *The Oil and Gas Journal* (Aug. 15, 1977): 36–37; "Cognac Rises into a Class of Its Own," 32.

33 Collipp interview by author; "Tallest Offshore Oil Platform," 22.

34 Paine interview; Shell Oil Company, *Cognac,* pamphlet, provided to author by Paine.

35 Paine interview.

36 "Why Oilmen Are So Cool to Offshore Leases," *Business Week,* Apr. 26, 1976, 72–73.

37 "Shell's New President Must Find More Oil," 20.

38 "Bookout Says OCS Search is Urgent, Industry Prepared," *The Oil and Gas Journal* (May 19, 1975): 153; Nanz, "The Offshore Imperative"; Shirley interview.

39 "'Be Prepared' Is the Motto of Clean Atlantic Associates," *Shell News* 4 (1978): 10–13; Shirley interview.

40. Shirley interview; "Spuddin' the Atlantic," *Shell News* 4 (1978): 6.

41 "Prospects Darken for Baltimore Canyon," *The Oil and Gas Journal* (Mar. 5, 1979): 72; "The Time to Start Looking Is Now," *Shell News* 6 (1984): 15; Threet interview by Beauboeuf.

42 "Georges Bank: First Step in the North Atlantic," *Shell News* 6 (1981): 2–4.

43 "Long Beach Builds Four Treasure Islands," *Shell News* 35, no. 1 (Jan.–Feb. 1967): 24–27.

44 Gramling, *Oil on the Edge,* 118–26.

45 Shell Oil Company, Western Exploration and Production Operations, *Project Beta,* pamphlet.

46 Carroll interview #1.

47 Ibid.

48 "Archimedean in Scope," *Shell News* 5 (1984): 31–33.

49 Ibid., 33.

50 "Oilmen Turn Cool in Alaska," *Business Week,* Sept. 1972.

51 "Why Oilmen Are So Cool to Offshore Leases," 72.

52 Ibid., 73.

53 Downey interview; Shell Oil Company, *Annual Report, 1976.*

54 Nanz interview; Shell Oil, *Annual Report—1977.*

55 "Shell: Alaska Holds 58% of Future U.S. Oil Finds," *The Oil and Gas Journal* (Nov. 20, 1978): 214; Little interview.

56 R. H. Nanz, Shell Oil Company, "'What We Need to Increase Domestic Oil and Gas Supplies" (Feb. 1977), provided to author by Nanz; "Shell Backs Offshore Cash-Bonus System," *The Oil and Gas Journal* (Apr. 29, 1974): 18.

57 "Caution Urged in Using New Offshore Bid System," *Oil & Gas Journal* (Feb. 25, 1980): 32.

58 "Interior's OCS Leasing Plan Advances," *Oil & Gas Journal* (Feb. 6, 1982): 70.

59 Nanz, "The Offshore Imperative"; Mike Forrest interview.

60 "The Time to Start Looking Is Now," 16.

61 Also see Pratt, *Prelude to Merger,* 114–16.

62 "High OCS Sale 71 Bids Top $2 Billion," *Oil & Gas Journal* (Oct. 18, 1982): 48–50.

63 "Diapir Basin High Bids Hit $877 Million," *Oil & Gas Journal* (Aug. 27, 1984): 38.

64 Quoted in Yergin, *The Prize,* 733.

65 Jim DeNike interview by author, July 13, 1998, Houston; "The Great Arctic Energy Rush," *International Business Week* (Jan. 24, 1983): 70–74.

66 Little interview.

67 Ibid; Forrest interview.

68 Shell, however, has never given up hope. In 2005, encouraged by soaring oil and gas prices, Shell Offshore Inc. purchased a large number of leases in the March 2005 Beaufort Sea OCS sale. And in the fall 2005 State of Alaska lease sale, the company purchased numerous nearshore and onshore leases in Bristol Bay. Alan Bailey, "Shell Declares Hand: Focused on Offshore Alaska, Committed to Using Cutting-Edge Technology," *Petroleum News* (Jan. 29, 2006), www.freerepublic.com/focus/ f-news/1566900/posts.

69 Holmes, "1970–1986 Lookback."

70 "Gulf Lease Sale Ranks Seventh Off U.S.," *The Oil and Gas Journal* (July 4, 1977): 33–34.

71 "Success Ratio High on June '77 Leases in Gulf," *The Oil and Gas Journal* (Oct. 30, 1978): 27–31.

72 "Development Slated for 14 Gulf of Mexico Fields," *Oil & Gas Journal* (Apr. 7, 1980): 46.

73 Holmes, "1970–1986 Lookback."

74 Blackburn interview; "Gulf Lease Sale Shatters Two Records," *Oil & Gas Journal* (Oct. 6, 1980): 34.

75 The block numbers for these code names were: Roberto—Matagorda 681; Hornet—Vermillion 221; Cougar—South Timbalier 300; Boxer—Green Canyon 19; Glenda—High Island A6; Wasp—East Cameron 300; Peccary—South Timbalier 292; Hobbit—Ship Shoal 259; Cheetah—Main Pass 310; Onyx—West Cameron 170; Persian—Ship Shoal 189. Holmes, "1970–1986 Lookback."

76 Holmes, "1970–1986 Lookback."

77 "At Issue: Land Access," *Shell News* 5 (1981): 18–19; Charles Frederick Lester, "The Search for Dialogue in the Administrative State: The Politics, Policy, and Law of Offshore Development" (Ph.D. dissertation, University of California, Berkeley, 1992), 91–93; Flowers interview.

78 Bookout interview #2.

79 Leighton Steward interview by author, May 11, 2004, Boerne, Tex.

80 Bookout interview #2.

81 Flowers interview; Broman interview.

82 J. Robinson West interview by author, Nov. 18, 2002, Washington, D.C.; Otteman interview.

83 "Interior's OCS Leasing Plan Advances," *Oil & Gas Journal* (Feb. 8, 1982): 70–71.

84 Bookout interview #2; "Ocean Drilling over a Mile Down: A Subject Deep but Not Dark," *Shell News* 2 (1983): 1–6; Carl Wickizer personal communication to author, June 27, 2001.

85 Carl Wickizer, "Out on the Horizon: Creating Means to Drill and Produce beyond Current Water Depth Limitations," presentation to Offshore Northern Seas Conference, Stavanger, Norway, Aug. 23, 1988.

86 "Revealing Secrets of the Deep," *Venture* 2 (1987): 1–3.

87 "Gulf of Mexico Exposure Totals $4.5 Billion," *Oil & Gas Journal* (June 6, 1983): 48.

88 "Gulf Deepwater Tracts Spark Bidding," *Oil & Gas Journal* (Apr. 30, 1984): 36.

89 Forrest interview.

90 Flowers interview; "*Shell America: A Sophisticated Ship, She Is*," *Shell News* 3 (1984): 1–9.

91 Thorpe, "Oil and Water," 143.

92 Ibid.

93 Broman interview.

94 Mike Forrest, "'Bright' Investments Paid Off," *AAPG Explorer* (July 2000), www.aapg.org/explorer/.

95 Ibid.

96 "Deep Water, Mobile Bay Tracts Spark Sale 98," *Oil & Gas Journal* (May 27, 1985): 46–47.

Chapter 7
Deepwater Treasures
in a New Era of Oil

1 "Shell, A Fallen Champ of Oil Industry, Tries to Regain its Footing," *Wall Street Journal,* Aug. 30, 1991, 1.

2 Texaco also acquired Getty Oil during this period but only after a monumental legal battle in which a Texas jury awarded Pennzoil, which had made an earlier deal to buy a large portion of Getty and sued Texaco over control, an unprecedented $11 billion judgment. Texaco was able to get the settlement reduced to $3 billion but was still forced into bankruptcy. See Steve Coll, *The Taking of Getty Oil.*

3 Donal Woutat, "Aggressive Shell on Hunt for Oil as Rivals Retreat," *Los Angeles Times,* Mar. 29, 1987, part 4, p. 1.

4 Toni Mack, "'It's Time to Take Risks,'" *Forbes,* Oct. 6, 1986, 126–28.

5 M. S. Robinson, quoted in Yergin, *The Prize,* 697.

6 Yergin, *The Prize,* 717.

7 Howarth, *A Century in Oil,* 348.

8 Quoted in Yergin, *The Prize,* 726.

9 Yergin, *The Prize,* 727.

10 "Royal Dutch Is Set to Swallow Shell Oil," *Business Week,* Feb. 6, 1984, 32.

11 "The Shell Buyout Becomes a Family Feud," *Business Week,* Apr. 16, 1984, 52; "Shell Oil's Man in the Middle," *Fortune,* May 28, 1984, 182–85.

12 "Why Royal Dutch/Shell Is Betting on the U.S.," *Business Week,* Feb. 20, 1984, 99.

13 Court of Chancery of the State of Delaware, *Smith v. Shell Petroleum, Inc.,* No. 8395, *Delaware Journal of Corporate Law,* spring 1991, 16 Del. J. Corp. L. 870.

14 "'It's Time to Take Risks,'" 128.

15 "Shell Buyout Becomes a Family Feud," 52.

16 Quoted in "Shell Oil's Man in the Middle," 184.

17 "Shell Oil's Man in the Middle," 184–85; Tom Stewart interview by author, June 30, 1999, Houston.

18 "Shell Oil's Man in the Middle," 182; "A Royal Shell Game," *Newsweek,* May 14, 1984, 54; "Down to the Wire at Shell Oil," *Business Week,* May 14, 1984, 41.

19 "Down to the Wire at Shell Oil," 42; *Smith v. Shell Petroleum, Inc.*

20 *Smith v. Shell Petroleum, Inc.;* "The Last Angry Man?" *Forbes,* June 17, 1985, 191–94; "Shell, A Fallen Champ of Oil Industry," 1; Associated Press, "Shell Stockholder Award Upheld," Apr. 23, 1992.

21 Quoted in Howarth, *A Century in Oil,* 346; "Shell, A Fallen Champ of Oil Industry," 1.

22 L. C. van Wachem to coordinators and independent division heads, Sept. 16, 1986, "Relationships with Shell Oil," copy provided to author by John Bookout.

23 "Launching a Superlative," *Venture* 1 (1989): 1.

24 Holmes, "1970–1986 Lookback."

25 "Rising above the Crowd," *Shell News* 57, no. 6 (1989): 28, 30.

26 Ibid., 30.

27 Ibid., 31.

28 Ibid.

29 Ibid.

30 "Bullwinkle Takes Shape," *Shell News* 55, no. 6 (1987): 5.

31 "Bullwinkle Lands on Its Feet," *Shell News* 56, no. 6 (1988): 2, 3.

32 Ibid., 3, 4.

33 "Rising above the Crowd," 30; "Bullwinkle Production Begins," *Venture* 6 (1989): 4.

34 "Bullwinkle Takes Shape," 6.

35 Holmes, "1970–1986 Lookback."

36 Forrest, "'Bright' Investments Paid Off."

37 "A Giant Step Outward," *Venture* 1 (1990): 1–5; Forrest, "'Bright' Investments Paid Off."

38 Richard Pattarozzi interview by author, May 15, 2000, Houston.

39 "A Giant Step Outward," 4.

40 "The Shape of Things to Come," *Shell News* 63, no. 3 (1994): 24.

41 Ibid., 23–24; "2,860 Feet under the Sea, a Record-Breaking Well," *New York Times,* Apr. 24, 1994, 9.

42 Pattarozzi interview; Thorpe, "Oil & Water," 144.

43 Pattarozzi interview; Bob Howard interview by author, May 18, 2000, Houston.

44 Rita Robison, "Bullwinkle's Big Brother," *Civil Engineering* (July 1995): 46; Pattarozzi interview.

45 "The Shape of Things to Come," 23.

46 Pattarozzi interview.

47 Jeff Ryser, "Hot Play in the Gulf," *Texas Business* (Aug. 1995): 33.

48 "Debottlenecking Removes Auger Production Constraints," *Oil & Gas Journal Online* (Nov. 11, 1996), www.ogj.com; "Auger: Moving into the Future," Supplement to *Hart's E&P* and *Oil and Gas Investor* (2001), 6.

49 Quoted in "Oil & Water," 144.

50 "Plumbing the Depths with Tahoe," *Shell News* 62, no. 4 (1994): 18.

51 Ibid., 19; "Shell Poised for Gulf Deepwater Installation," *Oil & Gas Journal Online* (Apr. 10, 1995), www.ogj.com; "Pulling Oil from Davy Jones' Locker," *Business Week,* Oct. 30, 1995, 74.

52 "Shell to Set Another World Water Depth Mark in Gulf," *Oil & Gas Journal Online* (June 5, 1995), www.ogj.com.

53 "BP Group Starts Up Troika Field in Gulf of Mexico," *Oil & Gas Journal Online* (Nov. 17, 1997), www.ogj.com.

54 "Auger: Moving into the Future," 46.

55 Forrest interview; "Shell's Mars Mission: A Deepwater Odyssey," Supplement to *Hart's E&P* and *Oil and Gas Investor* (1999): 6–9.

56 Forrest, "'Bright' Investments Paid Off"; "Launch Pad into the Deep," *Houston Chronicle,* Aug. 17, 1997, 2E.

57 "Shell's Mars Mission," 6–9; "Launch Pad into the Deep," 7.

58 Broman interview.

59 Ibid.; Forrest, "'Bright' Investments Paid Off.'"

60 "Launch Pad into the Deep," 2E; Pattarozzi interview.

61 "Creating a New Frontier," *Shell News* 63, no. 4 (1995): 15.

62 "Launch Pad into the Deep," 2E.

63 Ibid.

64 "Creating a New Frontier," 17–19.

65 Pattarozzi interview.

66 Ibid.; "Launch Pad into the Deep," 2E.

67 "Mission: Make it Bigger, but Don't Stop Drilling or Production," *Oil & Gas Investor* (1999): 44–49.

68 Pattarozzi interview; "Creating a New Frontier," 20; Peacock quoted in "Launch Pad into the Deep," 1E.

69 "Shell's Mars Mission," 12.

70 "The Cloning of Mars," *Shell News* 64, no. 1 (1996): 8–13.

71 Ibid., 10.

72 "Amoco, Shell Slate Another TLP for Eastern Gulf of Mexico," *Oil & Gas Journal* 95, no. 5 (Feb. 3, 1997): 26; Shell Oil, *Shell in the U.S.: 1999 Annual Review*, 16; "Shell Oil Goes Deep in Gulf," *Houston Chronicle*, Apr. 9, 1999, 1C, 8C; "Shell Sets Record at Princess," *Petroleum News On-Line*, Jan. 1, 2004, www.petroleumnews.com.

73 "Oil Firms Building Gulf Pipelines," *Houston Chronicle*, Aug. 23, 1996, 1C, 8C.

74 James R. McCaul, "Floating Production Systems Provide a Capability to Produce Deepwater and Remote Fields," *Oil & Gas Journal Online* (June 11, 2001), www.ogj.com.

75 "Shell Announces Two Additional U.S. Deepwater Successes," *Shell News* (Oct. 2000): 19.

76 Pattarozzi interview.

Epilogue
The Globalization of Shell Oil

1 "He's Changing the Drill at Shell," *Business Week*, Oct. 21, 1996, 35; "The Undersea World of Shell Oil," *Business Week*, May 15, 1995, 74.

2 Lane Sloan interview by Sam Morton, Aug. 31, 2000, Houston.

3 "Oil Industry's Shining Star Falls," *Houston Chronicle*, Aug. 11, 1991, 1; Little interview.

4 "Shell, a Fallen Champ of Oil Industry, Tries to Regain Its Footing," *The Wall Street Journal*, Aug. 30, 1991, 1; Phil Carroll interview #2 by author, Joseph Pratt, and Sam Morton, June 12, 1998, Houston,.

5 "Oil Industry's Shining Star Falls," 1.

6 Merrell Seggerman interview by author, Mar. 24, 1999, Houston.

7 The most stunning and imaginative step taken by Shell in restructuring its assets, however, was the announcement in late 1992 that the Mexican national oil company, Pemex, would purchase one-half of Shell's refining assets at the Houston Deer Park Manufacturing Complex. Shell Oil, *Annual Report, 1992*, 14.

8 "Royal Dutch Shell Group Europe's Second Largest IS Spender, Repositioning Role of IS, New IS Group Proposed," *Information Week* (Dec. 11, 1995): 45.

9 "He's Changing the Drill at Shell," 35.

10 See Thomas Frank, *One Market under God*, for a good analysis of the hucksterism of these and other management consultants during the "new economy" euphoria of the 1990s. For a good discussion of similar managerial

reforms at another oil company, see Pratt, *Prelude to Merger*, chapter 2.

11 Joel Kurtzman, "Is Your Company Off Course? Now You Can Find Out Why?" *Fortune*, Feb. 11, 1997, 129.

12 "America's Most Admired," *Shell News* 63, no. 2 (1994): 23.

13 "A First: Amoco, Shell to Form U.S. Joint Company," *Oil & Gas Investor* 93, no. 42 (Oct. 16, 1995): 36; Pratt, *Prelude to Merger*, 109.

14 "Shell, Mobil Agree to Combine California E&P," *Oil & Gas Journal* 95, no. 23 (June 9, 1997): 24.

15 Quoted in Coll, *The Taking of Getty Oil*, 337.

16 Shell Oil Company, *Annual Report, 1998*, 2. Carroll resigned to assume the presidency of Fluor. A month after the U.S. invasion of Iraq in 2003, Carroll took control of Iraq's oil production for the U.S. government and prevented efforts by neo-conservatives in the Bush administration to privatize Iraqi oil. See Greg Palast, "OPEC on the March: Why Iraq Still Sells Its Oil à la Cartel," *Harper's*, Apr. 2005, 74–76.

17 "Going Global," *Shell News* (summer 1998): 8–9.

18 "Shell Gets Rich by Beating Risk," *Fortune*, Aug. 26, 1991, 79; "Learning across a Living Company: The Shell Companies' Experiences," *Organizational Dynamics* 27, no. 2 (autumn, 1998): 61–69.

19 "Shell's Crystal Ball," *Financial World* (Apr. 16, 1991): 58.

20 "'The Opportunities are Enormous,'" *Forbes*, Nov. 9, 1992, 92–94; "A Tale of Two Strategies," *Forbes*, Aug. 17, 1992, 48–50.

21 "Slow Payoff," *Forbes*, Feb. 27, 1995, 64.

22 "Why Is the World's Most Profitable Company Turning Itself Inside Out?" *Fortune*, Aug. 4, 1997, 122.

23 "How Do We Stand?" *Shell World* (June/July 2000): 38–39.

24 "Why Is the World's Most Profitable Company," 125.

25 "The World According to Shell," *Houston Chronicle*, June 30, 2000, 1D, 7D.

26 Shell International Ltd., "Shell Restructures Portfolio," *Shell News Review* 12 (Dec. 1998): 8–9.

27 "Royal Dutch/Shell 4th-Quarter Profit More Than Doubles as Oil Prices Surge" (Feb. 10, 2000), www .bloomberg.com.

28 "Gulf Oil Fields Acquired from Shell," *Houston Chronicle*, Apr. 30, 1999, 1C; Shell Oil, *Annual Report, 1998*, 10; "Oil Companies Set to Team Up in Gulf,"

Houston Chronicle, July 20, 1999; "Shell Units Will Move Bases Here," *Houston Chronicle,* Mar. 12, 1999, 1C.

29 "SEPCo Charts a Course toward the Future," *Shell News* (Feb. 2000): 9.

30 "Ultradeepwater Play Paces Gulf, North America," *Oil & Gas Journal Online* (Nov. 5, 2001), www.ogj.com.

31 "At Shell, Strategy and Structure Fueled Troubles," *The Wall Street Journal,* Mar. 12, 2004, 1; Ferruh Demirmen, "Shell's Reserves Revision: A Critical Look," *Oil & Gas Journal* (Apr. 5, 2004): 43–46.

32 "Industry Wants Clarity on Reserves, Says SEC Reserve Rules out of Date," *Petroleum News Online* (Apr. 18, 2004), *www.petroleumnews.com;* "SEC Approves Technology Check for Gulf Proven Reserves," *Houston Chronicle,* Apr. 14, 2004, www.chron.com.

33 Shell Oil, "The Blueprint for Shell in the U.S," *Shell News* (Aug. 2000): insert.

34 "Refining Shell," *Continental* (Aug. 2003): 36–39.

35 Deffeyes, *Beyond Oil,* 9.

36 On the debate over peak oil, see Colin J. Campbell, *The Coming Oil Crisis;* Deffeyes, *Hubbert's Peak.* Campbell predicts the peak for the world outside the Middle East and North Africa in 2006 and the world including those regions in 2016. The U.S. Geological Survey, on the other hand, offers more conservative estimates of 2023 and 2040, respectively. Tim Appenzeller, "The End of Cheap Oil?" *National Geographic,* June 2004, 90 Also see Bob Williams, "Debate Over Peak-Oil Issue Boiling Over," 18–29.

Selected Bibliography

Appenzeller, Tim. "The End of Cheap Oil?" *National Geographic.*
June 2004, 81–109.

Austin, Diane E. "Women's Work and Lives in Offshore Oil."
In *Markets and Market Liberalization: Ethnographic Reflections,* edited by Norbert Dannhaeuser and Cynthia Werner,
163–204. Oxford: Elsevier, 2006.

Bamberg, James. "OLI and OIL: BP in the U.S. in Theory and Practice." In *Foreign Multinationals in the United States: Management
and Performance,* edited by Geoffrey Jones and Lina Gálvez-
Muñoz, 169–87. London: Routledge, 2002.

Beaton, Kendall. *Enterprise in Oil: A History of Shell in the United
States.* New York: Appleton-Century-Crofts, 1957.

Blair, John M. *The Control of Oil.* New York: Vintage Books, 1978.

Boulard, Gary. *Huey Long Invades New Orleans: The Siege of a City,
1934–36.* Gretna, La.: Pelican Publishing Company, 1998.

Bryce, Robert. *Cronies: Oil, the Bushes, and the Rise of Texas, America's
Superstate.* New York: Public Affairs, 2004.

Burleson, Clyde W. *Deep Challenge! The True Epic Story of Our Quest
for Energy beneath the Sea.* Houston: Gulf Publishing, 1999.

Campbell, Colin J. *The Coming Oil Crisis.* Brentwood, England, 1997.

Castells, Manuel. *The Rise of the Network Society.* Oxford: Blackwell,
1996.

Coll, Steve. *The Taking of Getty Oil: The Full Story of the Most
Spectacular—& Catastrophic—Takeover of All Time.* New York:
Atheneum, 1987.

Collipp, Bruce G. "Offshore Industry 1950–1965: Invention." In *A
Half Century of Maritime Technology 1943–1993,* edited by Harry
Benford and William A. Fox. Jersey City, N.J.: Society of
Naval Architects and Marine Engineers, 1993.

Committee on Assessment of Safety of OCS Activities, *Safety and
Offshore Oil.* Washington, DC: National Academy Press, 1981.

Cortada, James. *The Digital Hand: How Computers Changed the Work
of American Manufacturing, Transportation, and Retail Industries.*
Oxford: Oxford University Press, 2004.

Deffeyes, Kenneth S. *Beyond Oil: The View from Hubbert's Peak.* New
York: Hill and Wang, 2005.

———. *Hubbert's Peak: The Impending World Oil Shortage.* Princeton:
Princeton University Press, 2001.

Deterding, Sir Henri, as told to Stanley Naylor. *An International Oilman.* London: Ivor Nicholson and Watson, Ltd., 1934.

Detro, Randall A. "Transportation in a Difficult Terrain: The Development of the Marsh Buggy." *Geoscience and Man* 19 (June 30, 1978): 93–99.

Doyle, Jack. *Riding the Dragon: Royal Dutch Shell & the Fossil Fire*. Washington, D.C: Environmental Health Fund, 2002.

Forbes, R. J., and D. R. O'Beirne. *The Technical Development of the Royal Dutch/Shell, 1890–1940*. Leiden: Brill, 1957.

Fox, Milden J., Jr. "The Impact of Work Assignments on Collective Bargaining in the Petroleum Refining Industry on the Texas Gulf Coast." Ph.D. dissertation. Texas A&M University, 1969.

Frank, Thomas. *One Market under God: Extreme Capitalism, Market Populism, and the End of Economic Democracy*. New York: Doubleday, 2000.

Gerretson, Frederik Carel. *History of the Royal Dutch*. 4 vols. Leiden: Brill, 1953–57.

Goodwin, Craufurd, ed. *Energy Policy in Perspective: Today's Problems, Yesterday's Solutions*. Washington, D.C.: The Brookings Institution, 1981.

Gordon, Robert. "'Shell No!' OCAW and the Labor-Environmental Alliance." *Environmental History* 3, no. 4 (October 1998): 460–87.

Gramling, Robert. *Oil on the Edge: Offshore Development, Conflict, Gridlock*. Albany: State University of New York Press, 1996.

Greene, William N. *Strategies of the Major Oil Companies*. Ann Arbor, Mich.: UMI Research Press, 1985.

Hallowell, Christopher. *Holding Back the Sea: The Struggle for America's Natural Legacy on the Gulf Coast*. New York: HarperCollins, 2001.

Harris, Kenneth. *The Wildcatter: A Portrait of Robert O. Anderson*. New York: Weidenfeld and Nicolson, 1987.

Henriques, Samuel. *Marcus Samuel: First Viscount Bearsted and Founder of "Shell" Transport and Trading Company, 1853–1927*. London: Barrie and Rockliff, 1960.

Hollick, Ann. *U.S. Foreign Policy and the Law of the Sea*. Princeton: Princeton University Press, 1981.

Horsnell, Paul. "Oil Company Histories." *The Journal of Energy Literature* 5, no. 2 (December 1999): 3–31.

Howarth, Stephen. *A Century in Oil: The "Shell" Transport and Trading Company, 1897–1991*. London: Weidenfeld and Nicolson, 1997.

Hughes, Thomas P. *Human-Built World: How to Think about Technology and Culture*. Chicago: University of Chicago Press, 2004.

Jamison, H. C., L. D. Brockett, and R. A. McIntosh. "Prudhoe Bay—A 10-Year Perspective." In *Giant Oil and Gas Fields of the Decade 1968–1978*, edited by Michel T. Halbouty, 289–313. Tulsa: American Association of Petroleum Geologists, 1980.

Jones, Charles S. *From the Rio Grande to the Arctic: The Story of the Richfield Oil Corporation*. Norman: University of Oklahoma Press, 1972.

Jones, Geoffrey, and Lina Gálvez-Muñoz, eds. *Foreign Multinationals in the United States: Management and Performance.* London: Routledge, 2002.

Kallman, Robert E., and Eugene D. Wheeler. *Coastal Crude in a Sea of Conflict.* San Luis Obispo, Calif.: Blake Printery & Publishing, 1984.

Kaufman, Burton I. *The Oil Cartel Case: A Documentary Study of Antitrust Activity in the Cold War Era.* Westport, Conn.: Greenwood Press, 1978.

Kreidler, Tai Deckner. "The Offshore Petroleum Industry: The Formative Years, 1945–1962." Ph.D. dissertation. Texas Tech University, 1997.

Laborde, Alden J. *My Life and Times.* New Orleans: Laborde Printing Company, 1996.

Lankford, Raymond L. "Marine Drilling." In *History of Oil Well Drilling,* edited by J. E. Brantly, 1358–1444. Houston: Gulf Publishing, 1971.

Lawyer, L. C., Charles C. Bates, and Robert B. Rice. *Geophysics in the Affairs of Mankind: A Personalized History of Exploration Geophysics.* Tulsa: Society of Exploration Geophysicists, 2001.

Lerner, Steve. *Diamond: A Struggle for Environmental Justice in Louisiana's Chemical Corridor.* Cambridge, Mass.: MIT Press, 2005.

Lester, Charles Frederick. "The Search for Dialogue in the Administrative State: The Politics, Policy, and Law of Offshore Development." Ph.D. dissertation. University of California, Berkeley, 1992.

Louisiana State Mineral Board, *Biennial Report,* various years, Special Collections, Hill Memorial Library, Louisiana State University, Baton Rouge, La.

Mayne, W. Harry. *50 Years of Geophysical Ideas.* Tulsa: Society of Exploration Geophysicists, 1989.

McCloy, John J., Nathan W. Pearson, and Beverly Matthews. *The Great Oil Spill: The Inside Report—Gulf Oil's Bribery and Political Chicanery.* New York: Chelsea House, 1976.

McTaggart, Ralph G. "Offshore Mobile Drilling Units." In *The Technology of Offshore Drilling, Completion and Production,* compiled by ETA Offshore Seminars, Inc., 24–29. Tulsa: Petroleum Publishing Company, 1976.

National Archives and Records Administration (NARA), College Park, Md., Record Group 48, Records of the U.S. Secretary of Interior; Record Group 59, Records of the U.S. State Department, decimal file.

Okonta, Ide. *Where Vultures Feast: Shell, Human Rights, and Oil.* London: Verso, 2003.

Olien, Diana Davids, and Roger M. Olien. *Oil in Texas: The Gusher Age, 1895–1945.* Austin: University of Texas Press, 2002.

Olien, Roger M., and Diana Davids Olien. *Oil and Ideology: The Cultural Creation of the American Petroleum Industry.* Chapel Hill: University of North Carolina Press, 2000.

Owen, Edgar Wesley. *Trek of the Oil Finders: A History of Explora-tion for Petroleum*. Tulsa: American Association of Petroleum Geologists, 1975.

Painter, David S. *Oil and the American Century: The Political Economy of U.S. Foreign Oil Policy, 1941–1954*. Baltimore: Johns Hopkins University Press, 1986.

Palast, Greg. "OPEC on the March: Why Iraq Still Sells Its Oil à la Cartel." *Harper's*. April 2005, 74–76.

Parks, James M. "Recollections of Shell's Bellaire E&P R&D Lab." *Oil Industry History* 4, no. 1 (2003): 118–29.

Pratt, Joseph A. *Prelude to Merger: A History of Amoco Corporation, 1973–1998*. Houston: Hart Publications, 2000.

Pratt, Joseph A., Tyler Priest, and Christopher Castaneda. *Offshore Pioneers: Brown & Root and the History of Offshore Oil and Gas*. Houston: Gulf Publishing, 1997.

Priest, Tyler. "The 'Americanization' of Shell Oil." In *Foreign Mul-tinationals in the United States: Management and Performance*, edited by Geoffrey Jones and Lina Gálvez-Muñoz, 188–205. London: Routledge, 2002.

Rauch, Jonathan. "The New Old Economy: Oil, Computers, and the Reinvention of the Earth." *The Atlantic Monthly*. January 2001, 35–49.

Reifel, M. D. "Offshore Blowouts and Fires." In *The Technology of Offshore Drilling, Completion and Production*, compiled by ETA Offshore Seminars, Inc., 239–57. Tulsa: Petroleum Publish-ing Company.

Roderick, Jack. *Crude Dreams: A Personal History of Oil & Politics in Alaska*. Fairbanks, Alaska: Epicenter Press, 1997.

Sampson, Anthony. *The Seven Sisters: The Great Oil Companies and the World They Shaped*. New York: Bantam, 1975.

Scranton, Philip, ed. *The Second Wave: Southern Industrialization from the 1940s to the 1970s*. Athens: The University of Georgia Press, 2001.

Shell Development Company. *Bellaire Research Center: The First Fifty Years, 1936–1986*. Houston: Shell Oil Company, 1986.

Shell Oil Company. *Annual Report*, various years.

———. *Annual Report Form 10-K*, various years.

———. *History of Shell Offshore Seismic Operations in the United States*. Houston: Shell Oil Company, 1986.

Sherrill, Robert. *The Oil Follies of 1970–1980*. New York: Anchor Press, 1983.

Simmons, Matthew R. *Twilight in the Desert: The Coming Saudi Oil Shock and the World Economy*. New York: John Wiley & Sons, 2005.

Solberg, Carl. *Oil Power: The Rise and Imminent Fall of an American Empire*. New York: Mentor, 1976.

Stoff, Michael B. *Oil, War, and American Security: The Search for a National Policy on Foreign Oil, 1941–1947*. New Haven, Conn.: Yale University Press, 1980.

Stuart, Charles A. *Geopressures*. Baton Rouge: Louisiana State University, supplement to Proceedings of the Second Symposium on Abnormal Subsurface Pressure, 1970.

Sweet, George Elliott. *The History of Geophysical Prospecting*. Los Angeles: The Science Press, 1966.

Tanzer, Michael, and Stephen Zorn. *Energy Update: Oil in the Late Twentieth Century*. New York: Monthly Review Press, 1985.

Thorpe, Helen. "Oil & Water." *Texas Monthly*. February 1996, 90–93, 140–45.

Tidwell, Mike. *Bayou Farewell: The Rich Life and Tragic Death of Louisiana's Cajun Coast*. New York: Vintage Books, 2003.

Vietor, Richard K. *Energy Policy in America since 1945: A Study of Business-Government Relations*. Cambridge, N.Y.: Cambridge University Press, 1984.

Wells, Barbara. *Shell at Deer Park: The Story of the First Fifty Years*. Houston: Shell Oil Company, 1979.

Wells, Wyatt. *Anti-trust and the Formation of the Postwar World*. New York: Columbia University Press, 2002.

Wilkins, Mira. *The History of Foreign Investment in the United States, 1914–1945*. Cambridge, Mass.: Harvard University Press, 2004.

Williamson, Harold F., Ralph L. Andreano, Arnold R. Daum, and Gilbert C. Klose. *The American Petroleum Industry, 1988–1959: The Age of Energy*. Evanston, Ill.: Northwestern University Press, 1963.

Worster, Donald. *Under Western Skies: Nature and History in the American West*. New York: Oxford University Press, 1992.

Yergin, Daniel. *The Prize: The Epic Quest for Oil, Money, and Power*. New York: Simon and Schuster, 1992.

Equilon Enterprises, 271, 276, 279
Esso, 187. *See also* Exxon
Eugene Island 330 field, *48*, 132–33, 135
Eureka, 97, 218
Eureka platform, 207, 239
Europa field/project, *48*, 259
expendable well approach, 100
exploration and production (E&P), Shell
  Oil. *See* Shell Oil Company, exploration
  and production (E&P) organization of
Exxon (Standard Oil of New Jersey), 22,
  234; and 1942 consent decree, 25;
  acquisition of deepwater acreage by,
  223; acquisition of new reserves by, in
  the 1930s, 15; diversification of, 172;
  earnings of, 161; and *Exxon Valdez* oil
  spill, 214, 267; and Hondo project,
  196, 201; and March 1974 federal OCS
  sale, 194; merger with Mobil, 273, 276;
  participation of, in Ursa project, 261;
  political attack on, 159; and upgrading design wave height for offshore
  platforms, 190. *See also* Humble Oil
  Company; Esso
Exxon-Mobil, 11, 15, 277–78. *See also*
  Exxon; Mobil
Exxon Valdez, 214, 267

Faisal, King of Saudi Arabia, 155–56
Fannin building (Houston), 140
Federal Energy Administration (FEA), 159
Federal Energy Agency. *See* Federal Energy
  Administration
Federal Power Commission, 109
Federal Trade Commission, U.S. (FTC),
  14–15, 25, 67, 279
Felix prospect/field. *See* Grand Isle 76 field
Ferris, Bob, 37, 53, *79*, 80
Fifth Circuit Court of Appeals, U.S., 152
Fina, 273
Financial World, 274
Fisher, Arv, *188*
Florida, 220
Flowers, Billy: on Bay Marchand discov-
  ery, 126; on bright spot seismic, 130,
  132; and campaign to open up federal
  offshore leasing, 212, 217–19; as
  division exploration manager in New
  Orleans, 103; leadership of, in applying
  geophysics offshore, 133–34, 183; leadership of, deepwater exploration and
  leasing, 222–23, 225, 228; as marine
  division exploration manager, 124–25;
  and South Pass 65 discovery, 127; and
  understanding of faulting in Gulf of
  Mexico petroleum geology, 128
Forbes magazine, 18, 41, 153
Foreign Affairs, 158
Forrest, Mike: on area-wide lease sales,
  225; on Cognac prospect, 192, 195; as
  discoverer and advocate of bright spots
  seismic interpretation, 128–32, 135–
  36; as general manager of exploration
  for Shell Offshore, 223; as president
  of Pecten International, 171; on Shell
  Oil's reserve replacement, 211
Fortune magazine, 21, 235, 270
Four Corners area, 37

Fourier field, *48*, 263
Frederick, Don, 225–26
Fromenthal, Logan, 155
Funk, Jim, 244, 254–55

Galleria shopping mall (Houston), 141
Galloway, A.J.: on Bouwe Dykstra, 58; and
  budgeting for offshore exploration,
  45; and creation of E&P Economics
  Department, 77; on Monty Spaght,
  107; on offshore leasing, 62; promotion
  of, to executive vice president, 75; and
  purchases of East Bay leases, 47–48;
  retirement of, 78; as Shell Oil vice
  president for E&P east of the Rockies,
  37; on South Pass 24 field, 51
Galveston, Tex., 69
Garcia, Marisol, 155
Garden Banks, Gulf of Mexico, 243
Geer, Ron, 87–88, 90–91, 95–96, *97*, 99,
  104
General Motors, 22
Geological Survey, U.S., 96, 151–52, 204,
  219, 221
Geology of Texas, The, 39
geology, petroleum: of carbonates, 40; of
  deepwater (turbidite), 217–18, 248,
  254–55; and evolution of geologic
  theory, 38–39; and Gulf of Mexico
  play areas, 135–36; and methods for
  predicting reserve volumes, 126–27;
  and origin and migration of oil, 119;
  of pinnacle reefs, 121; of Prudhoe Bay,
  113, 116; of salt domes, 31, 41, 45, 47,
  92–93, 167, 217
Georges Bank, 202–204
Geophysical Research Corporation (GRC),
  31
geophysical research laboratory, Shell
  Oil, 42
geopressure technology, 64–65, 71
George Richardson, 258
Gershinowitz, Dr. Harold, 23
Getty Oil, 233–34, 254–55
Giliasso, 33
Ginsburg, Robert, 40
Glasscock Drilling Company, 69
Glenda prospect, 216
Global Marine, 69, 99, 243
Glomar II, 99, 111, 115
Godber, Frederick, 15
Godfrey, Dan, 196–97, 255–57
Golden Gate Bridge, 207
Golden Lane, Mexico, 203
Goldman Sachs, 233–34
Goldstone, Frank, 42
Graham, Bruce, 141
Grand Bahama terminal, 162
Grand Isle 76 field, *48*
Grand Isle, La., 90
Grand Lake, La., *32*
Great Depression, 15, 17, 247, 267
Great Britain, 15
Green Canyon, Gulf of Mexico, 222, 226,
  237, 252–53, 262
Grolla, Dick
Gulf Intercostal Canal, 66
Gulf Marine Fabricators, 240

National Environmental Policy Act (1969),
145, 205
National Ocean Industries Association
(NOIA), 183, 202
National Petroleum Council (NPC), 182
National Science Foundation (NSF), 82
nationalizations, 230
naval petroleum reserves, 112
Naval Technical Mission to Europe, U.S.,
106
net present-value approach, 77–79, 234
Netherlands, the, 15, 162, 279
New Orleans, La., 60–61, 79, 90, 269, 277
New York City, 15, 141–44, 269
New York Stock Exchange (NYSE), 18–19
New Zealand, 168
Niger Delta, 169
Nigeria, 164, 275, 278
Niobe, 125
Nixon, Richard M., 137, 156, 158–59, 168,
192
Norco refinery (Louisiana), 66–67, 101,
107, 175–76, 267
North Sea: offshore oil in, 187, 191, 201,
222, 231, 245, 251–52, 267, 275
North Slope, Alaska, 2, 7, 106, 112–17,
138. *See also* Prudhoe Bay
Not-In-My-Backyard syndrome, 209
Nova Scotia, 167, 173

O'Brien, Jerry, 75, 92–93, 126
Occidental Petroleum, 267, 277
Ocean Driller, 98, 100, 103
Ocean Industry magazine, 127
Ocean Drilling and Exploration Company
(ODECO), 58–60, 69, 83, 98
Offshore Company, The, 69
Offshore Co. v. Robison (1959), 152
Offshore Division, Shell Oil, 192, 218–19
Offshore East Division, Shell Oil, 221
Offshore Energy Center (OEC): Hall of
Fame, 264
Offshore Engineer magazine, 199
Offshore Operators, 128
Offshore Operators Committee, 55, 152
Offshore Production Division, Shell Oil,
184
Offshore Technology Conference (OTC),
188–89; Distinguished Achievement
Award, 99, 101, 120, 189, 238, 263
Oil and Gas Journal, 52, 123, 127
Oman, 278
one-atmosphere chamber. *See* Shell-
Lockheed system
One Shell Plaza (Houston), 141–44, *142*,
172, 279
One Shell Square (New Orleans), 184–86,
185
Onyx prospect, 216
Ora Mae, 47
Oregano field/project, 253
Organization of Petroleum Exporting
Countries (OPEC): cartel undermined,
231; control of world oil by, 2, 138,
163, 172; 1973 embargo by, 1, 123,
156, 180, 191, 209
Orr, Ken, 252
Otteman, Lloyd, 88, 194–95, 219–20

Oudt, Freddy, 52, 61, 92
Outer Continental Shelf Lands Act (1953),
55, 145; Amendments, 204, 210; or-
ders, 151-52; Section 8 of, 67
Owens, Chris, 61
Oxford University, 78

Pacesetter II, 194
Paine, Sam, 79, 80, 183–84, 195–96, 198,
200–201
Papua New Guinea, 163
Paraguay, 171
paramount rights, constitutional doctrine
of, 54
Parks, James, 39
Party 88, 45–46, 93–94
Pattarozzi, Rich, 245, 248–50, 256, 258,
264
Patterson, Maurice, 189–90
Patton, 75
Patton, General George, 75
Peacock, Steve, 259
Pearson, E.V., 79
Peccary prospect, 216
Pecten companies, Shell Oil, 168–70
Pecten International, 170–71, 233
Pecten, The, 48
Pei, I.M., 141
Pennzoil, 6
Pennzoil Oil and Gas Operators. *See* POGO
Pennzoil Place (Houston), 141
Pennzoil-Quaker State, 280
Permian Basin (Texas), 42, 87, 270
Persian prospect, 216
Perner, Ray, 88
Peru, 168
Peter Kiewit & Sons, 240
Petersen, Bill, 98, 251
Peterson, Reed, 121
Petrobrás, 187
Petroleum Club, New Orleans, 61
petroleum systems analysis, 120
Petronas, 169
petrophysics, 44
Petty-Ray Geophysical, 43
Phaedra, 125
Philippines, 171
Phillipi, Ted, 119
Phillips Petroleum, 126, 273
Picaroon prospect (Brazos A19 field),
215–16
Picou, Ed, Jr., 61, 93
pile-drivers, underwater, 239
Pine prospect, 134–35. *See also* South
Marsh Island 130 field
Piper Alpha platform, 267
pirogues, 32
Pirsig, Gerry, 120–23
pipelines, offshore, 65–67, 101, 252, 259,
262
Pittman, John, 56, 58, 71, 101
Placid Oil, 223, 245
Plaquemines Parish, La., 57
platforms, offshore: blowouts and ac-
cidents on, 137, 145–48, 150–51, 267;
and cycle time, 257; and earthquake
resistant designs, 205; fixed, steel-
jacket design of, 50, 69, 180, 185,

Russell, Don, 75, 84, 127, 144, 171, 174, 184

Sabah Shell Petroleum, 169
Sacramento Basin, Calif., 132
Saigon, 168
Samuel, Marcus, 12
San Antonio, Tex., 1
San Francisco, Calif., 207, 267
San Pedro Bay, Calif., 205–206
Santa Barbara, County of, 205
Santa Barbara channel, 83, 98–99, 118, 196, 204; 1969 blowout, 117, 137, 145, 192, 205, 207
Santa Fe Drilling Company, 115
Santa Fe Synder, 277
Saro-Wiwa, Ken, 275
Saudi Arabia, 1–2, 29, 155–56, 231, 271; Shell Oil petrochemical complex in, 138, 176
Savit, Carl, 130
Scaife, Bill, 134
Schlumberger, 80
Schoenberger, George, 81
Scott, George C., 75
Sea Gem, 145
Sears Tower (Chicago), 227, 237
Second World War, 15–16, 23, 25, 29, 47, 56, 106, 124, 173
Secretary of State, U.S., 156
Securities and Exchange Commission (SEC), U.S., 278
seismic exploration, 120–123; 3-D, 8, 246, 254, 274, 278; impact of digital computing on, 3, 106, 123, 125–26; offshore, 30, 41–47, 80, 93–94, 125, 126, 193, 217–18, 223-25, 245; in wetlands, 31–33, 32
semi-submersible drilling vessels, 8, 98–99, 115, 182, 243; invention of, 81–87, 90–91, 100, 102–103, 264
Senate, U.S., 14, 159
Senate Foreign Relations Committee, U.S., 268
Senate Interior Committee, U.S., 160
Senate Permanent Subcommittee on Investigations, 159
Senge, Peter, 269
Seriff, Aaron, 131
Serrano field/project, 253
Seven Sisters, 160
Shah of Iran, 210
Shatto, Howard, 49, 96–99
Shearson/American Express, 234
Shell America, 223–24, 224
Shell Camrex (Societe Shell Camerounaise de Recherches et d'Exploitation), 169
Shell Canada, 113, 116, 119–22, 138, 140, 164, 167, 173
Shell Caribbean Petroleum Company, 14, 21
Shell Centre (London), 107, 276
Shell Chemical Company, 23, 108, 143, 175, 229, 272; olefins plants, 175–77; matrix organization, 176
Shell Deepwater, Inc., 245
Shell Development Company, 22–25, 27, 77, 124, 190, 246

Shell Eastern Petroleum Products (New York), 15
Shell Exploration & Production Company (SEPCo), 253, 263, 277
Shell Explorer, 167
Shell-Lockheed system, 186
Shell Midstream Enterprises, 262
Shell Museum, 279
Shell News, 32–34, 49, 116, 156
Shell Offshore, Inc., 135, 216, 223, 247, 249–50
Shell Oil Company, 2–3; in Alaska, 110–118, 207–209, 212–15, 229, 266; and alternative energy, 2, 9, 137; Americanization of, 4, 11, 16–21, 25, 28; annual meeting, 235; anti-apartheid boycott of, 268; autonomy from Royal Dutch/Shell Group, 20–21, 38, 273; barge drilling, 34, 49; board of directors, 45, 192–94, 229, 234, 236–37, 244–45, 249; branded as yellow peril, 14, 18, 25; in Canada, 113; and coal mining, 2, 9; corporate history project, 279; corporate strategy, 172–78; creation of, 16–17; creation of separate operating companies, 270; contribution to global rebuilding of Group, 21–22; and deepwater Gulf of Mexico, 215–29, 243–66; Denver Area E&P office, 76, 121–22, 140; and development of mobile drilling, 81–99; dissolution of, 272, 280; divestments, 268, 272, 276; and downstream businesses, 9, 105, 107–109, 144, 175–77, 228–31, 237, 265, 271–72; and enhanced and heavy oil recovery, 2, 9, 178–80; 265–66; and environmental damage offshore, 145–149; Exploration & Production (E&P) organization, 4, 9, 30, 36–37, 77, 105–106, 110, 116, 118, 140, 154, 180–81, 184; financial performance and budget allocations, 108, 110, 159–60, 175, 228–29, 266–68, 272, 274; geological research, 38–41, 92-93, 126–27; geophysical research and development, 42, 120–21, 125, 131–32; globalization of, 265–66, 276–79; and Group buyout of minority shareholders, 227, 232–37; Houston Area E&P office, 37, 69, 121, 125, 128, 140; Information & Computer services, 139, 144; international exploration and development, 167–71; joint ventures, 270–72, 279; and lease bidding process, 193; long-range plan, 247; management reforms, 269–70; marine seismic exploration, 31–32, 92–94, 217–18, 223–24, 254–55; Midland Area E&P office, 37, 122, 125, 140; and minority shareholders, 4–5, 14, 19, 108, 164–66, 171; New Orleans Area E&P office, 37, 52–53, 56, 60–61, 64, 74, 97, 100, 110, 121, 124, 127, 129, 131–32, 140, 168, 182, 199; and offshore Atlantic, 202–206; and offshore California, 204-205; offshore Gulf of Mexico developments, 1960s, 124–136; offshore Gulf of Mexico developments,

Texaco (Texas Oil Company), 15; acquisition of Getty Oil, 233–34; breakthrough by, in wetlands drilling, 33; downstream alliances with Shell Oil, 271–73, 279–80; and June 1968 federal OCS sale (Texas), 129; leadership of, in gasoline sales 144, 153; lease position of, on Louisiana coastal plain, 31; and Louisiana State Lease 340, 38; outbid by Shell Oil for South Pass 28 lease, 80; and South Pass 24 field, 51

Texas A&M University, 124

Texas Commerce Tower (Houston), 141

Texas Miocene, 136, 215

Texas Railroad Commission, 156

Texas, State of: and tidelands controversy, 54; as swing oil producer, 156

Thirty Seconds over Tokyo, 16

Thomasson, Ray, 132

three-dimensional seismic exploration. *See* Seismic exploration: 3-D

Threet, Jack, 169–70, 183, 203, 211, 219, 222–23, 228

THUMS group, 204

Tichy, Noel, 269, 275

tidelands controversy, 54–55, 63

torsion balance, 33, 38, 41, 47

Total, 273

Tracy, Spencer, 16

Trans-Alaska Pipeline, 160

Transco, 134

Transco Tower (Houston), 141

transformation, Shell Oil's, 269–70, 275

transportation, offshore, 66

Treaty of Friendship, Commerce and Navigation, United States and The Netherlands, 68

Trinidad, 162

Troika field/project, *48*, 253

Trosclair, Lucius, 149–50, 152

true amplitude recovery, 126

Truman, Harry S., 26, 54, 67

Two Shell Plaza (Houston), 142

Tulane University, 269

Tulsa University, 43

turbidite sands, 217–18, 248, 254–55.

See also geology, petroleum; Gulf of Mexico

Union Oil Company of California, 14–15, 82, 117, 145, 201, 204, 213

Union Oil Company of Delaware, 14

United States v. Timken Roller Bearing Co., 26

UNIVAC computer, 124, 140

University of California–Berkeley, 84

University of Oklahoma, 182

University of Texas, 173

University of Zurich, 119

Uppencamp, George, 239

Ursa: prospect/field, *48*, 226; TLP/project, 259, 261–62

Utah, 45

Vallejo, Calif., 207

van de Vijver, Walter, 263, 278

van Everdingen, A.F. ("Toni"), 43, 77

van Wachem, Lodewijk, 232–33, 236

Vancouver Island (Canada), 167

Venezuela, 21, 73, 78, 164

Venice, La, 57

Ventura field, 76

Viater, Keith, 56, 151

Vibroseis, 126

Viosca Knoll, Gulf of Mexico, 252, 261

Vietnam, 168

Vietnam War, 137, 168

Vining, Alden, 149

VLCCs (Very Large Crude Carriers), 162

Voiland, Gene, 225

voluntary oil allocations, 158

Wall Street, 18

Wall Street Journal, 82, 135, 227

Ward, Skip, 190–91

Warrington, Ralph, 207

Washington, D.C., 159

Wasp prospect, 216

Wassan field, 15, 277

Watergate scandal, 137, 159–60

Waterton field, 119

Watt, James, 219–21

Watts, Sir Philip, 278

Weeks Island field, 30, *34*, 35, 44, *48*, 49, 88

Weiner Bob, 132

Weiss, Seymour, 61

West, J. Robinson, 219

West Africa, 277

West Cameron 192 field, *48*, 65

West Delta 30 field, *48*, 71

West Delta 143 platform, 252, 259

West Delta 133-A platform, *103*

West Delta 134 field, *48*

Western Geophysical, 130

Westhollow Research Center, 144

West Jumping Pound field, 119

West Louisiana Miocene, 136, 215

well logs, 80; continuous velocity, 42; technical advances in, 44

White Castle field, *48*, 92

Wickizer, Carl, 186, 221, 225, 246

Wide Bay, Alaska, 110

Williams, Art, 88

Williston Basin, 37, 40, 165

Wilmington Canyon, 221–22

Wilson, James E. ("Jim"), 124–25, 127–28

Wilson, James Lee, 40, 124

Win-or-Lose Corporation, 38

women's rights movement, U.S., 137, 154

Wood River refinery (St. Louis), 107, 276

work environment, offshore, 56

World Petroleum Congress, Fifth (1959), 75, 79

Yergin, Daniel, 73, 158, 172, 231–32

Yom Kippur War, 155–56

Zane Barnes, 243–44

ISBN-13: 978-1-58544-568-4
ISBN-10: 1-58544-568-1